OAK SEED DISPERSAL

A Study in Plant-Animal Interactions

MICHAEL A. STEELE

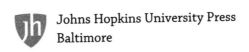
Johns Hopkins University Press
Baltimore

© 2021 Johns Hopkins University Press
All rights reserved. Published 2021
Printed in the United States of America on acid-free paper
9 8 7 6 5 4 3 2 1

Johns Hopkins University Press
2715 North Charles Street
Baltimore, Maryland 21218-4363
www.press.jhu.edu

Library of Congress Cataloging-in-Publication Data

Names: Steele, Michael A., author.
Title: Oak seed dispersal : a study in plant-animal interactions /
    Michael A. Steele.
Description: Baltimore : Johns Hopkins University Press, 2021. |
    Includes bibliographical references and index.
Identifiers: LCCN 2019057286 | ISBN 9781421439013 (hardcover) |
    ISBN 9781421439020 (ebook)
Subjects: LCSH: Oak—Seeds—Dispersal. | Seed dispersal by animals. |
    Acorns.
Classification: LCC QK929 .S83 2021 | DDC 581.4/67—dc23
LC record available at https://lccn.loc.gov/2019057286

A catalog record for this book is available from the British Library.

The illustrations that open each chapter are by Tad Theimer.

*Special discounts are available for bulk purchases of this book. For more
information, please contact Special Sales at specialsales@press.jhu.edu.*

Johns Hopkins University Press uses environmentally friendly book
materials, including recycled text paper that is composed of at least
30 percent post-consumer waste, whenever possible

To Margaret (Mooge) Ganse Steele, my best friend and most discerning critic, and Michael, Tyler, and Emily, the other three most important dimensions of my world

For after some weeks of close scrutiny I cannot avoid the conclusion that our modern oak woods sooner or later spring up from an acorn, not where it has fallen from the tree, for that is the exception, but where it has been dropped or placed by an animal.

HENRY DAVID THOREAU, *Faith in a Seed: The Dispersion of Seeds and Other Late Natural History Writings*, 1993, 129–130

CONTENTS

FOREWORD

Oaks (genus *Quercus*) comprise a large, widely distributed, and diverse group of plants. But why has Michael Steele written a book entirely dedicated to seed dispersal of a single genus? The answer is that, from the perspective of seed dispersal, oaks are one of the most complex and interesting groups of plants in the world. Oaks are dispersed almost entirely by the actions of nut-caching animals, primarily rodents and jays. These animals bury nuts in shallow, widely scattered sites, and, if not reclaimed by an animal, the acorns can germinate and produce new plants. These are old and complicated coevolved relationships between plants and dispersers. These complexities arise from the genetic diversity of oaks (the genus has five distinct sections), the broad variety of habitats (e.g., tropical, north temperate, and semi-arid), and the wide geographic range (North and Central America, Europe, and Asia) over which oaks are found. In these different environments, oaks have interacted with a wide variety of seed-dispersing animals. Further, across this range, there is a wide variety of seed predators (e.g., weevils, boars, and acorn woodpeckers) that act as parasites of the mutualism. These animals and physical environments exert strong selective pressures on seed morphology, nutritional content, and germination behavior.

Most nuts that are dispersed by scatter-hoarding animals are large and nutritious. This is necessary to attract potential dispersers. For example, chestnuts, walnuts, hickory nuts, beech nuts, Brazil nuts, almonds, hazel nuts, and many others are very attractive to animals. But acorns take a different approach. They are large and contain carbohydrates, fats, and other nutrients to attract animals, but they also contain tannins, which are bitter and, at high concentrations, toxic. Oaks appear to be attracting and repelling their potential mutualists at the same time. This schizophrenic strategy has its advantages, which Steele describes in great detail.

Despite the tannins, there is a great deal of competition for the acorns. Even before the acorns mature, weevils and other insects lay eggs inside the acorns where their larvae develop. When rodents and jays gather the acorns they often discriminate against infested acorns. Once cached, animals steal each other's acorns and move them to new sites. This cache pilfering can become rampant when other foods in the environment are rare; therefore, many acorns can be moved between a series of cache sites with implications for longer dispersal distance. To compete better in the competition of stored acorns, many species of animals have evolved accurate and extensive spatial memories and others have evolved olfactory abilities that increase their ability to pilfer caches. These adaptations give individuals an advantage when searching for buried acorns, increasing their access to a larger share of the stored food resource. But the rampant pilfering has far-reaching implications for the evolution of scatterhoarding—If animals do not get their own acorns back, how did this behavior evolve?

The acorns of some species of the white oak group attempt to avoid being eaten by animals by germinating in the autumn just days after being cached. The seedling quickly establishes a root and then translocates much of the nutrients from the cotyledons into the growing root. The shoot of the future oak does not elongate but remains underground until spring. It seems odd to germinate just as winter approaches, but by doing so the seedling avoids the predation that is likely to come when a rodent or jay retrieves the acorn. Several species of squirrels have adapted to fall-germinating acorns by excising the embryo from the acorn before they bury it. Interestingly, these squirrels do not bother excising the embryos of oak that germinate in the spring.

Masting plays an important role in this story. Masting is the synchronized production of large crops of acorns across a broad geographic range of a species, alternating with one or more years of little or no acorn production. This is an adaption to increase acorn survival and dispersibility. Populations of some specialist seed consumers crash following non-mast years so that more acorns survive following the mast years. This works pretty well, but when crops of acorns fluctuate dramatically across years, it can send shock waves through the animal community. Acorn mast is an important resource in the community for a wide variety of animals far beyond those that serve as agents of acorn dispersal. Deer, bear, woodpeckers, turkeys, mice, and many other animals direct their attention to the acorn mast during mast years. People are also sometimes affected by masting. When mouse populations increase in mast years, so do the populations of ticks and the incidence of Lyme disease. Few people realize that these far-reaching community dynamics are rooted in the caching behavior of rodents and jays.

Michael Steele is uniquely situated to tell this story. He has studied the caching behavior of rodents and other animals for more than 30 years. He is a co-author of *Squirrels of the World* (Johns Hopkins University Press 2012) and *North American Tree Squirrels* (Smithsonian Institution Press 2001) and, of course, squirrels play an important role in this story. He has published more than 100 papers, many on oak dispersal. His descriptions of oaks, squirrels, and their interactions are detailed and authoritative. This book should serve the scientific community well, as it provides a comprehensive analysis of this important forest ecosystem process.

Stephen B. Vander Wall
Reno, Nevada

PREFACE

One might ask, Why devote an entire book to the subject of oak dispersal? The answer to this question is simple and, at the same time, quite complex. The process of seed dispersal, in which seeds are moved away from their parent source, is an absolutely critical stage in the life cycle of most flowering plants. That is the simple part. The dispersal of oaks is also complex because the process can have far-reaching effects on a species' biology—it is often interconnected with many aspects of a species' reproductive traits, as well as numerous aspects of its ecology and evolution. An understanding of the dispersal process can reveal an entire story about the species. This is particularly the case in the oaks, in which the dispersal of the acorn is tied to numerous acorn and tree characteristics, as well as the behavior and ecology of the animals that feed on and move these acorns to their final destination. Even forest structure, composition, and forest genetics often follow directly from the dispersal process, but also, in turn, influence it. Herein, I attempt a synthesis of my lab's research over the past three decades on the interactions between oaks and their seed consumers (rodents, birds, and insects) and their ultimate influence on acorn dispersal—a key stage in the oak life cycle.

My specific goal is to synthesize findings from my research, that of my colleagues, and that of other scientists, on the oak dispersal system, a potential model system, that should stimulate new insights on how the underlying mechanisms of animal-mediated dispersal drive numerous ecological and evolutionary processes in oak forest ecosystems (e.g., forest structure, seed masting, the evolution of seed characteristics, and predator/prey interactions), many of which feed back into the dispersal process. I intend this synthesis to demonstrate the multidisciplinary approaches, including studies of plant-animal interactions, cognitive behavior, plant ecology, seed chemistry, seedling morphology, biogeography, spatial ecology, and evolution, all necessary for understanding the dispersal process and its influence on forest ecosystem function. As such, it is my hope that this synthesis will offer a transformative perspective on the mechanisms driving terrestrial ecosystem processes and, in turn, some aspects of ecological theory.

Although single authored, this book was not possible without the help of countless others. First and foremost, I dedicate this final product to Margaret, my spouse and closest friend of more than 40 years, and my three children, Michael, Tyler, and Emily, all of whom guide me in nearly every decision I make. They have not only tolerated my constant obsessions with science, natural history, and field work, but they have also encouraged it, even immersed themselves in it. In the process, they have accompanied me to some of the most remote places on the planet. No one has a better support network, and for that, I am eternally grateful.

Drs. Tad Theimer and Stephen Vander Wall, two colleagues and friends—whom I hold in the highest regard—helped put the final touches on this book. To both, I extend a special salute. The foreword, by Stephen Vander Wall, professor of biology, University of Nevada, Reno, clearly illustrates why he is revered as *the* leading authority on food hoarding in animals and animal-mediated seed dispersal, among other topics. The exquisite artwork that opens each chapter in the book was created by Tad Theimer, professor of biology, Northern Arizona University. Yes, Tad, a well-established biologist, with major contributions to the study of seed dispersal, is also an accomplished artist. As humble as he is, it was only after a lovely overnight stay at Tad's home that I discovered this special talent hidden in his study. And, when I began the book, I capitalized on that memory. For each image in this book, Tad first researched the content of each corresponding chapter, then produced a sketch, and afterward a final drawing. The image for Chapter 14 was only finalized after considerable protest by the author and some good humor exchanged between the two of us. I also extend my sincere appreciation to private landowners, Mr. George Vanesky and his family, William and Mary Martin, Floyd Balliet and his family, the extended Rinehimer family, and Hawk Mountain Sanctuary for permission to use their properties for field studies and long-term monitoring of oak mast. Their patience and support have allowed me to litter their forests with seed collectors and other equipment for nearly two decades, all in the name of science.

My most sincere gratitude goes to my research associate, Shealyn Marino. First hired as a research technician on external funding, Shea has been my dedicated and meticulous research partner for more than 15 years, and, if I have my way, many more years to come. In addition to overseeing much of the long-term mast and small-mammal, live-trapping surveys discussed in this book, she has been an integral contributor to countless other studies. She is also my right hand in the lab as she regularly works with me to oversee a small army of undergraduates in their independent research, senior projects, and summer research positions. Shea and several students, especially Tyler Brzozowski, Ryan Giberson, Brendon Kelly, Kelvin Mejia, and Morgan Novakovich also provided valuable assistance with finalizing the figures and literature cited in this book.

My collaborators number in the dozens and student researchers in the hundreds. For fear of overlooking anyone, I just extend my deepest gratitude to everyone. Although single authored, this book has been shaped, as my approach to science, by some key colleagues. To my graduate advisor, Dr. Peter D. Weigl, I extend my deepest gratitude for encouraging my curiosity and independence but also instilling the life-long message that my research is only as good as the "big picture" it paints. This book is in some small way a tribute to that message. Another long-time colleague, collaborator and friend, Dr. Robert Swihart, welcomed me into his research team at the Department of Forestry and Natural Resources at Purdue University. That experience, extending now for more than 18 years, has allowed me to grow substantially as a scientist and to explore a number of aspects of oak dispersal that I would not otherwise have addressed.

Several colleagues read and commented on all or part of the book, including my friends and colleagues Debra Chapman, and Drs. William Biggers, Michal Bodziewicz, Kenneth Klemow, Jeffrey Stratford, Stephen Vander Wall, and an anonymous reviewer. I thank them for their time and tremendous effort for helping to catch errors and transform my rough ideas into a more coherent presentation. I accept full responsibility for any remaining errors and flaws.

Time, a fulltime research associate, nine postdoctoral associates, well over 200 undergraduate research technicians, various equipment and supplies, and, of course, a constant supply of acorns all cost money. In my 30 years at Wilkes University, I have been fortunate to receive a steady flow of the necessary dollars to conduct my collaborative research across at least eight states in the United States, as well as Mexico, Costa Rica, China, England, Poland, and Spain. Particularly important for my research have been funds from the US National Science Foundation (e.g., DEB-9442602, DBI-9978807, DEB-0642504), especially the most recent OPUS award (DEB-15556707), which directly supported the preparation of this book. I also recognize funds from the US Fulbright Foundation for my research in Mexico, Howard Hughes Medical Institute, The Pennsylvania Wild Resource Conservation Fund, Poland's National Science Foundation, the Department of Biology, Wilkes University, a Bullard Fellowship from Harvard University, and the H. Fenner Endowment of Wilkes University. This last fund has partially supported my position for the past 15 years, providing a slight reduction in my teaching load, support for my research associate, Shealyn Marino, and funds for supplies, equipment, and student technicians.

I recognize several individuals whose direct support made this book possible. My program officer for the OPUS grant from the National Science Foundation (DEB), Dr. Douglas Levey, a leading authority on plant-animal interactions and seed dispersal, was a tremendous help with securing the award and guiding me through it. Everyone should have such a champion. And, at Johns Hopkins University Press, I extend my deepest gratitude to Vincent J. Burke, my editor for this book, for two previous books published by JHU Press, and for another at Smithsonian Institution Press. His quick wit and keen editorial eye always helped to guide me through the challenging tasks of writing these books. Most recently I have had the wonderful experience of working with Tiffany Gasbarrini, Senior Science Acquisitions Editor, and Esther P. Rodriguez, Editorial Assistant at JHU Press. I thank them both for their diligence, good humor, and patience and for constantly making this author's experience with the editorial process a truly positive one. I also extend my most sincere thanks to Liz Radojkovic the professional copy editor who carefully and systematically polished and standardized my otherwise rough ideas into something that is clean and far more presentable. Liz also added good humor and a positive tone to what otherwise can be a most difficult task for authors.

Finally, to you, the reader, whether you open this book as a budding naturalist just wanting to learn something about oak-animal interactions, or, the professional ecologist with a keen interest in seed dispersal, I believe there is something for you. My goal has been to share one small aspect of how the natural world works. Indeed, we must continue to pursue an intimate understanding of the natural world if we are ever to protect it for future generations.

OAK SEED DISPERSAL

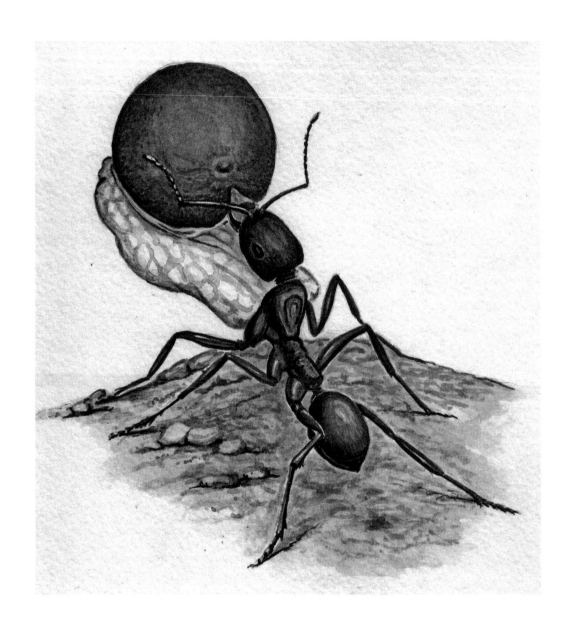

INTRODUCTION

The dispersal of the acorn turns out to be one of the most significant stories of ecology and evolution, and it often takes place just beyond the edge of our own backyards. It is a process that impacts the behavior and ecology of the animals that feed on oak acorns and, at the same time, shapes forest structure and composition. Found across the globe, the many species of oaks are distributed from the northwest corner of South America, through Central and North America, and across Eurasia. The highest species diversity occurs in Mexico and southeast Asia.

This book, the culmination of three decades of research on three continents, focuses on the interactions between oaks and rodents, birds, and insects. It reveals the oak dispersal system in a way that provides new insights on how animal-mediated dispersal drives ecological and evolutionary processes in forest ecosystems.

The ecological system I describe is replete with species interactions and feedback between trophic levels. The dispersal process sometimes involves symbioses, strong competitive interactions, predation, and a host of other complex interactions. To tell this story fully, I relied on multidisciplinary studies, my own and those conducted by others, of plant-animal interactions, cognitive behavior, plant ecology, seed chemistry, seedling morphology, biogeography, spatial ecology, and evolutionary ecology. Each of these approaches was needed to uncover the intricacies of the dispersal process and its influence on forest ecosystem function. It is my hope that this book offers a transformative perspective on the mechanisms driving terrestrial ecosystem processes and, in turn, some aspects of ecological theory.

In this chapter, I briefly review why seed dispersal is important, what is involved in the process, the various modes of seed dispersal, the fundamental theoretical foundation for the study of seed dispersal, and where the study of seed dispersal is headed. I then move into a more detailed account of animal-mediated seed dispersal, reviewing various mechanisms of seed dispersal by animals, highlighting a range of plant and animal taxa involved in this relationship, as well as various biomes and systems in which animal-mediated dispersal is essential. I then close this chapter with a brief introduction to the importance of scatterhoarding for the oaks, thereby foreshadowing the chapters to follow.

THE BIOLOGY OF DISPERSAL

Dispersal is a fundamental biological process for nearly all organisms. It involves the movement of an individual organism (offspring or propagules) away from their place of birth (natal dispersal), or the movement of one individual from one breeding population to another (migration or breeding dispersal) (Ronce 2007). The former of these is especially critical for offspring, as they find new resources and mates following the dispersal event. And, by definition, dispersal results in gene flow (Ronce 2007), so, while it is often critical for the individual, it also influences population genetics. Ronce (2007, p. 232) defines dispersal quite simply as "any movement of individuals or propagules with potential consequences for gene flow across space." Any quick electronic search produces an enormous number of publications on the subject, and perusal of these titles reveal the tremendous diversity of biological processes influenced in one way or another by the dispersal process.

Whereas dispersal is important for nearly all organisms, it is an especially critical part of the life cycle of any sessile organism, such as a plant. It is necessary for most flowering plants to have their seeds, fruits, or nuts dispersed from their origin, where the probability of survival and seedling establishment is often slim. And today, with increasing worldwide fragmentation of habitat, climate change, and the loss of biodiversity, the subject of dispersal has become even more relevant, as we attempt to understand how anthropogenic changes are modifying or impeding this important natural process.

MODES OF SEED DISPERSAL

The adaptations for seed dispersal seem potentially endless. In fact, a visual reflection on the many modes of seed dispersal bring to mind the popularized quote by early British geneticist and evolutionary biologist, J. B. S. Haldane, regarding the tremendous diversity of beetles on earth, in which he referenced "God's inordinate fondness for beetles" (Gould 1993). That admiration could apply equally well to the vast diversity of adaptations and strategies plants employ for moving their seeds.

Evidenced by publications as far back as 1785 with Holmberger's "On the Dispersal of Plants over the World" (cited in van der Pijl 1972), botanists have long recognized the tremendous diversity of strategies or syndromes that plants employ to move their seeds. And, van der Pijl's (1972) detailed account of these strategies, although descriptive, still serves as a cornerstone for understanding the basic biology of seed dispersal, the various plant and seed/diaspore traits used to achieve dispersal, and the specific terminology used to characterize it. Here I borrow from the terminology of van der Pijl (1972) to briefly review the various strategies involved in plant dispersal. The order in which I present this terminology is arbitrary and my own. These dispersal mechanisms (strategies) are organized around the ecological process involved in moving the seed. My discussion is selectively brief and intended only to point the reader toward the diversity of strategies plants employ in moving their seeds. I also seek to provide a context for the specific mode of dispersal in the oaks involving the direct movement of the fruits (acorns) by primarily birds and mammals in a process known as "scatterhoarding."

Autochory refers to dispersal that is accomplished by the plant itself rather than by any other physical or living dispersal agent. Autochory can take several forms, three of which I discuss here: barochory, ballochory (or ballistic dispersal), and passive autochory. Barochory is sometimes mistaken simply as movement of the seed due to its weight, which according to some authors includes genera of plants that produce relatively large fruits, such as coconuts (*Cocos* spp.), walnuts (*Juglans* spp.), and even the acorns of oaks (*Quercus* spp.). However, as emphasized by van der Pijl (1972) and my colleague Dr. Stephen Vander Wall (personal communication), this is a misleading designation because, while the rounded form of these fruit or nuts may move the seed short distances from the parent, especially on hillsides, these diaspores ultimately depend on additional dispersal agents to make it to their final site of seedling establishment—water in the case of *Cocos* and scatterhoarding animals in the case of *Juglans* and *Quercus*. These latter two genera are clearly dispersed by animals that move and cache their fruits. The dropping of the seeds to the ground or to water is better considered just presentation of the diaspore to the first agent of dispersal (Stephen Vander Wall, personal communication).

Ballochory involves the actual release of seeds due to a bursting (explosion) of the fruit as a result of building turgor pressure. The specific mechanisms by which this is accomplished are quite varied across plant genera and typically result in only short dispersal events just beyond the parent plant (see review by van der Pijl 1972). Common plants that show this kind of abrupt dehiscence of the fruit and ejection of the seed include species of *Geranium* and *Impatiens*

1.1. Examples of ballochory (ballistic dispersal). Shown is **(a)** the flower of the Himalayan balsam (*Impatiens glandulifera*) adjacent mature seed pods. The build-up of turgor pressure inside the seed causes them to readily explode upon contact with any passing object. (*Chiew Loo*) **(b)** Intact and "exploded" seed pod of *Impatiens capensis*, also called spotted touch-me-not or common jewelweed. (*Herman Geithoorn*) **(c)** The fruit of the sandbox tree (*Hura crepitans*), also called the dynamite tree or possumwood, is found in the tropical forests of the Americas. When the large fruit dries it eventually explodes, dispersing its seeds more than 45 m (Swaine and Beer 1977). (*Ahmad Fuad Morad*) SEE COLOR PLATE

(Fig. 1.1). One of the most dramatic examples of ballochory occurs in the sandbox tree (*Hura crepitans*), an evergreen found in the tropics of the Americas. The species produces fruit about 5 cm in diameter that resemble a small squash with as many as 16 carpels that all separate when the fruit violently explodes dispersing the seed in each carpel up to 45 m (van der Pijl 1972; Swaine and Beer 1977), although even greater distances have been cited.

Finally, passive autochory can involve the deposition of seeds as a result of movement of the plant as it grows along the ground or the actual movement of diaspores themselves as a result of changes in humidity that cause trichomes (bristles) on the seed or diaspore to move it short distances from the parent (van der Pijl 1972).

All other forms of seed dispersal that depend on other mechanisms of dispersal—either physical mediums (e.g., water, wind) or animals—are collectively referred to as allochory. Allochory includes movement by water (hydrochory), wind (anemochory), or animals (zoochory), the last of which involves a tremendously diverse array of strategies.

Hydrochory can be as simple as the force a rain droplet has on the spore of a fungus or liverwort, causing them to pop off the structure in which they are held. Although, in a sense, a ballistic form of dispersal, it is caused by the rain and thus considered separate from autochory (van der Pijl 1972). Some species also depend on rain wash to move seeds. Others depend on the movement of water currents in streams and rivers. At the other end of the extreme are plants that rely on ocean currents for dispersal, such as coconut of *Cocos* spp., which are well suited for floating for long periods. Long assumed to represent an adaptation primarily for long-distance dispersal, the Atoll Ecosystem Hypothesis suggests otherwise for one species, *C. nucifera* (Harries and Clement 2013). While the diaspores of many plants that are moved through fresh water and ocean systems rely on their buoyancy for dispersal, others do not. Van der Pijl (1972) describes a number of plant taxa that do not float but are suspended in the water column by a slimy covering or hairlike projections that allow them to be moved by water currents. Thus, while hydrochory accounts for the dispersal of a limited range of taxa, it represents a diversity of adaptations for moving diaspores through this medium.

Wind dispersal (anemochory), in contrast, is far more widespread (van der Pijl 1972), but like hydrochory, is a passive mechanism that depends on the physical currents produced by the medium. Adaptations of wind-dispersed diaspores are diverse, from the samaras of the maples (Acer spp.), the enormous wings of the *Centrolobium robustum* of the American tropics (van der Pijl 1972), or the many plants like dandelion (*Taraxacum officinale*) and milkweed (*Asclepias* spp.) that depend on other ornate structures to harvest the wind (Fig. 1.2). Indeed, many plant species are masters at harvesting air currents. Moreover, in the past two decades, our fundamental understanding of wind dispersal has been advanced considerably by theoretical models that, in many situations, can now predict the fate and consequences of wind-dispersed seeds (Nathan et al. 2002, 2011). Such models, for example, can demonstrate how forest structure can alter patterns of uplift, which in turn contribute to critical long-distance dispersal events (Nathan et al. 2002). These advances in the study of wind dispersal contrast significantly with those in other studies of seed dispersal (e.g., animal-mediated dispersal), which have been considerably slower to develop due, in part, to the diversity of specific mechanisms by which other modes of dispersal are achieved and by the lack of continuity in these various processes (but see Cousens et al. 2010; Lichti et al. 2017).

1.2. Examples of wind dispersed seeds. Shown are **(a)** samaras of silver maple (*Acer saccharinum*), native to eastern and central United States and southeastern Canada. Mature seeds are blown off the parent tree and drift to the ground in a spiral motion. The drifting seeds can travel up to 50 m. (*Michael A. Steele*) **(b)** Native to the Americas and Australia, the seeds of the fireweed plant, also called American burnweed or pilewort (*Erechtites hieraciifolius*), are shown here prior to dispersal. Mature seeds easily detach when exposed to the wind. (*Michael A. Steele*) **(c)** Seeds of the common milkweed (*Asclepias syriaca*) are dispersed once the pod has opened and each lightweight seed with its cluster of hairlike projections are carried by the wind. (*Michal Sansfacon*) SEE COLOR PLATE

Animal-mediated dispersal, or zoochory, is arguably far more widespread than most other modes of seed dispersal. It involves a tremendous diversity of adaptations in which plants depend on animals to move their seeds. Seed dispersal by animals is carried out by a number of insects, other arthropods, and many species of vertebrates, including fish, some reptiles, birds, and mammals. Birds and mammals represent a disproportionately greater contribution to the process. Animals move seeds by several means, including epizoochory in which fruits can attach to the body of animals, especially mammals; endozoochory in which animals eat seeds or fruit and then either regurgitate or defecate them; and the process of scatterhoarding in which animals, primarily birds and mammals, disperse, store, and under some circumstances fail to recover these seed stores, resulting in their germination and establishment (Herrera 2002). This latter mechanism of animal-mediated dispersal is a central focus of this book, as this is the primary mechanism by which rodents and birds move the oaks.

Epizoochory, which is relatively rare, occurring in less than 5% of plants, involves the attachment of seeds or fruits to birds and, more commonly, mammals by means of burrs, hooks, or sticky, gum-like fluid (van der Pijl 1972; Sorensen 1986; Fig. 1.3). More common in ground-level herbaceous plants, dispersal by this means is accidental, in that the animal is often not aware that the diaspore has become attached. However, the result is often significant in that the diaspore can be dispersed a considerable distance before it is detected and removed or arrives at its final resting place when the animal dies (Sorensen 1986). Although the final deposition is often random, the primary advantage achieved by this means of dispersal is likely the high probability of long-distance dispersal. It is also important to note that this form of animal-mediated dispersal, unlike the others, does not involve a mutualism between animal and plant.

Frugivory is by far the most common means of endozoochory. It is the consumption of the fruits surrounding seeds, usually by vertebrates, often followed

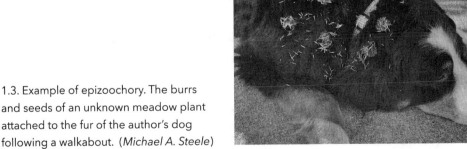

1.3. Example of epizoochory. The burrs and seeds of an unknown meadow plant attached to the fur of the author's dog following a walkabout. (*Michael A. Steele*)

1.4. Examples of frugivory in which animals consume fruits and then either defecate or regurgitate the seeds some distance from the parent. Shown are (a) the southern cassowary (*Casuarius casuarius*), which is a well-known seed disperser across its range in northeastern Australia and New Guinea. *C. casuarius* can grow to just under 2 m in height, and has the potential to disperse seeds more than a kilometer from their source (Weber and Woodrow 2004) *(Daintree Discovery Centre)* (b) Seeds sprouting from a pile of cassowary droppings. *(Druce Horton)* (c) A green Honeycreeper (*Chlorophanes spiza*), shown here eating seeds of *Miconia longifolia*, is native to Costa Rica and other parts of central America. *(Jeffrey Stratford)* (d) The orangutan (*Pongo pygmaeus*), found in the rain forests of Borneo and Sumatra, often drops or spits seeds as they are eating, or defecates seeds after swallowing fruits. *(Matt Trevillion)* SEE COLOR PLATE by the deposition of seeds at some distance from the parent plant. This can result by spitting out the seeds as is often done by the many rodents and primates, such as the orangutan (*Pongo pygmaeus*), or by consumption of the seeds along with the fruit, followed by defecation or regurgitation, as in the case of the cassowary (*Casuarius casuarius*), which is known to be an important disperser of numerous fruiting species where it occurs in northeastern Australia and New Guinea (Fig. 1.4). Frugivory occurs in many forest types but is most common in the tropical rain forests, where a majority of woody plant species (up to 90%) depend on this mode of seed dispersal (Jordano 2000). Many vertebrate frugivores appear to exhibit morphological or physiological adaptations for dispersing fruit although few dispersers maintain an exclusive diet of fruit. Thus, early suggestions for strong coevolutionary relationships between plants and frugivores (Snow 1971; Howe 1993) has not been widely supported (Levey and Benkman 1999).

A specific type of frugivory by ants—myrmecochory—deserves special mention. Not only does it merit its own name, but it involves a specific adaptation to attract ants (elaiosome), and it accounts for the dispersal of an estimated 11,000 species of angiosperms (in approximately 334 genera of plants) (Lengyel et al. 2010). Myrmecochory occurs nearly worldwide and in a variety of plant species. By comparing angiosperm lineages with global patterns of myrmecochory, Lengyel et al. (2010) concluded that seed dispersal by ants may have evolved independently in as many as 147 occasions, but in a minimum of at least 101 angiosperm lineages. Many cases of myrmecochory involve the elaiosome, an energy-laden seed attachment, comprised mostly of lipid, which the ant consumes after moving the seed from its source (Fig. 1.5). The study of myrmecochory is thus a field unto its own. A quick search over the past 10 years from the time of this writing revealed approximately 185 publications on the subject. It is indeed a rich area of research that new students may consider a fruitful subject for exploration.

Scatterhoarding is a specific form of animal-mediated dispersal in which animals disperse and cache seeds and nuts in individual widely spaced caches for future use. Often the behavior involves storage of propagules in shallow underground cache sites that are ideal for storage but also coincidentally optimal for germination and establishment when animals fail to recover these items. Performed most often by birds and mammals, the behavior is now widely recognized as a primary avenue of seed dispersal for many seed and nut producing species, such as the oaks (Vander Wall 1990). Our understanding of the basic natural history of scatterhoarding behavior grew over several decades (see reviews by Vander Wall 1990 and Brodin 2010), eventually culminating in the theoretical prediction that the evolution of the behavior requires that the individual scatter-hoarder must maintain a recovery advantage over competitors that could potentially pilfer caches (Andersson and Krebs 1978). This was

(a)

(b)

1.5. Ants and seeds. Shown are **(a)** the *Messor wasmanni* ants native to the Greek Island of Kefalonia. The Messor genus consists of more than 100 species, most known as harvester ants, feeding primarily on seeds and mushrooms. Harvester ants are best known for their extensive seed granaries in which seeds are stored in dry conditions to prevent germination. As a result, these ants are primarily seed predators. *(Colin Millington)* **(b)** Numerous other ant species feed on seeds with an elaiosome, a high energy attachment that the ants remove and eat after dispersing the seeds. Seed dispersal by ants (*myrmecochory*) has evolved independently in numerous lineages of flowering plants (see text). *(Robert Klips)* SEE COLOR PLATE

followed in turn by several key studies that demonstrated how this is likely achieved by both mammals (Stapanian and Smith 1978) and birds (Tomback 1982; Vander Wall 1982). These systematic studies on scatterhoarding behavior were eventually accompanied by a growing number of studies demonstrating the consistent dispersal of propagules by scatterhoarders and the evidence for a coevolutionary relationship between scatterhoarders and trees (Lanner 1996; Vander Wall 2001). As the dispersal of the oaks depends almost exclusively on scatterhoarding of acorns by mammals and birds, this behavior is a central theme throughout this book. Subsequent chapters focus on the details of mammalian and avian scatter-hoarding behavior and how it influences the fate of cached acorns and the likelihood of acorn dispersal and oak seedling establishment.

THE BREADTH AND DIVERSITY OF ANIMAL-MEDIATED SEED DISPERSAL

From myrmecochory to scatterhoarding, animal-mediated seed dispersal involves a diversity of interactions in what seems like an unlimited number of conditions and circumstances. Indeed, many dispersal events seem to be quite fortuitous and unpredictable but essential for the species involved. Whereas many frugivores, for example, disperse plant seeds by actively seeking the plant's fruit, the seeds of many other plants can be accidentally dispersed by herbivores. Janzen (1984) suggested that smaller seeds are likely to be dispersed when larger herbivores simply consume vegetation, and in the process inadvertently ingest and pass seeds through their digestive track. Although numerous authors have verified this hypothesis, in one of the more extensive studies to test this, Myers et al. (2004) reported that during the course of one year, white-tailed deer (*Odocoileus virginianus*) passed viable seeds of 70 plant species in their feces. This included the diaspores of a diversity of herbs, shrubs, and trees and those that exhibit the full range of dispersal modes discussed previously. Myers et al. (2004) further argued that deer may be particularly important agents of long-distance dispersal for many of these seed species because of the animal's extensive range sizes and long gut retention times. Other authors speculate that numerous other herbivorous megafauna, extinct since the Pleistocene, may have served a similar critical role in seed dispersal in many parts of the tropics (Janzen and Martin 1982; Corlett 2013; Jara-Guerrero et al. 2017). This likely means in today's forests, some plant species are dispersal limited while others may depend on less optimal solutions for dispersal (Jansen et al. 2012).

Another example is a study of the dispersal of a keystone plant species (crowberry, *Empetrum nigrum*) in the alpine tundra of southern Norway (Steyaert et al. 2018). The authors followed patterns of seed deposition in and around decomposing carcasses of reindeer (*Rangifer tarandus*) following a lightning kill of an entire herd of 323 individuals. Following the event, reindeer carcasses were regularly visited by scavengers, such as ravens (*Corvus corax*) and hooded crows (*C. cornix*), as well as several species of meso-carnivore and rodents. The birds and meso-carnivores consistently dropped feces near the carcasses with the former frequently depositing viable seeds of crowberry. The authors maintain that these avian scavengers in effect directed seeds toward these "decomposition islands" that serve as optimal sites for seedling establishment (Steyaert et al. 2018).

Yet another example of a potentially under-appreciated mechanism of animal-mediated seed dispersal is caliochory—the accidental transport of seeds in plant material and mud when birds construct their nest (Warren et al. 2017). The authors collected viable seeds from 144 plant species in the nests of 23 bird species and with the use of nest boxes, experimentally excluding seeds that may have arrived by wind or fecal deposition. The 144 plant species included species that are typically recognized by one of the seven classic dispersal modes, with anemochory (wind) being the most common. Warren et al. (2017) also found that 43% of the plant species found viable were cleistogamous (self-fertilizing), suggesting that this may be an especially important mode of dispersal for seeds with this genetically limiting means of reproduction. Although the question is still open as to whether these dispersed seeds successfully establish, it seems quite possible that caliochory represents a viable means of seed dispersal for numerous species.

The breadth and diversity of the mechanisms by which animals move seeds also raises the question of how these interactions may vary geographically. In one of the first, more systematic and rigorous evaluations of the geographic variation of animal-mediated seed dispersal, Vander Wall and Moore (2016) evaluated the patterns of animal-mediated dispersal across North America, north of Mexico. Based on data from 197 sites, these authors assigned greater than 123,000 plant records of 12,424 plant species to dispersal mode (non-mutualistic, frugivory, scatterhoarding, or myrmecochory) and, in the process, identified 14 specific modes of animal-mediated dispersal. Many of the 14 modes of dispersal involved both primary dispersal followed by a secondary mode of dispersal involving an animal mutualist. Among the 20,426 records of animal-mediated dispersal, 14.6% of these accounts involved both a primary and a secondary stage of dispersal. The majority of primary dispersal events (75%) were the result of frugivory by birds and/or mammals. Myrmecochory surprisingly accounted for 23.7% of the records in at least one stage of the dispersal process. Vander Wall and Moore (2016) reported that the highest frequency of plants that depend on animal mutualisms for dispersal is most strongly correlated with precipitation and range of elevation and, as a result, associated with the southeastern region of North America. This study is a major step forward in helping us understand the environmental conditions that drive the evolution of plant-animal mutualisms involved in seed dispersal.

The above discussion illustrates the vast diversity of mechanisms by which seeds are dispersed. This variation illustrates how difficult it is to identify the common selective pressures that likely gave rise to such variation. Perhaps it is no surprise then that a general theoretical and unifying framework for seed dispersal continues to remain out of reach (but see Schupp et al. 2017). Although this is partially true today, early studies of seed dispersal followed directly from early theoretical predictions. In the search that followed, numerous themes and concepts helped to better characterize the ecology and evolution of seed dispersal, but despite these accomplishments, a common unifying structure is still on the horizon. I briefly review some of these central ideas and themes that only in part unify today's general body of knowledge on seed dispersal.

BASIC THEORY AND CENTRAL THEMES UNDERLYING THE STUDY OF SEED DISPERSAL

Janzen-Connell (J-C) Model: The Seed / Seedling Shadow

Few single ideas in the history of ecology—and many other areas of science—have had more of a long-standing and guiding inspiration on a field of study than the Janzen-Connell (J-C) model has had on the study of seed dispersal. Some even place the J-C model up there with Robert MacArthur's and Edward Wilson's theory of island biogeography (1967) or Evelyn Hutchinson seminal 1959 paper on the fundamental importance of energy in structuring natural communities. Named for independent contributions by Daniel H. Janzen (1970) and Joseph H. Connell (1971), predictions of this conceptual model were simple, compelling, and long-lasting in their influence on the field of ecology. The J-C model is the first conceptual hypothesis to address the question of why seed dispersal is important in plants. It basically predicted that in order to survive, seeds must be dispersed away from parents to avoid areas near the parent where density-dependent mortality is highest. It offered a unifying theme on how the ecology and evolution of plants was tied to the need for each species to move its seeds—whether it be by self, water, wind, or animals.

The J-C model, based on Janzen's (1970) observations in central and South America and those of Connell (1971) in Australia, is that most adult trees of a given species are far more evenly distributed than would follow from the distribution of their seeds, which were typically clustered around the parent plant. This general pattern, they deduced, was due to two factors: (1) it supported that seed densities, following their final dispersal destination, should decline with distance from the parent source with the highest densities occurring under the parent tree, and (2) there should be a disproportionately higher effect of enemies, such as seed predators and pathogens, near the parent source where seed and seedling densities are highest (Fig. 1.6). As a result, seeds dispersed farther from their source should experience a higher probability of establishment. This made considerable intuitive sense, since higher densities of seeds would attract seed predators and allow easy transfer of pathogens. Seedlings, if they were to establish under the parent tree, should also face intense competition for light, water, and nutrients with the mother and siblings. In contrast, seeds at lower densities, farther from the parent, perhaps near the parent of another species, should experience higher survival and higher rates of seedling establishment.

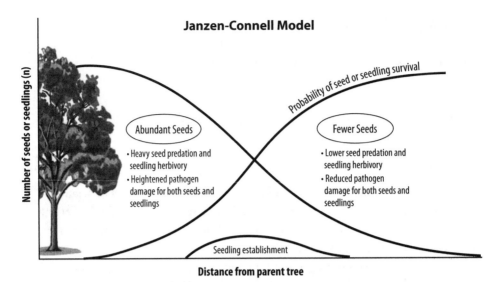

Janzen-Connell Model

Number of seeds or seedlings (n)

Probability of seed or seedling survival

Abundant Seeds
- Heavy seed predation and seedling herbivory
- Heightened pathogen damage for both seeds and seedlings

Fewer Seeds
- Lower seed predation and seedling herbivory
- Reduced pathogen damage for both seeds and seedlings

Seedling establishment

Distance from parent tree

1.6. Janzen-Connell model showing as the distance from the parent tree increases, the abundance of seeds drops and the probability of seed and seedling survival increases. (Adapted from Janzen 1970)

The J-C model, as basic as it was, offered a universal explanation for (1) the rather even distribution of adult trees of each species observed by Janzen and Connell (but see Hubbell 1980), (2) the countless mechanisms of seed dispersal that appeared to ensure successful movement of the seeds away from the parent plant, and (3) even a possible explanation for why so many tree species evolved in the tropics (Terborgh et al. 2002). In science, when a relatively simple concept can potentially explain such a broad range of phenomena, it is likely to prevail.

And prevail it did—the J-C model, intuitive and compelling, inspired nearly 50 years of research on seed dispersal. But as inspirational as it was, many of the studies to immediately follow never directly tested the underlying hypotheses put forward by Janzen and Connell. Many of these studies centered primarily on the behavior of frugivory, the most common mode of seed dispersal, in the tropics where Janzen's and Connell's research originated. Evidenced by numerous early reviews and comprehensive edited volumes (Howe and Smallwood 1982; Estrada and Fleming 1986; Fleming and Estrada 1993; Howe 1993; Jordano 2000; Levey et al. 2002), most investigations centered on specific aspects of the frugivore-fruit interactions (e.g., the nutrient composition of fruit, fruit displays, strategies of fruit digestion, and so on) and far less on seed fates and demographic or fitness consequence of seed movement and seedling establishment (Levey and Benkman 1999; Jordano 2000). Although these efforts were certainly important, in that much was learned about the ecology and evolution of these interactions (see reviews by Stiles 2000; Jordano 2000; Wilson and Traveset 2000), any hopes of a unifying theory to follow from this intensive focus on only frugivory were eventually dashed (Levey and Benkman 1999; Jordano 2000). What was needed most was a broader range of investigations on the process of seed dispersal and, most important, some direct tests of the J-C model.

By 2000, however, nearly 30 years after the introduction of the J-C model, two major breakthrough studies provided direct tests of the hypotheses central to the model (Harms et al. 2000; Packer and Clay 2000). In the first, Harms et al. (2000) followed patterns of seed fall in numerous species of tropical trees in Panama and compared this with the distribution and survival of thousands of seedlings of the same species. As predicted, relative mortality was highest near parent sources of all species and establishment consistently occurred where sibling densities were lowest. In a more detailed study of just one species, black cherry (*Prunus serotina*) in a temperate forest, Packer and Clay (2000) found the same predicted patterns of distribution and mortality and that the cause was a soil fungus that easily spread when seedling densities were highest.

In the years to follow, the study of seed dispersal gained traction as its focus broadened well beyond the study of frugivory. Between 2000 and 2015, four international conferences, resulting in several special publications from these meetings (e.g., Fenner 2000; Levey et al. 2002; Forget et al. 2005; Dennis et al. 2007; see special issue in *Acta Oecologia* 2011) and greater than 2000 additional publications (Schupp et al. 2010) have steadily strengthened our understanding of the process of seed dispersal. During this time, numerous unifying concepts and themes have emerged. I briefly list some of these concepts that have recently strengthened our perspective and moved us closer to a unified theory on seed dispersal (but see Lichti et al. 2017; Schupp et al. 2017).

🐾 *Rare long-distance dispersal (LDD) events* are often important in many seed dispersal systems. They frequently drive the success of the disperser-plant mutualism (Clark 1998; Nathan 2006; Levey et al. 2008) but are often the most difficult events to document (Jordano 2017). Statistical methods that account for a range of dispersal distances (including both short and long distance) are needed to more accurately predict patterns of tree movement (Clark et al. 1999).

🐾 *The conditional nature of the disperser-plant mutualistic relationship* can cause the seed consumer to shift from seed predator to seed disperser under different environmental conditions (e.g., seed availability) (Theimer 2005).

꙳ Seed dispersal is often *context dependent* with respect to site variability (Schupp 2007). Thus, a detailed understanding of the habitat mosaic is needed to understand the dispersal success.

꙳ Some dispersal agents may move seeds to sites disproportionately more suitable for seedling establishment, a process referred to as *directed dispersal* (Howe and Smallwood 1982; Wenny and Levey 1998; Hirsch et al. 2012).

꙳ *Landscape structure, habitat fragmentation*, and *resulting habitat corridors* can significantly influence the outcome of animal-mediated seed dispersal events (Levey et al. 2005).

꙳ *Other mutualistic interactions* (e.g., mycorrhizal plants [i.e., plants with symbiotic relationship with soil fungi]) may influence the outcome of many seed dispersal systems and should therefore be considered and integrated into models of seed dispersal (Theimer and Gehring 2007; Correia et al. 2018).

꙳ In any study of seed dispersal, efforts should be made to understand different conditions contributing to variation in *seed dispersal effectiveness*, which can be assessed by summarizing both *quantitative and qualitative effectiveness of seed dispersal* (Schupp 1993; Schupp et al. 2017).

꙳. The seed dispersal process often involves multiple steps between the parent tree and the final point of establishment, leading to *a clear distinction between primary, secondary, and, even, multiple stages of dispersal* (Vander Wall and Longland 2004; Forget et al. 2005, and references therein; Jansen et al. 2012).

꙳ In many systems, final seed fates may even require the linkage of two or more modes of seed dispersal, a process known as *diplochory* (Vander Wall and Longland 2004).

꙳ *Predators of scatter-hoarding birds and mammals may drive the seed dispersal process* by both influencing the placement of seeds in high-risk sites where pilferage rates are reduced and preying on some individuals after the caching process. (Steele et al. 2011, 2014, 2015, this volume).

꙳ *Plant responses to climate change is often tied closely to patterns and mechanisms of seed dispersal.* Dispersal may either benefit or hinder responses to climate change (Johnson et al. 2019; Naoe et al. 2019; Snell et al. 2019).

The above represents only a cursory look at some of the key breakthroughs in the study of seed dispersal in the past few decades, especially those involving animal-mediated dispersal. Despite these critical advances there is still a need for a broader theoretical framework that serves to unify these findings. In contrast to studies on animal-mediated dispersal, those on wind dispersal have received a century of attention with especially rapid advances in the past few decades and now some successful predictive models (Nathan et al. 2011). As several authors have noted, the most significant challenges in modelling animal-mediated dispersal is to capture how seed fate follows from the animal's behavior and how this behavior is linked to environmental context (Cousens 2010; Lichti et al. 2017) and to better understand how communities of animals collectively contribute to seed fate (Muller-Landau et al. 2008).

At a broader level, Schupp et al. (2017) recently presented a general framework for assessing several types of mutualism, including plant-seed disperser interactions. They emphasize the importance of measuring both quality and quantity of the effect of one mutualist on another to assess the overall effectiveness of the interaction. However, they also emphasize the historical bias of many studies that measure effectiveness of such interactions on just one of the mutualists. Clearly mutualisms such as animal-mediated dispersal is a two-way process.

In an attempt to synthesize much of the research to date on seed dispersal by scatter-hoarding rodents, my colleagues and I, under the direction of Dr. Nathanael Lichti, have offered the predictive framework for modeling the outcome of these interactions (Lichti et al. 2017). The review identified six stages in the process of seed dispersal by scatterhoarders, nine intermediate variables (motivation, food value, costs of secondary compounds, handling time, missed opportunity costs, metabolic costs, seed perishability, pilferage risk, and predation risks) that link other characteristics, such as animal and seed traits with behavioral decisions that drive the outcome of final seed fates. The goal was to review what is known about these various stages in the dispersal process, to identify areas of research that are still much needed, and to ultimately place this information into a structured predictive framework. Although it is only a first step toward modeling one aspect of one specific type of animal-mediated dispersal, it may serve to spark other related endeavors.

Much research is yet needed before we can synthesize the early J-C models with a broader predictive framework on seed dispersal. The pages that follow are not directed at integrating the diversity of studies on scatterhoarder-mediated seed dispersal toward a theoretical framework. Instead, I take a different approach. My goal is to focus exclusively on a specific system—the oak dispersal system—to demonstrate how so many aspects of this one system influence the dispersal process, and how this process, in turn, defines so much of the oak ecosystem. By focusing exclusively on the dispersal of oaks, it is my hope that I can provide some insight regarding the complexities of animal-mediated dispersal in this one system. Indeed, it is far more complex than we ever imagined or what was ever expected in the few decades following the seminal predictions of Janzen and Connell. Moreover, our understanding of the dispersal process in oaks is far from complete; however, it does represent a system in which there is an interesting story to tell, one that I hope inspires future research not just in this system but others like it.

THE DISTRIBUTION, DIVERSITY, AND EVOLUTION OF THE OAKS

2

INTRODUCTION

I now focus on the characteristics that distinguish the oaks (genus *Quercus*), their general taxonomy, and their distribution across five of the seven continents, which includes extensive regions of North America, Europe, and Asia, as well as limited areas of northern South America and Africa. With species found in temperate, subtropical and even tropical biomes, the oaks represent a tremendously diverse group of flowering plants with a significant role in many terrestrial ecosystems. My emphasis is on North America, where *Quercus* is the most diverse genus of native trees, with well over 90 species found north of Mexico and greater than 150 species distributed throughout Mexico and Central America, many of which are endemic.

I briefly discuss the evolutionary history of the oaks with reference to recent studies on the phylogeny and biogeography of oaks in both North America and Eurasia. I then close with a brief overview of geographic and taxonomic variation of key acorn characteristics (e.g., germination schedules and acorn size), which, as I discuss in future chapters, directly influence oak dispersal.

WHAT ARE THE OAKS? WHERE ARE THEY FOUND?

The oaks belong to the genus *Quercus*, one of the most widespread of flowering plants found in the northern hemisphere (Nixon 2006). The genus *Quercus* belongs to the family Fagaceae, which includes two subfamilies: Fagoideae, which includes the beeches of the genus *Fagus*, and the Quer-

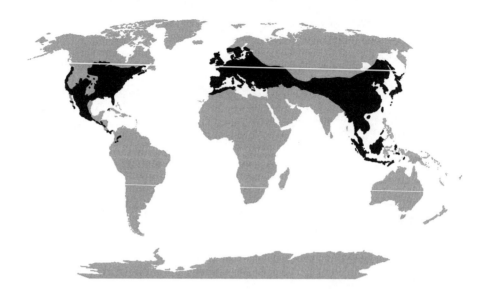

Figure 2.1. Global distribution of all oaks (genus *Quercus*). This includes all species of both subgenera. (*Distribution in North, Central, and South America follows from Nixon 2006*)

coideae, which includes the genera *Castanea* (chestnuts), *Castanopsis* (the chinquapin), *Lithocarpus* (the stone oaks), the newly designated genus *Notholithocarpus* (the tanoaks) (Manos et al. 2008), *Chrysolepis* (also called chinquapin), and a few other genera found in the tropics. Members of this genus are distributed across the globe from the Americas through Europe and Asia (Fig. 2.1), although different taxa occur in different regions.

The *Lithocarpus*, which include anywhere from 100 to more than 300 species (Flora of China 1994; Nixon 1997; Kremer et al. 2012) are found only in Asia. One species from North America, formerly considered a stone oak, is now distinguished by the genus *Notholithocarpus*. All species of *Lithocarpus* are evergreen trees or shrubs that produce an acorn-like fruit with a hard, woody shell; hence, their common name—the stone oaks.

Within the genus *Quercus* there are typically two subgenera recognized: *Quercus* (recognized by some as *Euquercus*) and *Cyclobalanopsis*. These two groups are also recognized by some authors as two separate genera (Chengjiu et al. 1999). The *Cyclobalanopsis* are exclusive to eastern and southeast Asia (including Borneo, China, Laos, Japan, Korea, Malaysia, Myanmar, Nepal, Taiwan, and Vietnam). Species of this subgenus are distinguished by caps with pronounced scalelike rings (Fig. 2.2).

Within the subgenus *Quercus* (= *Euquercus*) five sections are frequently recognized (Table 2.1). The two accounting for the largest number of oak species are

2.2. Shown are acorns of *Q. myagii* found on the Japanese Island of Okinawa. The scaled caps are typical of acorns of approximately 150 species in the subgenus *Cyclobalanopsis*, also called the ring-cupped acorns. Acorns of *Q. salicina* on the main Island of Okinawa weigh approximately 7 g, but can reach 14 g on Iriomote Island of the Okinawa Perfecture, located between Okinawa and Taiwan (Takuya Shumada, personal communication). (*Yuya Watari*) SEE COLOR PLATE

section *Quercus*, also referred to as the white oaks (previously *Leucobalanus* or *Lepidobalanus*), and section *Lobatae*, the red oaks (previously *Erythrobalanus*). Species belonging to the section *Quercus* produce acorns that mature in six months, and an acorn cotyledon that is typically low in tannin levels. The inside of the acorn shell lacks hairlike follicles and the style (the stalk that extends from the ovary to the stigma in the flower of the oak) is short relative to those of *Lobatae*. Members of the white oak section are found throughout the Americas, Europe, North Africa, and Asia, which is nearly the entire range of the genus.

Section *Lobatae* accounts for a large number of species as well, but all are found exclusively in North, Central, and South America. The red oaks are distinguished by acorns that take approximately 18 months for maturation, flowers with longer styles than that which are found in species of white oaks, and the presence of hairlike follicles on the inside of the acorn pericarps. Red oak species are found in Canada, throughout the United States, Mexico, Central America, and the northwestern tip of South America. Adult red oaks and white oaks are typically, but not always, distinguished by leaf morphology with red oaks having pointed lobes with a bristle-like projection at the end of each lobe and white oaks exhibiting rounded lobes without any projection (Fig. 2.3). However, many red oak species, especially in Mexico, exhibit rounded lobes but still exhibit the distinctive projections at the end of these lobes.

Another important difference between the red oaks and white oaks concerns their germination schedules. In Chapters 5 and 6, I review how the germination schedule in many parts of the *Quercus'* distribution is a key factor influencing caching decisions by rodents and thus the dispersal and establishment of oaks. White oaks germinate immediately, without any need for stratification,

Table 2.1. Overview of the genus *Quercus*, in which two subgenera are often recognized: *Quercus* and *Cyclobalanopsis*. Within the subgenus, *Quercus*, five sections are recognized. Worldwide estimates of species diversity vary considerably, but diversity is highest in the red oaks (section *Lobatae*), white oaks (section *Quercus*), and the *Cyclobalanopsis*. *Protobalanus* includes only six species (Nixon 2006). Only the white oaks are distributed worldwide. The red oaks and *Protobalanus* are found only in the Americas, and *Cyclobalanopsis* is restricted to Asia. See commentary in text on potential revisions to taxonomy.

SUBGENUS	SECTION	CHARACTERISTICS	LOCATION
Quercus = Euquercus	*Quercus* (the white oaks)	Short styles; acorns mature in 6 months; acorn cotyledon typically lower in tannin; inside of acorn shell lacks hairlike follicles	North and Central America, Europe, Asia, North Africa
	Lobatae (the red oaks)	Long styles; acorns mature in 18 months; acorn cotyledon higher in tannin; inside of acorn shell has hairlike follicles	North, Central, and South America
	Protobalanus (the intermediate oaks)	Short styles; acorns mature in 18 months; acorn cotyledon higher in tannin; inside of acorn shell of some species have hairlike follicles, while others do not	Southern California, SW Arizona, and NW Mexico (Baja California)
	Mesobalanus (similar to section *Quercus*)	Long styles; acorns mature in 6 months; acorn cotyledon higher in tannin; inside of acorn shell lacks hairlike follicles	Europe and Asia
	Cerris	Long styles; acorns mature in 18 months; acorn cotyledon higher in tannin; inside of acorn shell lacks hairlike follicles	Europe and Asia
Cyclobalanopsis (ring cupped oaks)		scales on acorns; many species clustered on branches	southeastern Asia

provided conditions are not too dry. Red oaks, in contrast, exhibit delayed germination requiring a period of cold stratification (or moisture stratification in more xeric conditions). In eastern deciduous forests, for example, red oaks such as the northern red oak (*Q. rubra*), pin oak (*Q. palustris*), and black oak (*Q. velutina*) require an extended period of cold that changes the permeability of the pericarp and the ability of the acorn to germinate. These differences in germination schedule have a dramatic effect on dispersal patterns, with the red oak highly preferred for caching and white oaks selectively eaten, especially in the presence of red oaks. In Chapters 5 and 6, I also review a number of

2.3. Shown are **(a)** northern red oak (*Quercus rubra*) tree shape (*Dr. Guillaume Droval*), and **(b)** leaf form. (*Eric Hunt*) **(c)** white oak (*Q. alba*) tree form (*Melinda Young Stuart*) and **(d)** leaf form. (*Scott Zona*)

studies—many from my lab—that demonstrate how germination influences dispersal and caching patterns.

In addition to the red oaks (section *Lobatae*) and white oaks (section *Quercus*), a third section of oaks, *Protobalanus* (the intermediate oaks), is found in the New World but is limited in distribution to the southwestern US (primarily California and Arizona) and northwestern Mexico (Baja California). Like the red oaks, all species of *Protobalanus* are endemic to the Americas (Manos 1997; Nixon 1997). Oaks in this section share some of the characteristics of red oak, except the style is often, but not always, shorter like that of the white oaks (Fig. 2.4). *Protobalanus* includes only about 5–6 species (Nixon 2006; Johnson et al. 2009).

Oaks in the section *Mesobalanus* share many of the characteristics of section *Quercus*, with the exception of long styles. Hence, these oaks are sometimes in-

2.4.Palmer's oak (*Quercus palmeri*), an example of oak species belonging to the section *Protobalanus*. Shown are **(a)** the typical tree and **(b)** leaves and acorns of Palmer's oak. (*Doug Wirtz*) SEE COLOR PLATE

cluded in the section *Quercus*. The section *Mesobalanus* includes a small number of species found in Spain, North Africa, Europe, and Asia (Fig. 2.5).

Finally, the section *Cerris* includes a relatively small number of species that are distributed from southwestern Europe and Northern Africa to eastern Asia, including Japan. Species in this section share similar characteristics of the red oaks. The section includes the cork oak (*Q. suber*), the sawtooth oak (*Q. acutissima*), the holm oak (*Q. ilex*), and the Chinese cork oak (*Q. variabilis*), all of which are highlighted in subsequent chapters (see Fig. 13.4).

Although the taxonomic review above captures the long-accepted view at the time of this writing, it is important to emphasize that the more recent phylogenetic studies suggest a taxonomic revision, based primarily on differences in pollen morphology. Summarized by Denk et al. (2017), this taxonomic revision identifies two monophyletic lineages comprised of two subgenera: the subgenus *Quercus* and the subgenus *Cerris*. The subgenus *Quercus* includes five sections

(*Protobalanus*, *Ponticae*, *Virentes*, *Quercus*, and *Lobatae*), and the subgenus *Cerris* includes three sections (*Cyclobalanopsis*, *Ilex*, and *Cerris*).

A BRIEF EVOLUTIONARY HISTORY OF THE OAKS

The family Fagaceae is reported to have appeared in the late Cretaceous period and was well represented by the early Tertiary (about 50 million years ago [mya]). The origin of *Quercus* is roughly estimated at about 45 mya, followed by a rapid diversification between the Eocene and the Oligocene, about 36 mya (Moore 1984, and references therein). However, as noted by several authors, the fossil record of the oaks is limited and complicated by numerous factors and is still in need of further clarification.

Recent high resolution phylogenetic and biogeographic analyses of oaks in North America, based on next-generation DNA sequencing coupled with fossil records and biogeographic data, indicate that oaks in the new world originated in northern temperate areas, then

2.5. The Algerian oak or Mirbeck's oak (section *Mesobalanus, Quercus canariensis*), a medium-sized deciduous (or semi-evergreen) oak tree found in North Africa, Portugal, and Spain. Shown are **(a)** adult tree. (*Geoff Nowak*) **(b)** Typical leaf and acorn. (*John Jennings*)
SEE COLOR PLATE

moved southward and diverged into two groups: the oaks now found in California and those of eastern North America (Hipp et al. 2018). These eastern oaks then established and diversified in Mexico and eventually in Central America about 10–20 million years ago (Hipp et al. 2018). Following their colonization in Mexico, a rapid radiation appears to have occurred in both the red oak and white oak lineages simultaneously in relation to leaf type (e.g., evergreen, deciduous) and climatic conditions (e.g., moisture conditions) (Hipp et al. 2018). A similar phylogenetic analysis at different spatial scales in many of the sites in the continental US show that at more localized sites, especially in the east, oaks diverged into different species potentially as a result of interspecific interactions but, at the same time, converged toward similar leaf types at these local levels (Cavender-Bares et al. 2018). These authors argue that diversification of species at the local level was a major factor driving the assembly of oak communities, including the high diversity of oaks in places such as the southeastern US.

Deng et al. (2018) have recently completed a molecular phylogenetic and biogeographic analyses of the *Cyclobalanopsis*, which they treat as a section of *Quercus*. They conclude that the diversification of these evergreen oaks resulted from major tectonic events involving the Eurasian, India, and Indo-China tectonic plates and several climatic events occurring after the late Eocene that resulted in tremendous topographic and ecological variation. And, in another recent genetic study, Plomion et al. (2018) sequenced the entire genome of the pedunculate oak (*Q. robur*). They demonstrated that this long-lived oak accumulates somatic mutations, including genes that render the oak potentially resistant to various pathogens and arthropod pests. These somatic mutations that occur during the life of the oak are believed to originate in stem cells that give rise to apical meristems of the oak.

DIVERSITY OF OAKS

According to one of the leading authorities on oak systematics, Kevin Nixon (2006, and references therein), the overall number of species in the western hemisphere is about 220 (but see Johnson et al. 2009, estimates as high as 250). This includes all species found from Canada, through the United States, Mexico, Central America, and the northwest corner of South America. Nixon (2006) cites the number of species in each country within this portion of the *Quercus* range: Canada (4), United States (91), Mexico (up to 165), Cuba (1), Belize (9), Guatemala (up to 26), El Salvador (up to 10), Honduras (up to 15), Nicaragua (14), Costa Rica (14), Panama (12), and Colombia (1). Within this region, hot spots in diversity include the southeastern US (Aizen and Patterson III 1990) and, above all else, the mountains of central and southern Mexico (Nixon 2006).

In the United States, approximately 30 species are found in the west, with

2.6. Regional diversity map of oak species of the Atlantic region. Numbers indicate number of oak species. Contours were drawn from a grid in which the number of oak species present in cells 125 × 125 km was recorded. *(From Aizen and Patterson 1990)*

the remainder occurring in the east. According to Johnson et al. (2009) only the "chinkapin" oak (*Q. muehlenbergii*) and the bur oak (*Q. macrocarpa*) occur in both the eastern and western regions. Johnson et al. (2009) recognizes three major regions in the western US with the greatest number of oaks: Texas, the southwestern US, and the western coastal region. In the eastern US, the highest diversity of oaks occurs in the southeast close to where Alabama, Florida, and Georgia intersect (Aizen and Patterson III 1990; Fig. 2.6).

In Mexico, the diversity of oaks follows from the diverse climatic conditions. Nixon (1993b) reports that the white oaks (section *Quercus*) in particular, which includes more than half of the oaks in Mexico, are more diverse than the red oaks (section *Lobatae*) and are distributed widely with respect to environmental conditions. Nixon reports that the section *Quercus* in Mexico is comprised of two groups of species (subsections), one of which is found in extremely dry environments.

Oaks take many forms and occur in a diversity of forest types, from shrub oaks that are often less than a meter in height to the iconic giant live oaks (*Q. virginiana*) found in the coastal plain of the southeastern US (see Chapter 13). As reviewed by Nixon (2006, and references therein), oaks are associated with a diversity of forest types, from temperate deciduous forests to savannah biomes that extend from temperate to tropical regions, a diversity of evergreen forests, tropical and semi-tropical montane forests, and cloud forests to chaparral (shrubland) ecosystems found in southern California and Baja California,

Mexico, as well as similar systems in the Mediterranean Basin. The diversity of ecosystems with which they are associated make the oaks extremely important from an ecological perspective. And, whereas some oaks are found in a number of ecological systems, it is important to appreciate that some species are dominant in all of the systems they occupy. One such example is the Garry oak (*Q. garryana*) of northwestern North America, which is found in open grasslands and savannahs, closed canopy forests of only Garry oak, and Douglas fir (*Pseudotsuga menziesii*) forests where Garry oak is an early successional species that depends on frequent disturbance by fire (see Chapter 14). As a result, the diversity of systems in which Garry oak is associated supports a number of vulnerable species, many of which are now threatened due to the loss of this oak.

VARIATION IN ACORN SIZE

Acorn size varies tremendously across oak species from the large acorns of *Q. germana*, the Mexican royal oak, found in the cloud forests of eastern and northeastern Mexico, which reaches the size of a golf ball (Fig. 2.7), to *Q. insignis*, found from southern Mexico through Central America to Panama, which produces acorns even twice that size. However, the acorns of most oak species range in size from approximately 1–2 g (willow oak, *Q. phellos*) up to 5 g (red oak, *Q. rubra*), although some can reach closer to 8 g such as the acorns of bur oak (*Q. macrocarpa*), which produces one of the largest acorns in the Americas, north of Mexico. Acorn size also varies within species, although sizes of acorns produced by individual trees usually show less variation.

Across geographic ranges, however, individual oak species show considerable variation in the size of acorns produced. Aizen and Woodcock (1992) examined biogeographic variation in acorn size in 32 oak species in eastern North America and found that many of these species showed a significant negative relationship between latitude and acorn size; that is, individual trees produce larger acorns at lower latitude than those in more northern latitudes. Koenig et al. (2009a) tested several hypotheses regarding the negative relationship between acorn size and latitude in bur oak and found strong evidence that abiotic factors, such as rainfall and temperature, likely contributed to this pattern. Interspecific comparisons in acorn size revealed that species that produce larger acorns have significantly larger geographic ranges and often occur at higher latitudes than small-seeded oaks (Aizen and Patterson III 1990).

As discussed throughout this book, acorn size is a critical factor influencing oak dispersal as well as germination and seedling establishment. However, conflicting selective pressures may often operate on acorn size. For example, blue jays (*Cyanocitta cristata*) consistently prefer smaller acorns of some red oak

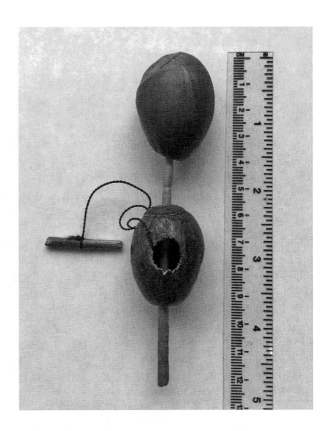

2.7. A spin toy made by a local guide, Miguel Flores, in the small enchanted town of Cuetzalan, Mexico, nestled in a cloud forest in the northeast corner of the State of Puebla, in central Mexico. The toy was constructed of acorns of *Q. germana* collected from a tree just behind Miguel's house, which he gladly showed to the author. *(Timothy Lavoie)*

species for dispersal, whereas rodents often selectively disperse larger acorns of these same species farther from their source. Similarly, Gómez (2004) found that larger acorns of the holm oak (*Q. ilex*) showed higher rates of both germination and seedling survival but these larger acorns were also highly preferred by post-dispersal seed predators. There is much yet to understand about how acorn size influences patterns of dispersal and regeneration, but for sure, this one characteristic has much to do with the ecological processes surrounding oak dispersal.

In the next chapter, I move to the life cycle of the oak, with a specific emphasis on acorn development, as this life stage directly relates to the interactions with the insects, birds, and mammals that feed on and disperse the oak fruit.

THE OAK
LIFE CYCLE 3

INTRODUCTION

Herein I briefly review the life cycle of the oak from flower and pollen production, to acorn maturation, primary dispersal of the acorn to germination in the cache, seedling establishment, and growth of the sapling and
the adult tree. An emphasis will be placed on the acorn, its anatomy, and
the markedly different germination strategies of red oak versus white oak
species. I will also review several discoveries by the author and associates
demonstrating differences in germination morphology in North American
and Asian oaks that suggest alternative adaptive strategies for tolerating
acorn pruning by rodents. I will emphasize why acorns, technically fruits,
serve the oak as a fruit—as much as a seed—in the oak's life cycle. I will
also review my lab's research showing how and why the acorns typically
house far more energy than is needed for germination and establishment
and how many oaks are well adapted for both tolerance and escape of seed
(acorn) predators. This discussion foreshadows more detailed investigations of acorn-animal interactions in Chapters 5–11.

THE MALE OAK FLOWER

To illustrate the oak life cycle, I review some basic biology and cite examples typical of oaks in the eastern deciduous forests of North America where the conditions influencing flower maturation, fertilization, and
acorn development are reasonably well studied. For many oaks, especially
those found in non-temperate forests of Asia, Mexico, and Central America, such details are not well understood. I begin this review of the life cycle

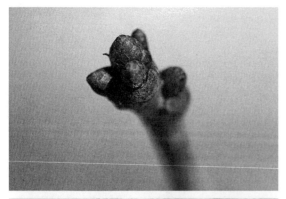

3.1. Mixed buds of the male flower of the white oak (*Q. alba*). These buds will develop at the terminal end of the branches a full year before maturing into pollen-producing catkins and primordial leaves. *(Suzanne Mrozak)* SEE COLOR PLATE

3.2. Catkins emerge from the developing male flowers and begin to release pollen on an adult southern red oak (*Q. falcata*) in northeastern Maryland on April 28, 2018. The stalks of the catkins are very long and contain many pollen-producing stamen, ideal for maximizing exposure to the wind. *(Michael A. Steele)* SEE COLOR PLATE

of the oak with the adult tree and its production of male and female flowers. The details of this flowering process provide important context for later discussions on acorn development and, in particular, aspects of acorn biology that directly relate to the dispersal process in oaks.

The oaks are monoecious, meaning each tree produces both male and female flowers. The trait of separate unisexual male and female flowers—regardless of whether they occur on the same tree or not—is referred to as an imperfect flower. Imperfect male flowers (staminate flowers) consist of the stamens, which produce and store the pollen grains in bulbous anthers. Male flowers are produced on elongated clusters called catkins or aments. The female flower (pistillate flowers) consists of the pistil, comprised of the stigma and style at the base of which is the ovary containing the ovules. Although many angiosperms have petals, in the oaks, both male and female flowers lack petals, though they do have sepals (Keator 1998).

The cycle begins with the early development of the male and female flowers—a full growing season before flowers mature and pollination begins (Sork et al. 1993a, and references therein). For example, for both the white and red oak this occurs the summer before pollination. The developing male flowers are usually located at the terminal end of the tree branches in what are called mixed buds (Fig. 3.1). Mixed buds house the male flowers that will eventually develop into the pollen producing catkins, as well as in primordial leaves that are packed around these developing male flowers. By the end of the next winter, following early development, these mixed buds begin to swell, and pollen develops in the male flowers. Later in spring—near early May in Northeastern Pennsylvania, for example—the mixed buds open and the catkins emerge, grow in length, and the anthers begin to shed pollen (Fig. 3.2). As described by Keator (1998), the catkin is a long stalk, containing numerous male flowers that originate from the stem. Each male flower on the catkin contains several stamens that, at maturity, extend into the air to maximize exposure of the pollen grains to wind. Inside the anther, pollen mother cells undergo meiosis to produce haploid pollen grains. As in other angiosperms, the haploid nucleus divides once by mitosis to form two nuclei, one of which directs the development of the pollen tube (the tube nucleus) and the other which produces the sperm to fertilize the egg (generative nucleus).

Maturation of the male flowers on the catkins and the release of pollen typically occur while leaves are still young and small thereby preventing interference of pollen movement into the air. In eastern deciduous forests, shedding of pollen typically occurs in early spring; however, year-to-year variation in precise timing of pollen release is related to weather (Keator 1998).

THE FEMALE OAK FLOWER

The female flowers are less prominent and found inside axillary buds located at the juncture of a leaf and a twig. Their arrangement can vary depending on the species. At the base of the female flower is a cluster of bracts that surround the ovary, which usually contains six ovules (Keator 1998). Following fertilization and acorn development, these bracts form the cap of the acorn. Extending from the

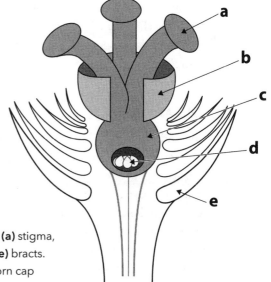

3.3. Diagram of the female oak flower showing **(a)** stigma, **(b)** location of sepals, **(c)** ovary, **(d)** ovule, and **(e)** bracts. The acorn develops from the ovary and the acorn cap from the bracts. *(Ryan Giberson)*

ovary are three styles that each extend to a stigma, the morphology of which varies by species. At the top of the ovary, surrounding the styles is a small cluster of six green sepals (Fig. 3.3).

The embryo sac inside each of the six ovules originates from a megaspore mother cell which, like its male counterpart, undergoes meiosis to form four haploid daughter cells, of which only one survives to form an embryo sac. This nucleus undergoes three mitotic divisions to produce eight nuclei. One of these nuclei becomes the egg and two become polar nuclei.

POLLINATION

The first flowering plants appeared in the fossil record about 165 million years ago (mya) in the early Cretaceous period and were dominant across the globe by 100 mya (Raven et al. 2017). The oaks are reported to have appeared by about 32–35 mya with direct ancestors of extant species appearing 23 mya (Jensen 2011). While most of the angiosperms went the direction of animal pollination—arguably a more efficient method—the oaks depend on the wind to deliver the pollen grain to the stigma of the female flower.

Oak pollination begins when the pollen grain is released from the anther and is carried to the stigma of the female flower. The pollen grain then germinates, and the tube nucleus directs the growth of a pollen tube that extends to the base of one of the ovules. The pollen tube then follows the micropyle—a narrow canal—to enter the embryo sac inside the ovule, releasing the two sperm that were produced from the generative nucleus. In oaks, the first pollen tube to reach one of the six ovules is often, but not always, the only one to fertilize and produce the new acorn (Mogensen 1975). One of the sperm in the pollen tube fertilizes the egg, which becomes the plant embryo. The other sperm fertilizes the two polar nuclei and become the endosperm that later serves as an energy source for the developing embryo.

THE ACORN

The endosperm in many flowering plants surrounds the embryo and provides an energy source for the developing seed. Because of the process of double fertilization, the endosperm is, in a sense, separate from the embryo itself. The young embryo develops two cotyledons—storage units that, in the case of oaks, are high in lipids and carbohydrates and usually low in protein. In the acorn, like in other nuts, the cotyledons are hypogeal, which means they remain as storage units below ground as the seedling germinates. In contrast, epigeal cotyledons, typical of many weeds, are the first part of the plant to emerge from the soil, shed their seed coats, and become the first photosynthetic leaves of the developing seedling. Plants that produce epigeal cotyledons must germinate and begin photosynthesis quickly and are susceptible to early pruning by herbivores. In contrast, hypogeal cotyledons, like those of the oak, take over energy storage from the endosperm early in development and then remain below ground. There they support the development of the seedling for an extended period, and buffer against the loss of a pruned seedling that can be replaced if initially lost (see below).

At the apical (bottom or distal) end of the acorn, still locked inside the pericarp until germination, is the radicle (the embryonic root) (Fig. 3.4). Behind the radicle tip is the hypocotyl, which is the stem of the germinating seedling, and the epicotyl, which will develop into the young leaves and eventually the aboveground seedling. The cotyledons are attached to the junction between the hypocotyl and epicotyl. The epicotyl and hypocotyl are derived from the plant embryo and, in effect, constitute the seed. The plumule represents the tip of the young stem.

The oak's strategy for reproduction is, in most cases, to invest considerable energy to support a single fertilized ovule. Such a significant investment in the energy of one seed is similar to that of other hardwood tree species (e.g., walnuts). Keator (1998) contrasts this k-strategy of high nutrient investment in a single ovule with the strategy employed by orchids in which the ovary of one plant contains countless ovules, none of which are endowed with an energy source for future development.

The acorn cap varies by species in structure and the extent to which it covers the body of the acorn. It remains attached to the acorn until it is shed. I have observed that if still attached after it is shed, the acorn is likely aborted or heavily damaged by insects. The adaptive significance of the cap, although not well understood, is suggested by some as a covering that protects against insect infestation (Keator 1998). I hypothesize below that the cap may interact with chemical gradients in the acorn to reduce insect damage.

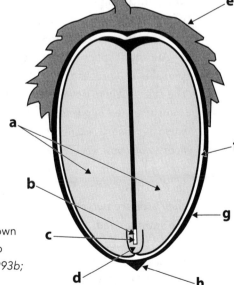

3.4. Anatomy of an acorn showing **(a)** cotyledons, **(b)** epicotyl, **(c)** hypocotyl, and **(d)** radicle. Also shown are the **(e)** cupule, **(f)** seed coat (testa), **(g)** pericarp (fruit wall), and **(h)** remains of style. *(From Nixon 1993b; Shealyn A. Marino)*

DEVELOPMENT OF THE ACORN
AND IMPLICATIONS FOR OAK-ANIMAL
INTERACTIONS

The previous discussion of the basic biology of the oak flowers, pollination, and acorn formation is a bit tedious but necessary. Beginning with pollination through fertilization and the development of the acorn—to even the germination of the acorn—the story quickly becomes directly relevant to the process of seed dispersal and seedling establishment in the oaks. Development of the pollen tube, for example, differs significantly between the red oaks (section *Lobatae*) and the white oaks (section *Quercus*). In the latter, the pollen tube grows quickly, one of the ovules is penetrated, fertilization is complete, and the acorn develops and matures in about six months—typically in the autumn following pollen production. For the red oaks the process is slower and at the end of the growing season, growth of the pollen tube as well as the development of the ovules ceases. And, as described by Keator (1998), the final stages of pollination do not resume until the following spring. The result means that it takes about 18 months from fertilization to the maturation of red oak acorns. Little is known about either the precise proximate control of this process or the ultimate factors underlying the origin of these differences between the red oaks and the white oaks. Keator (1998) suggests that possibly environmental unpredictability led to the evolution of this delay in fertilization in the red oaks. These differences in development of the pollen tube and the final stages of fertilization in the two oak groups translate into profound effects on the ecology of oak-animal interactions. For example, the extent to which weather (e.g., late spring frost) influences initial flower development and pollination will translate into mast failure of the white oaks in the same year and crop failure of red oaks the following year for trees in the same forest (see Chapter 4).

Another aspect of acorn development that can directly influence acorn dispersal is the process of ovule abortion, or more specifically, the failure of it. In the oaks, early in development, there are six ovules, five of which typically abort soon after the egg in one is fertilized (Mogensen 1975). However, this is not always the case. In some situations, perhaps more often than we now even realize, acorns of some species will occasionally produce two or even three developing embryos (seeds). As Dr. Amy McEuen and I reported, at least 14 species of oaks in North America, on occasion, produce acorns that are multi-seeded; that is, they produce two or more embryos in a single acorn (McEuen and Steele 2005). We summarized, from our own observations and those of at least 10 other authors, the occurrence of multiple-seeded acorns in 14 species of oaks: eight red oak and six white oak species. This included five species found in Mexico or Central America, seven species from North America (north of Mexico), and two species introduced to the United States, one from Europe, and one from Asia (McEuen and Steele 2005).

Multiple-seeded acorns are relevant to oak-rodent interactions because squirrels of several genera often excise the seed embryo (the hypocotyl, epicotyl, and radicle) at the apical end of the acorn. This is primarily performed on the early germinating acorns of white oak species. By scraping a hole with their incisors in the apical end of the acorn, squirrels are able to arrest germination in nondormant white oak acorns. This prevents germination and kills the acorn, but, for the squirrel, it extends storage time from a few weeks up to six months (see Chapter 6). However, when acorns produce more than one embryo, the additional embryos consistently develop in another part of the acorn—anywhere from the side to the basal end of the acorn. We hypothesized that these multi-seeded acorns may provide an advantage when squirrels perform embryo excision at the apical end of the fruit (McEuen and Steele 2005). Both embryo-excised acorns found in the field and those in which we simulated embryo excision in the lab often successfully germinated as long as they had a second embryo hidden in the acorn.

Although the production of multiple-seeded acorns is an advantage to the oak when it sustains damage to the apical embryo, it does not necessarily follow that this is a direct adaptation to the squirrel's behavior of embryo excision. Or is it? If this were the case, we would expect that multiple-seeded acorns should only be found in nondormant white oak species. Yet, we found the trait to occur in both white and red oaks. When one also considers the effects of pre-dispersal insect predation and the behavior of radicle clipping of germinating acorns by numerous rodents, and the occasional embryo excision when red oaks begin to germinate in squirrel caches in the spring (Steele et al. 2001b), it becomes clear that multi-seeded acorns may be an advantage for both white oaks and red oaks under varying circumstances. Clearly more research is needed on this relationship between seed predators and the oaks and why embryo excision across the globe is performed nearly exclusively by squirrels. The only other rodent that we have observed performing any behavior close to this is the Central American agouti (*Dasyprocta punctata*), which severs the apical tip along with the embryo.

Another characteristic of the acorn that influences patterns of seed dispersal and the probability of caching and seedling establishment is the size of the fruit. Acorn size follows directly from cotyledon size. Across species, acorns can vary in size from less than 1 g (e.g., pin oak, *Q. palustris*) up to 14 g (e.g., *Quercus miyagii* = *salicina*). Even within a species, acorn size can vary tremendously; for example, northern red oak (*Q. rubra*) acorns can range in size from 2 g to 6 g although size is generally similar in the same tree. From the oaks' perspective it seems logical to conclude that the size of the cotyledons lends itself directly to a greater energy source for the seedling and a higher probability of seedling survival.

Although partly true, there appears to be much more to seed size than just feeding the seedling. First, as I discuss in detail in Chapter 6, larger acorns are more likely to be cached and dispersed farther—at least by rodents—than smaller acorns. Eastern gray squirrels selectively disperse larger acorns farther and, under many circumstances, trade-off greater predation risks when selecting cache sites for these acorns of higher value (Steele et al. 2014). This advantage of larger acorn size for dispersal, however, should not be assumed for all oak species. For example, smaller seeded oak species, as well as smaller acorns of the same species, are often selectively dispersed by jays (Bartlow et al. 2011).

Acorn size also appears to significantly influence the fruit's ability to tolerate partial seed damage by seed predators. Both insects and vertebrates often inflict only partial damage to acorns, especially when acorn crops are heavy. This partial damage is also consistently directed at the basal end of the acorn, away from the embryo. Chemical gradients in the cotyledon of some acorns render the basal end of the acorn higher in lipid and lower in tannin, and thus higher in energy, more palatable, and easier to digest (see Chapter 6). To the extent that these chemical gradients are a general pattern found in many oak species worldwide, it follows that larger cotyledon size may allow some oaks to tolerate partial damage to the seed predator. To test this, we simulated the kind of damage typically inflicted by rodents, birds, and insect larvae on six species of oaks in the Central Hardwoods Region of North America (Bartlow et al. 2018a; see Chapter 6). We found that some acorns of all six species (25–60% of acorns) were able to sustain more than 50% cotyledon damage (removal) and still germinate and establish. Within each species, larger acorns sustained damage better than smaller acorns. These simulations suggest that seed size plays an important role in resistance to insect and rodent damage (Yi et al. 2015). Similar studies in Asia, led by my colleague, Xianfeng Yi, showed that partial rodent damage of *Q. mongolica* acorns occurs more often during a mast crop, and that these damaged acorns frequently germinate and show enhanced root and shoot growth compared to whole, intact acorns (Yang and Yi 2012). Similarly, Yi and Yang (2010b) suggested that larger *Q. aliena* acorns may serve to satiate weevils, especially in years when mast crops are lower.

Beyond cotyledon size, other aspects of the acorn's anatomy also may play a role in helping to escape rodent embryo excision. For example, the position of the plumule in at least one species of oak (*Q. mongolica*) appears to be far deeper in the center of the fruit than that of at least five other early germinating oaks in Asia and, as far as I am aware, the white oaks in the Americas. The position of the plumule appears to allow this species to escape any attempt of embryo excision by squirrels in its range (Zhang et al. 2014b). Finally, it has been suggested that pericarp thickness varies with the sites weevils typically select for ovipositing (Yi and Yang 2010b). These authors report that in *Q. variabilis* pericarp thickness is strongly negatively correlated with weevil oviposition sites, weevil larvae in the cotyledon, and holes in the pericarp where weevils have exited the cotyledon. This suggests that a thicker pericarp in the apical portion of the acorn may discourage weevil predation near the plumule (seed).

Survival of the acorn to germination and seedling establishment is indeed the most precarious of the life stages of the oak. Even if the acorn escapes pre-dispersal insect predation and fungal damage, the probability of dispersal and establishment is often quite low. I have observed nearly 100% of sound acorns not making it to germination in most years of low mast, due almost entirely to rodent predation. In fact, in one experiment during a low mast year, I documented predation of 8000 acorns by rodents in less than 48 hours. Even in a high mast year, or in experiments in which we simulated high mast, we have observed <5% of acorns producing seedlings that made it through dispersal, establishment, and the first year of growth.

For the reasons outlined in Chapter 1, even if an acorn successfully germinates under the parent tree, the likelihood of it surviving much beyond the first year is bleak. This is why dispersal to a site more suitable for initial establishment and growth is so critical. In addition, if buried or cached, the acorn is far more likely to germinate and produce a seedling (Davis et al. 1991). In a recent study under the direction of my colleague, Dr. Harmony Dalgleish, several of us quantified germination success of red oak, white oak, American chestnut (*Castanea dentata*), and hybrid chestnut (*C. dentata* × *C. mollissima*) on the ground surface, under leaf litter, and below soil surface in artificial caches at several sites in the eastern US. All species experienced significantly higher rates of germination and seedling establishment in artificial caches than on the soil surface. Although caching was absolutely necessary for germination and seedling survival of red oak, it often increased probability in the other two species as well (Sawaya et al. 2018; see Chapter 8).

ACORN DORMANCY AND GERMINATION SCHEDULES: IMPLICATIONS FOR ACORN PREDATION AND DISPERSAL

When it comes to acorn germination, especially as it relates to acorn predation and dispersal by scatterhoarders, all acorns are not the same. One of the most obvious differences in germination schedules is evident in the red oaks (section *Lobatae*) compared with the white oaks (section *Quercus*). Nearly all white oak species germinate immediately following maturation, provided that conditions are suitable. In most cases, this simply means that they have been dropped from the tree and are able to absorb moisture, begin radicle development, and extend the radicle into the soil. Such is the case in more northerly regions, where species such as *Q. alba* and *Q. macrocarpa* germinate after seed fall when they are under leaf litter in close association with the soil. However, for some white oak species in more southernly portions of their range (Florida and Central Mexico), acorns may even germinate while still attached to the tree, thus gaining a head start before the acorn drops to the ground. Often this is a race to ensure establishment of the radicle in the soil to overcome drought conditions that would otherwise kill the seed (acorn).

In contrast to white oak acorns, those of red oaks exhibit delayed germination. In northern latitudes, acorns of red oak species (e.g., red oak, black oak [*Q. velutina*], and pin oak [*Q. palustris*]) overwinter before germinating. Germination in these species requires cold stratification—extended exposure of the embryo to low temperatures—before the seed (embryo) begins the germination process. Typically, temperatures <4°C (39°F) for a few months will break dormancy, which means most red oak species overwinter before germinating. Even in temperate, montane forests of Mexico, stratification is necessary for most red oak species, although there, it is often reduced in length and severity. In these more southernly regions, stratification may take a different form.

This dramatic difference in germination schedule in the red oaks versus the white oaks has a profound influence on the caching decisions of the rodents that disperse these oaks. Germination in nondormant white oaks, such as *Q. alba*, result in the rapid transfer of energy from a highly utilizable form of nutrients in the cotyledon to a taproot high in cellulose that is no longer of any value to the squirrel (Fox 1982; Steele and Smallwood 2002). Gray squirrels selectively cache acorns of red oak species over those of white oaks because of the reduced perishability (higher storability) in the former (Hadj-Chikh et al. 1996; see Chapter 6). In fact, we found that despite other chemical differences between red oak and white oak species (higher tannin and lipid levels in red oaks), it is the reduced perishability due to delayed germination that specifically drives the selective caching of red oak acorns over those of white oak (Hadj-Chikh et al. 1996). And, as described above, on those occasions when squirrels cache white oaks, they often first excise the embryo, arresting early germination and extending the storage capacity of these otherwise perishable foods. Whereas many rodents selectively cache dormant red oaks over the more perishable white oaks, it is only the squirrels of a few genera in North America and Asia that are genetically equipped for the behavior of embryo excision (see Chapter 6).

The ability of squirrels to detect dormant versus nondormant acorns likely follows from both the physical and biochemical changes in the cotyledon and the pericarp that ultimately result as the acorn begins to break dormancy (Vogt 1974; Bonner and Vozzo 1987; Steele et al. 2001b; Johnson et al. 2009; see review by Sundaram 2016). In Steele et al. (2001b), for example, we reported that when presented with artificial acorns in which white oak cotyledon was sealed inside the pericarp of a red oak acorn, gray squirrels selectively cached these acorns.

And when the pericarps of red oak acorns were first soaked in acetone, potentially removing any plant wax or other chemicals on the pericarp, squirrels selectively ate these acorns. These results led us to suggest that squirrels detect dormancy or nondormancy based on the condition of the pericarp and that when squirrels are able to detect odors from the kernel, they perceive the acorn as no longer dormant. More recently, my colleague Dr. Mekala Sundaram (2016) used light microscopy, scanning electron microscopy, and chemical analyses (e.g., gas chromatography-mass spectrometry) to evaluate the structure of the pericarp of both dormant and nondormant acorns. She found that dormant acorns were coated with a plant wax that appeared to peel at the time of germination. She also found that dormant acorns coated with compounds from the kernel or chemicals associated with germination (acetaldehyde) were consumed by squirrels.

These results suggest that although there are a number of biochemical changes that occur when dormancy breaks (Vogt 1974; Bonner and Vozzo 1987; Sundaram 2016), it is simple changes in the physical structure of the pericarp that now allow moisture to enter the kernel and for seed predators to detect that the seed is now ready to germinate or be eaten. Although it is assumed by some that dormancy in red oak species is a fundamental plant trait that evolved during harsh environmental conditions (Keator 1998), it is also plausible that this strategy allowed dormant red oaks to escape rodent seed predation. Recent research suggests that seed hardness in many plant species evolved specifically to prevent detection by mammalian seed predators (Paulsen et al. 2013, 2014). Certainly, this opens a new discussion and an avenue of research regarding the evolution of rodent-oak interactions and the possibility that dormancy in oaks arose as result of strong selection by seed predators. At the very least, seed dormancy serves as a central issue in determining patterns of seed predation and dispersal in extant species of oaks across the globe.

GERMINATION AND SEEDLING ESTABLISHMENT

Once moisture makes it across the pericarp, the seed begins to germinate. The first evidence of this is the protrusion of the radicle from the apical end of the acorn, which then quickly extends into the soil where it is able to extract water and nutrients. The development of the radicle also represents a rapid transfer of energy from the highly nutritious and edible cotyledon to mostly cellulose in the radicle, which is no longer of any use to seed predators—a characteristic that many authors argue likely evolved for this very reason (Lewis 1911; Barnett 1977; Fox 1982; Vander Wall 1990; Steele et al. 2005; Steele 2008). This transfer of energy into the radicle is particularly rapid in the white oaks (Q. alba), which often experience intense seed predation especially in the presence of red oaks or American chestnuts, both of which exhibit dormancy (Lichti et al. 2014). However,

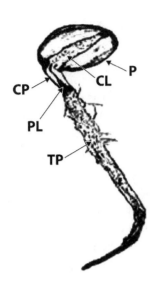

3.5. Anatomy of germinating acorn of *Quercus montana*. Shown is a germinated acorn with cotyledon (**CL**), cotyledonary petiole (**CP**), plumule (**PL**, embryonic stage of the epicotyl), taproot (**TP**), and pericarp (**P**). *(From Yi et al. 2013b)*

the rapid formation of the radicle also allows the acorn to ensure water uptake and thereby thwart the effects of temporal and spatial periods of drought, so a full understanding of the adaptive significance of this trait is still open to question.

Sometime after the radicle extends into the soil, the epicotyl pushes upward and at the same time develops the first leaves of the aboveground shoot and the establishing seedling. The young leaves soon produce chlorophyll and are able to photosynthesize. Although it is often assumed that the acorn is an essential energy supply well into the first year of the seedling's life, there is considerable evidence that this perspective may be exaggerated, or in many cases simply not true. In addition to supporting the young seedling, the acorn serves a number of functions related to tolerance of partial seed predation and promotion of dispersal (Bartlow et al. 2018a). In fact, there is now strong evidence that oak seedlings may survive loss of the acorn shortly after germination and establishment of the radicle and the budding epicotyl (Yi et al. 2019).

As white oak species rapidly germinate and extend their radicles into the soil, rodents (Yi et al. 2012b, 2013a) and jays (Bossema 1979) will frequently counter this by pruning the acorn and leaving the radicle and epicotyl intact (Yi et al. 2013a). It appears that successful removal of the acorn is only possible because of an anatomical structure, the cotyledonary petiole—an extension of the two cotyledons that pushes the plumule beyond the apical end of the acorn (Fig. 3.5). This separation of the acorn from the plumule—the point at which the epicotyl and radicle diverge—offers an ideal location where seed predators can easily snip off the acorn and leave the seedling undisturbed. First described in detail by Lewis (1911) in germinating acorns of *Q. virginiana*, the cotyledonary petiole of this species is particularly exaggerated. Germination begins first with fusion and the extension of the petioles of both cotyledons for several centimeters before the radicle grows more than a few centimeters. This is then followed by rapid growth of the radicle, followed by extensive thickening of the radicle, and only then, six weeks after initial germination, growth and emergence of the epicotyl. The acorn becomes well separated from the seedling, the radicle becomes both an energy and water source, and the epicotyl, soon thereafter, becomes a long-term energy source.

At this point it appears the acorn can be harvested by seed predators without killing the seedling. Both lab and field experiments with seedlings of chestnut oak (*Q. montana*) and other white oak species demonstrate that these seedlings are able to consistently survive acorn pruning (Yi et al. 2012b, 2013a). The cotyledonary petioles in some white oak species can extend close to 3 cm in length or farther (Lewis 1911; see Chapter 7). And, whereas the cotyledonary petiole is substantially larger in many white oak species, compared with red oaks, some red oaks exhibit the structure as well. Interestingly, we have observed similar acorn pruning in northern red oaks during their germination (Yi et al. 2013b).

The final fate of the seedling depends to a large extent on where it finds itself. Competition for water and light and herbivory by insects and/or vertebrates, such as white-tailed deer (*Odocoileus virginianus*), can easily seal its fate. Although few studies have followed seedling fates, those that have find seedling mortality rate can, under some circumstances, even exceed that of acorns (Davis et al. 1991). Seedlings, however, have the ability to "die back" due to water stress, spring frosts, lack of suitable light, and other stressors, but then resprout the following year from dormant buds anywhere along the stems or at the protected root collar (Johnson et al. 2009). Although the oaks generally do not form seed banks in the soil, this regenerative ability of seedlings allows the oaks to form seedling banks.

THE SAPLING AND THE ADULT TREE

Given that my review focuses primarily on acorn dispersal and seedling establishment, I close this chapter with only a brief reference to the sapling and adult stages of the oak life cycle. If the seedling survives a few years and makes it to the sapling stage, its probability of mortality declines considerably. Likewise, the adult tree, once established, maintains providence. The time to first acorn production varies with species and with the various stressors faced by the seedling and sapling stages. For example, I am aware of a Garry oak (*Q. garryana*) on Vancouver Island that first produced acorns at eight years and a red oak that took approximately 15 years until the first acorns appeared. Both of these trees were maintained at ideal growing conditions. In contrast, I have also followed a white oak growing in less than optimal conditions (high soil moisture) remain less than 2 m in height and never producing acorns before age 17.

For additional information on the sapling and adult oaks, see the exhaustive review by Johnson et al. (2009) on "The Ecology and Silviculture of Oaks." This review provides detailed treatment of the regeneration ecology of oaks, including population dynamics, site productivity, effects of disturbance on regeneration, as well as data on annual growth and phenology in sapling and adult trees. Although much of Johnson et al. (2009) is dedicated to various silviculture practices that maximize growth and regeneration potential for the primary benefit of humans, the volume accomplishes this in the context of an extensive review of the basic research on sapling and adult oak biology as well.

MASTING IN OAKS 4

INTRODUCTION

What is masting or mast seeding? What causes it? How does it relate to oak dispersal? And, how does it affect oak forest ecology? Masting is highly variable seed production across years that is highly synchronized among trees of the same species. I open this chapter with an explanation of the well-known characteristic of masting in oaks, the likely evolutionary explanations for why it occurs, and some of the proposed mechanisms for how acorn production is often synchronized widely among individual trees. I then review a few of the animal species in eastern deciduous forests of North America and elsewhere that depend on oak mast for food. I follow with an overview of a 16-year study in which my students, collaborators, and I have monitored acorn production and insect seed predators, as well as small populations in oak forests of Pennsylvania. This study demonstrates the importance of oak mast as a keystone resource—a food source that profoundly influences a vast array of ecological interactions between acorn consumers, predators, parasites, and pathogens in today's forests. I highlight the importance of acorns as a periodic energy pulse that drives ecosystem function. And, I briefly review the seminal discoveries by Richard Ostfeld and associates that demonstrate how oak mast influences the spatial and temporal variation of Lyme disease. Finally, the importance of oak mast as a keystone resource is placed in the context of the catastrophic loss of the American chestnut nearly a century ago.

WHAT IS MASTING?

One of the most notable characteristics of oaks is that they produce large seed crops in some years, and in others, barely an acorn can be found. This is most evident to those that study such "tree behavior," but you do not need to be a scientist to witness this. Ask anyone with a mature oak in their yard. However, this pattern of seed production is not limited to individual trees but occurs over large geographic areas. In other words, the process of masting—the production of episodic seed crops in some years, followed by one or more years of low mast or mast failure—is widely synchronized across individual trees of the same species and even across oak species, especially those of the same section of *Quercus*. This synchronization of the acorn crops of individual trees often occurs across an extensive area. Koenig and Knops (2005), for example, report that in the blue oak (*Q. douglasii*) of California, crop synchronization frequently occurs across the entire 20,000 km² range of the species. Although common in the oaks, masting, or mast fruiting in trees, is not unique to oaks and occurs across the globe in temperate, tropical, and boreal forests in a number of species (Kelly 1994; Koenig et al. 2015, and references therein).

Over the past 20 years, the study of masting behavior has received considerable attention. From those that have contributed to the subject, we have learned quite a bit on why it occurs (i.e., the evolutionary factors driving the phenomenon), how masting occurs (i.e., the proximate mechanisms controlling the phenomenon), and the ecological impact of the phenomenon, which can include a diversity of cascading effects in forest ecosystems. Although many questions remain unanswered, a fascinating picture has emerged. One suggests that mast seed production is a major force driving ecological processes in plant communities. In the oaks, mast production is key for acorn dispersal and seedling establishment, but it also results in a number of cascading effects in oak forest systems that influence a range of ecological processes that feed back into the dispersal process.

WHY DO OAKS MAST?

To ask why oaks mast is to ask what is the adaptive significance of the masting phenomenon? In other words, does masting lead to greater oak reproductive success in the form of acorn survival and germination, seedling establishment, and ultimately, another reproducing oak tree? Masting in plants in general is thought to result from one or more of three advantages: predator satiation, dispersal efficiency, or pollination efficiency (Koenig et al. 2015). The latter of these, proposed by Kelly and Sork (2002), suggests that wind pollinated species, especially those that rely on outcrossing (e.g., oaks), should produce flowers and pollinate all at the same time to maximize pollination efficiency. When the energy is available to produce large crops of flowers, pollination efficiency ensures large mast production in oaks and may help to synchronize seed crops over extensive geographic areas. However, the likely ultimate (evolutionary) explanation for masting is captured in the predator satiation hypothesis (PSH). Proposed by Janzen (1971), the PSH suggests that masting allows a plant species to satiate seed predators in high mast years, thereby allowing many seeds to escape predation and survive to germinate and establish. Low mast years, in contrast, would be expected to also benefit the plant by reducing seed predator populations through starvation. Despite its popularity, however, there is a catch when applying the PSH to the oaks. Whereas satiation of seed predators alone would facilitate some plants' reproductive success (e.g., those that are wind dispersed), this is unlikely the case for plants (e.g., oaks) that depend on animals to move their seeds. This led to the dispersal effectiveness hypothesis (DEH), basically an addendum to the PSH, as a possible third explanation.

Table 4.1 summarizes the majority of studies on oaks that have attempted to address the evolutionary basis of masting in oaks either via the PSH or the DEH. Also included is a more recent and limited hypothesis based on seed size (SSH) that does not fully account for variation in crop sizes. The SSH posits that larger individual acorns of any oak species will result in greater tolerance of partial seed predation by pre-dispersal seed predators such as insects. The ability of acorns to tolerate partial seed damage by insects (and vertebrates) is an important strategy in oaks, facilitated primarily by chemical gradients in the cotyledon that render the basal half of the acorn higher in energy and more palatable than the apical half, containing the embryo (see Chapter 6). The result is partial acorn consumption and subsequent survival and germination of some of these partially eaten acorns, especially when acorns are abundant. The SSH suggests that larger individual acorns, which likely come at a cost of producing smaller acorn crops, may increase tolerance to predator damage. We also know that larger acorns are more likely to be dispersed farther and possibly cached in sites more suitable for germination and establishment (Steele et al. 2014; see Chapter 6). Hence the SSH holds some merit as a secondary strategy of oaks nested within the broader context of masting.

The other two hypotheses (the PSH and the DEH), often treated independently throughout the literature, should be considered hand in hand for the oaks. Satiation of seed predators alone is not likely in most cases sufficient for ensuring oak regeneration, if acorns are not dispersed and cached as well. Hence, Table 4.1 illustrates some of the inherent flaws in the published literature to date. First, the majority of studies in oaks focus specifically on predator satiation of primarily insects. Of the 22 studies highlighted in the table, only four test the DEH and only seven consider the effect of acorn abundance on

Table 4.1. Overview of hypotheses to explain potential satiation effects on the predators and dispersers of acorns. Shown are various topics covered and whether data are significant (S) or not significant (NS) in support of the predator satiation hypothesis (PSH), seed size hypothesis (SSH), and the dispersal effectiveness hypothesis (DEH).

AUTHOR AND PUBLICATION YEAR	SPECIES	LOCATION	PREDATION	DISPERSAL	GERM.	ESTAB.	INSECTS	RODENTS	BIRDS	PSH	SSH	DEH
Bartlow et al. 2018a	6 spp.	US/Asia	X		X	X	X	X	X	S		S
Bogdziewicz et al. 2018d	Q. ilex	Spain	X				X			S/NS depending on habitat		
Crawley and Long 1995	Q. robur	England	X		X	X	X	X	X	S/NS		
Espelta et al. 2008	Q. humilis, Q. ilex	Spain	X		X	X	X			S		S
Espelta et al. 2009b	Q. humilis, Q. ilex	Spain	X		X	X	X			S		NS
Fukumoto and Kajimura 2011	Q. serrata, Q. variabilis	Japan					X			NS		
Higaki 2016	Q. acutissima	Japan	X				X			S		
Koenig et al. 1994	several species	Calif., US	X					X	X	S		
Maeto and Ozaki 2003	Q. crispula = mongolica	Japan	X				X			NS		
Moran and Clark 2012b	Q. rubra, Q. velutina, Q. falcata, Q. coccinea	S US		X						S limited to indiv of each spp.		

Moore et al. 2007	Q. rubra, Q. palustris	C and E US	X	X	X				S, NS	S, NS	S, NS
Perea et al. 2011c	Q. pyrenaica	Spain	X	X	X				S		
Pérez-Ramos and Maranon 2008	Q. suber, Q. canariensis	S Spain	X	X	X	X	X		S	S	
Schnurr et al. 2002	Q. rubra	NE US	X	X	X	X			S/NS		
Sork et al. 1993a	Q. alba, Q. rubra, Q. velutina	Mid US	X	X					S		
Tong et al. 2017	Castanopsis sclerophylla	China	X	X	X				S at tree level		
Xiao and Zhang 2006	Lithocarpus harlandii	SW China	X	X	X	X			S		
Xiao et al. 2017	Q. serrata	SW China	X	X					S, NS		
Xiao et al. 2005b	Castonopsis fargesii	SW China	X	X	X	X			S, NS	NS	
Xiao et al. 2007	Quercus variabilis, Q. serrata, Cyclobalanopsis glauca	SW China	X	X	X	X			S		
Yang and Yi 2012	Q. mongolica	China	X	X	X	X			S		
Yi and Yang 2010b	Q. aliena	China	X	X	X				NS		

either rodent abundance or rodent or jay behavior (acorn dispersal). Regardless, it is clear from independent studies focusing on the impact of acorns as a pulsed resource (see extensive discussion below), that rodent populations usually spike sometime later following acorn masting events and crash following mast failures. Given the role of rodents in moving acorns, it might follow that these observations therefore support both the PSH and the DEH.

There is also increasing evidence that weevils (*Curculio* spp.)—the most significant insect seed predator of acorns—may not be directly influenced by variation in crop size as predicted by the PSH. Fukumoto and Kajimura (2011), for example, found that neither populations of weevils, nor that of the tortricid moth (a generalist insect seed predator) were related to acorn crop size. For weevils, they attributed this to the ability of *Curculio* to vary its length of diapause in the soil. In other words, weevils may be able to time their development to avoid mast crop failures. Similarly, my colleagues and I analyzed patterns of weevil infestation in three oak species over a 17-year period and found evidence that in at least two white oak species (white oak and chestnut oak), weevils were able to avoid the effects of satiation by aggregating and increasing reproduction at seed-rich trees (see below) (Bogdziewicz et al. 2018c).It is entirely possible that while the PSH may apply to vertebrate acorn predators such as rodents, it may not have the same effect on the insects.

Although the seed size hypothesis (SSH) is included in Table 4.1, it is somewhat peripheral to the other two hypotheses and by itself generally unlikely to explain defense against seed predators in most cases. This is because while seed size operates at the seed and tree level, it should always be considered in the context of acorn abundance as well (i.e., the population level). Nevertheless, larger acorns often allow the individual acorn to better survive partial acorn consumption by insects such as weevils. Yet, because the tree is limited in the energy available to produce both high numbers of acorns and large acorns, these traits should all be considered simultaneously when evaluating the overall strategy. It should also be noted that in small-seeded oaks, smaller individual acorns are often less preferred by insects (Espelta et al. 2009a). Perhaps in the oaks, larger seed size is directed at the insects and masting at the dispersal agents.

Many authors believe the most plausible explanation for masting in oaks is captured in the PSH, assuming increased survival of seed predators also translates into an increase in dispersal (Crawley and Long 1995; Koenig and Knops 2005; Xiao et al. 2013b). However, the evidence to support the PSH/DEH in oaks is not unequivocal. Based on a number of studies from across the globe (Table 4.1), a closer look suggests that the PSH may apply to oaks only under specific situations. Crawley and Long (1995), one of the first to test this hypothesis, examined variation in acorn crops of *Q. robur* over a 15-year period and assessed patterns of seedling recruitment in response to acorn abundance. They found that oak establishment does correspond with acorn crops at some sites (e.g., where seed predators are fewer) but not others where combined rates of seed predation are high, establishment sites are limited, and leaf damage due to herbivory by rabbits was high. Crawley's and Long's results, therefore, suggest that support for PSH/DEH is spatially variable and highly context dependent.

Research from my lab, those of my collaborators, and other studies also point to several situations in which higher mast abundance may not always translate into higher rates of seedling establishment. In contrast, low mast years universally result in high rates of acorn predation and virtually little or no acorn survival (Steele et al. 2001a; Moore et al. 2007); the opposite does not always occur in high mast years. During bumper years, acorn dispersal and scatterhoarding are common but can differ in the manner in which acorns are dispersed. In Moore et al. (2007), for example, we show that when acorns are abundant, dispersal distances are significantly shorter, which is likely to limit rates of successful establishment. In other words, during heavy mast years, rodents appear to trade-off better cache sites with the time to cache more seeds. They invest far less time per acorn in caching, which may overall contribute to lower rates of seedling establishment.

Second, although we observed higher rates of seed caching in mast years in today's forests, many of these emerging seedlings are lost to herbivory by the end of the first growing season (see Chapters 6 and 7). Thus, the abundance of herbivores, which is often independent of acorn crops, can significantly affect seedling establishment even if acorns escape cache recovery.

Third, acorn abundance may influence the recruitment success differently in different species. As I discuss in Chapters 6 and 7, early germinating acorns of white oak species are at a significant disadvantage if they mast at the same time as red oaks, which are more highly preferred for caching because of their delayed germination and reduced perishability (Hadj-Chikh et al. 1996; Smallwood et al. 2001).

Finally, Schnurr et al. (2002) showed that different species of masting trees will have a variable effect on the behavior of different rodent species. As emphasized by Moore et al. (2007), large sample sizes across numerous sites are necessary in order to understand how acorn abundance (masting) and related factors interact to influence oak dispersal and establishment. High acorn abundance does not always translate into extensive oak regeneration, but it is the only situation in which regeneration is likely to occur.

Throughout this book, I return to the concepts of the PSH-DEH several times as they are central guiding principles for understanding the origin of masting in the oaks.

HOW DO OAKS MAST?

In the past two decades, considerable attention has been devoted to understanding the proximate mechanisms underlying the masting phenomenon in oaks and other woody species. Although a complete explanation is still on the horizon, the established investigators that have pursued it have made considerable progress. For the general reader, one of the best earlier reviews on the subject (Koenig and Knops 2005) highlighted the possible drivers of both crop variability and synchronization of seed crops. The most recent and extensive studies that have further explored these proximate mechanisms of seed production can be attributed to Dr. Walt Koenig (and members of his laboratory at Cornell University), as well as Dr. Michal Bogdziewicz (Adam Mickiewicz University, Poland), Dr. Elizabeth Crone (Tufts University), and several others. Here I briefly review some of these mechanisms which, in most cases, must be considered jointly to explain masting behavior, especially in the oaks.

Masting in oaks, as in many other tree species, occurs at two levels. It occurs at the level of the individual tree that produces high crops in some years, followed by years of little or no acorn production. However, masting also occurs at the population level, as crops are synchronized across individuals over an extensive geographic area. Indeed, for predator satiation to be successful, synchronization must occur at the population level. Thus, to understand masting, it is important to understand mechanisms at both of these levels (Crone and Rapp 2014; Pesendorfer et al. 2016a; Bogdziewicz et al. 2017c). In the individual tree, it is hypothesized that the energy needed to produce numerous flowers and a heavy acorn crop is costly for the tree, limiting energy for growth and preventing heavy crops in back-to-back years (Sork et al. 1993a).

But how are crops synchronized? One of the first hypotheses to potentially explain this, the resource tracking hypothesis, suggested that widespread weather patterns influenced energy availability in trees and at the same time synchronized crop production; however, as Koenig and Knops (2005) explain, this idea was not supported by a number of studies.

The resource budget hypothesis, alternatively, suggested that individual trees must acquire or mobilize a minimum level of energy to turn on reproduction and that other mechanisms synchronize seed production across individuals (Isagi et al. 1997; Satake and Iwasa 2000). One set of factors that has been proposed to aid in synchronization is pollen limitation, which can be controlled by one of two mechanisms: pollen coupling or the pollination Moran effect (Pearse et al. 2015; Bogdziewicz et al. 2018a). The pollen coupling hypothesis suggests that pollination success, and thus fruit production, depends on the annual production of flowers by trees, which is determined by both weather and the trees' stored resources (Bogdziewicz et al. 2017a). In contrast, the Moran effect suggests that weather alone, which can be correlated over a large geographic area, drives the success of pollination. This second hypothesis suggests that acorn production in any given year is dependent on weather and less so on the density of pollen.

Pesendorfer et al. (2016a) developed a model to explain the 35-year pattern of acorn production in California's valley oak (*Q. lobata*). They found that patterns of acorn production in this species followed from resource availability in individual trees coupled with pollen limitation resulting from variation in flowering phenology (Koenig et al. 2012, 2015). In other words, phenological synchronization of flowers corresponded with higher acorn production. As another example of a study that tested this, Bogdziewicz et al. (2017a), working with a 19-year data set of beech (*Fagus sylvatica*) and two oak species (*Q. robur* and *Q. petraea*) in Poland, found that pollen abundance best correlated with seed production in beech, whereas acorn production in the oak species appeared to be associated with weather conditions that often resulted in short periods of pollination and good seed set, thus supporting the pollination hypothesis or Moran effect (Bogdziewicz et al. 2017a). Although numerous other studies have explored the mechanisms that drive patterns of acorn production (e.g., Koenig et al. 2012, 2015; Pesendorfer 2016a), overall there is no consistency in the mechanisms that appear to contribute to synchronization (Crone and Rapp 2014; Pearse et al. 2015, 2016). In one of the first experiments of its type, Pearse et al. (2015) supplemented female flowers of *Q. lobata* with pollen and found that in one year, but not the other, this manipulation resulted in a significant increase in acorn production. They also followed female flowers over several years and found that many flowers were aborted even when pollen was available, demonstrating that factors other than pollen limitation are influencing acorn crop size.

A specific alternative to pollen limitation that might synchronize acorn crops over considerable geographic areas is captured in the environmental veto hypothesis. This is the idea that a major environmental event can result in correlated reproductive failure in one year and then synchronize resource budgets among trees, thus resulting in a subsequent masting event (Bogdziewicz et al. 2018a). Although it is known that acorn crops are positively correlated with spring temperatures (Sork and Bramble 1993; Koenig and Knops 2015), which directly relates to flowering rates (Koenig and Knops 2015; Bogdziewicz et al. 2017a, 2017b), an environmental veto is a more severe event that universally shuts down reproduction in a given year. A late spring frost, for example, can kill oak flowers over a large geographic area and prevent subsequent acorn production (Bogdziewicz et al. 2018a; Steele, personal observation).

To test this, my colleagues, Michal Bogdziewicz and Elizabeth Crone, developed a mathematical model to determine if such an environmental veto could explain acorn masting patterns. They calibrated the model with a 16-year data set on masting in red oak (*Q. rubra*) and white oak (*Q. alba*) from my lab (see below). The simulations demonstrated that assumptions regarding resource budgets coupled with occasional widespread reproductive failure were the key factors necessary to reproduce the patterns of masting observed at these study areas. This was accomplished without incorporating assumptions regarding pollen limitation.

In a similar analysis of acorn production in the holm oak (*Q. ilex*) on the Iberian Peninsula, Bogdziewicz et al. (2017b) found evidence supporting the Moran effect but it was regulated by environmental vetoes related to flowering phenology (Koenig et al. 2012, 2015) and acorn abscission due to spring drought. Bogdziewicz et al. (2017b) concluded that in this species, masting weather vetoes best explained synchronization of masting rather than any model based on simple weather cues or pollination limitation.

The proximate cues controlling masting in oaks is indeed a complicated one. Although numerous studies have explored these phenomena, there is no general agreement on a single set of factors that drive synchronization of acorn crops. This is most likely due to the highly variable environmental conditions under which oaks grow and the many ways weather may interact with the oaks' ability to store and mobilize resources and influence pollination (Bogdziewicz et al. 2017b).

OAK MAST: A MAJOR FOOD RESOURCE FOR FOREST SPECIES

In forests where they are common, oaks produce a key nutritional resource for a diversity of animals. Van Dersal (1940) cited approximately 186 mammal and bird species that regularly depend on oaks (mostly acorns) for food in North America, north of Mexico (Goodrum 1959). Include in this list the many species of insects (Moffett 1989) and other animals that prey on these insects and vertebrates, and an entire ecological network begins to emerge. All of this follows from the consumption of acorns.

Many of the larger acorn-consuming species in North American forests (e.g., turkey [*Meleagris gallopavo*] and black bear [*Ursus americanus*]) function exclusively as seed predators, grinding the acorn in small pieces before digesting it, and therefore never dispersing the fruits. The black bear masticates the fruit whereas the turkey swallows the nut whole and grinds it in the crop. Vaughan (2002) reviewed 24 studies on the autumn diet of black bears across their range in North America. Twelve of these studies cited acorns as the most dominant food item, ranging as high as 76% of the autumn diet. Several studies cited by Vaughan (2002) also demonstrate that activity patterns, home range sizes, denning behavior, reproduction, and mortality of adults and cubs are all linked to acorn abundance, at least in parts of the species' range, such as the Appalachian Mountains of the southeastern US.

Acorns can comprise 75% of the turkey's diet in autumn, winter, or spring (Steffen et al. 2002). While certainly a key dietary item that provides a high fat, high energy resource that is associated with turkey body condition, the influence of acorn availability on turkey survival, reproduction, and behavior is less clear. One reason for this may be the turkey's ability to shift to other food items in the absence of acorns (Steffen et al. 2002). A third game species in the eastern US that also depends on acorn crops is the white-tailed deer (*Odocoileus virginianus*). This species is reported to consume as much as 1 kg of acorns/individual per day during autumn seed fall (McShea and Schwede 1993; Feldhamer 2002). Although white-tailed deer shift diets with food availability, body mass and condition of fawns and the number of does entering the breeding season are significantly correlated with annual acorn production (Feldhamer 2002). Likewise, antler condition, home range size, and patterns of habitat use of white-tailed deer are also directly tied to acorn abundance (McShea and Schwede 1993; Feldhamer 2002). In Europe, the wild boar (*Sus scrofa*) is an especially important consumer of acorns that also functions exclusively as a seed predator. Suselbeek et al. (2014a) demonstrated the tremendous ability of this acorn competitor to recover acorns from both larderhoards and scatterhoards of granivorous rodents or jays, indicating just how important this acorn predator may be for limiting oak establishment across its range in parts of Eurasia and North Africa.

A limited number of bird species, especially the jays and other members of the family Corvidae regularly consume acorns. Some of these species are potential agents of oak dispersal via their scatter-hoarding behavior. Most of these species are discussed at length in Chapter 10 so here they receive limited attention. Pesendorfer et al. (2016b) reviewed the published evidence of the close mutualism between the corvids and the oaks, highlighting acorn dispersal by five species of scrub jay (*Aphelocoma* spp.), the Steller's jay (*Cyanocitta stelleri*), and the blue jay (*C. cristata*) in North America, and the Eurasian jay (*Garrulus glandarius*) in Europe and much of Asia. Each of these species are associated with specific oaks and, as discussed later, tend to disperse oak species that produce smaller acorns. In eastern North America, other avian species, such as the common grackle (*Quiscalus quiscula*) (Steele et al. 1993), the tufted titmouse (*Baeolophus bicolor*), the red-bellied woodpecker (*Melanerpes carolinus*), and white-breasted nuthatches (*Sitta carolinensis*) (Richardson et al. 2013), all consume acorns but are generally thought to have a limited effect on dispersal. One avian species known to rely heavily on acorn mast is the acorn woodpecker (*Melanerpes formicivorus*) found in the western US, southward through Mexico and Central America, and even into South America. This species regularly stores acorns above ground level in trees and other wooden structures in which it excavates individual holes for storage of each acorn. Because of this particular behavior, this species does not contribute to oak dispersal.

Among the mammals, it is the granivorous rodents that feed heavily on oaks and at the same time contribute to oak dispersal. In the deciduous forests of the eastern US, for example, both the white-footed mouse (*Peromyscus leucopus*) and the deer mouse (*P. maniculatus*) are significant consumers of oak mast as are other species of these new world mice where they co-occur with the oaks. Similarly, the eastern chipmunk (*Tamias striatus*) also depends heavily on acorn mast throughout these forests. All three of these rodents first scatterhoard acorns and, as a result, often contribute to dispersal of the oaks. However, in many cases these scatterhoards are often recovered by cache owners and then moved to larderhoards where the acorns are either eaten or die. The tree squirrels (*Sciurus* spp.) are also scatter-hoarding granivores but unlike these other species, they only scatterhoard acorns and thus serve as a primary agent of oak dispersal. In much of the eastern half of North America, it is the tree squirrels, especial the eastern gray squirrel (*S. carolinensis*) and fox squirrel (*S. niger*) that are keystone dispersal agents.

In Europe, N. Africa, and Asia, the geographic range of the oaks is occupied by *Apodemus* spp. of the family Muridae (the family of true rats and mice). These mice are considered critical agents of oak dispersal throughout this region, although they tend to disperse seeds relatively short distances. Many of the genera of squirrels found in Asia (e.g., *Sciurotamias* and *Callosciursus*), other than *Sciurus*, appear to be closely associated with the oaks of this region, sometimes dispersing acorns and other times functioning nearly exclusively as acorn predators (see Chapters 5–9). The Eurasian red squirrel (*Sciurus vulgaris*), distributed throughout conifer and deciduous forests of Europe and Asia, limit their consumption of acorns and are not known to regularly disperse the oaks (Steele and Wauters 2016).

Not only do invertebrates feed on the acorns but an entire insect community can make their home inside acorns (Moffett 1989). Numerous species of moths, beetles, flies, and other species depend on acorns at least during one or more stages of development. Their impact as potentially devastating pre-dispersal seed predators is discussed in subsequent chapters.

The great majority of animals that feed on acorns function exclusively as acorn predators. And, the few that disperse and cache acorns do so with intensions of recovering those caches. This means the vast quantities of acorns produced by oaks never make it to initial germination. Nevertheless, scatterhoarding behavior by the relatively few species that engage in it is the primary means by which the oaks are dispersed. The one important exception to this is the possibility that many seed predators often only consume the top (basal) half of the acorn due to various physical and chemical characteristics of the fruit, which in some cases may allow the remaining apical half to germinate and establish (see Chapters 5 and 6). Overall, though, because of the intense predation of acorns by such a diverse array of seed predators, it is only during years of high mast production that successful germination and seedling establishment is likely to occur. This is why an understanding of the phenomenon of masting is so essential to understanding the dispersal process.

MONITORING PATTERNS OF OAK MASTING

I now shift the discussion to a brief review of my lab's efforts to follow acorn masting in forests of eastern Pennsylvania. My team first monitored acorn production in natural forest stands from 2000 to 2002 at three sites (see Chapter 7). At the time, my associates and I were interested in assessing natural acorn availability so we could then compare patterns of acorn abundance with patterns of acorn dispersal and predation by rodents (Moore et al. 2007). Attempts to determine acorn abundance in forested sites, however, require quite an effort, at least in eastern deciduous forests. Compare for example the technique used by Koenig and Knops (2013) when studying California oaks. In their system, where oak trees often stand alone in open grasslands, two investigators simply conduct 15-second acorn counts using binoculars to scan the canopy and determine relative abundance. For Koenig and his team, this technique required a careful, systematic but modest effort, that is also highly reliable and accurate. We were jealous.

Try as we did to view the canopy of oaks in our more densely forested stands, it was often difficult to even see acorns, and, if we could, it was often impossible to determine to which tree acorns were attached because of the overlapping branches in the canopy. Other investigators have overcome this problem by randomly placing wooden, framed quadrants under each oak shortly after seed fall and counting acorns (Wolff 1996). After visiting Dr. Jerry Wolff's sites at the Mountain Lake Biological Station (in southwestern Virginia) in 1995 and discussing this approach, I was first convinced of its efficacy and, above all else, the time that it saved. A few years later, however, when my lab was conducting a seed dispersal experiment in which we selectively provisioned small mammals with tagged acorns, we soon learned why this approach was not likely to be effective. It was a low mast year, and at one site, small mammals removed and consumed more than 8000 acorns in just 48 hours. Although this occurred during a low mast year and the rates of removal were more than 10 times faster than in years of high acorn abundance, it still gave me pause—rapid removal or consumption of seeds on the ground were likely to introduce too much inaccuracy. But it was my postdoc at the time, Dr. Thomas Contreras, who convinced me that we needed to employ a more systematic approach.

4.1. Acorn collection baskets monitored at three sites in northeastern Pennsylvania for 17 years. At each site, 15 trees of three oak species (n = 45 trees per site) were monitored for acorn production. Shown are **(a)** under each tree's canopy we place two 12-gallon plastic bins mounted on stakes approximately 1 m above ground, and **(b)** to prevent seed bounce and predation by birds and rodents, a piece of metal chicken wire placed in the top of the bin. *(Shealyn A. Marino)Xianfeng* SEE COLOR PLATE

Tom's solution, modified from other techniques cited throughout literature, was to attach a plastic bin (Fig. 4.1a) bolted to a wooden plate (8″ × 8″) that was attached to the top of a 4′-long, 2″ × 2″-wooden stake that was driven vertically into the ground. The bin thus stood a few feet above the ground directly below the canopy of the tree (Fig. 4.1a). A piece of mesh chicken wire was also placed in the bin at approximately half its depth (Fig. 4.1b). This allowed acorns to fall through the wire mesh and into the bottom of the container. The position of the bin and the mesh together prevented access by birds and small mammals that might otherwise harvest the acorns from the containers. In 2001, we began our survey and placed two such seed collectors under each of 15 trees of each of three species of oak at three sites: two in northeastern PA in Luzerne County and a third in Schuylkill County at Hawk Mountain Sanctuary. At two sites (V and HM) we monitored red oak, white oak, and chestnut oak (*Q. montana*) and at one site (S) we monitored red oak, white oak, and black oak (*Q. velutina*). Tree species selection was based on relative abundance at each site and the total trees selected at each site were all within approximately a 5-ha area. Each tree was permanently tagged, and the two bins were carefully positioned under the canopy so that the acorns that fell in each container originated from the same tree (Fig. 4.1a). Each year, beginning in early September, we visited each of the 135 trees every seven to 10 days, recording the number of acorns per tree and collecting and storing the acorns (4°C) for subsequent analysis of insect damage. This was continued usually until the end of November, a few weeks beyond recovery of the last acorn and the cessation of seed drop. Upon death of a tree, it was replaced with another.

Despite my initial skepticism on the effort involved, after three years I was impressed with records we had produced, their reliability in assessing both individual tree and population-level production, as well as the ability to assess pre-dispersal predation by insects for each tree. Our collections represented only an index of acorn abundance but they seemed quite consistent with patterns of production—so, it seemed like a good idea to continue. And continue we did, year after year. As many an undergraduate technician in my laboratory would attest, in addition to relocating metal-tagged acorns in the forest, the monitoring of acorn bins and the careful dissection of acorns to assess weevil and other insect damage were simple rites of passage in the Steele lab. At the time of the publication of this book we were entering our 19th year of acorn collections in one study and year 27 in another.

Shown in Figure 4.2 is a condensed summary of acorn production for two species of oaks (red oak and white oak) across three study areas and a third species (chestnut oak) collected at two of the sites. Each point represents the mean number of acorns collected per tree species, per site. Several conclusions are evident from a quick review of Figure 4.2. First, during the 16-year period, masting events occurred on only four occasions with a fifth major event appearing to unfold in 2018 (not shown). Only two of these events (2001 and 2005) were significant masting events, whereas the other two events (2010 and 2012) produced only half as many acorns in comparison. The 2001 masting event was followed by an independent masting of red oaks in 2002 at only one site (V), whereas acorn production for all other species and sites was low that year. Among the four masting events, three showed considerable synchronization across study sites in 2001, 2005, and 2012. Although two of the study areas (V and S) were only 10 km apart, the third site (HM) was located approximately 65 km due south of these two sites, thus suggesting considerable geographic synchronization.

Over the 16 years, there was also some evidence for synchronization across oak species, which is especially interesting given that these data include a red oak species and two white oak species. The timing differences in flowering and acorn development in white oaks (section *Quercus*) and red oak (section *Lobatae*) would be expected to reduce the likelihood of synchronization, so this pattern seems interesting. In one of their many extensive analyses of oak masting pat-

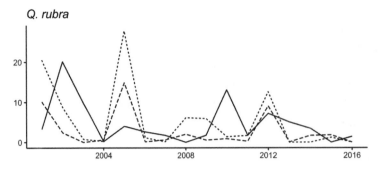

Q. rubra

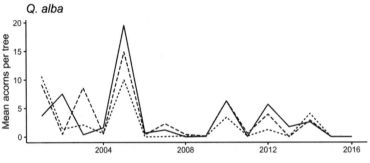

Q. alba

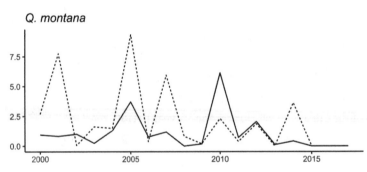

Q. montana

4.2. Acorn production per site per year at three study areas in Luzerne and Schuylkill Counties, Pennsylvania. Shown are the mean number of acorns per tree, per site for red oaks (*Q. rubra*), white oaks (*Q. alba*), and chestnut oaks (*Q. montana*). Solid, dashed, and dotted lines indicate each of the three study areas. Acorns were collected in two seed collectors under each of 135 trees (15 trees of each species at each study site) *(Michael A. Steele, unpublished data)*
SEE COLOR PLATE

terns, Koenig and Knops (2002) analyzed 80 data sets from 24 studies in North America, Japan, and Europe. They concluded that these data generally supported predictions of the PSH with one notable exception: the lack of synchronization of oak species within a forest community (e.g., white oaks versus red oaks)—an outcome that may prevent the reduction in seed predator numbers. As I discuss in Chapters 3, 5, and 6, the nondormancy and rapid germination of white oak acorns, which renders their acorns more perishable and less storable to rodents than those of red oaks, would favor white oaks masting independent of red oaks. In contrast, synchronized masting of the two oak groups would favor the red oaks. With perishable white oak acorns to eat, the red oaks would be selectively dispersed and cached. Colleagues and I are now testing several predictions to follow from contrasting masting patterns in these two groups of oaks.

We collected mast data not only to determine patterns of acorn production but
to also assess how acorn production related to patterns of acorn predation by
insects and rodents, as well as rodent dispersal of acorns. Measuring acorn mast
by itself is interesting and important, especially for better understanding the
mechanisms driving masting, but I knew even at the time we started these surveys that acorn production had been measured in oak stands throughout the
world and several investigators were well on their way to uncovering some of
the mysteries of oak masting.

To evaluate the impact of the insect acorn predators, every acorn collected
in the seed collectors was returned to the lab and systematically measured, dissected, and inspected for larval insects (Fig. 4.3). Although numerous species of
insects infest acorns (see Chapter 12), by far the most common are the weevils
(*Curculio* spp.). Late in the summer, adult weevils emerge from the soil following
a lengthy diapause of one to three years, then they climb nearby oaks, mate, and
soon thereafter oviposit in developing acorns. Inside the acorns, eggs hatch and
develop into larvae that feed on the oak cotyledon. When the acorn then drops
from the tree, the larvae bore a hole in the acorn pericarp, exit the acorn, and
then burrow in the soil to complete the life cycle.

With 17 years of data on weevil infestation and acorn production at these
three sites, several of my colleagues and I were able to test the predator satiation hypothesis for weevils in this system
(Bogdziewicz 2018b). In this study, we first
tested for evidence of weevil satiation in high
crop years; that is, we measured whether increased acorn abundance resulted in a greater
proportion of acorns escaping weevil predation. We also tested for evidence of starvation and that poor mast years reduce weevil
populations. This was measured by testing
if poor mast years were followed by years of
high acorn survival because of a reduction
in weevil numbers. We found that in two of
the three oak species studied (white oak and
chestnut oak), masting did not lead to a reduction in seed predation. This was due to
two factors: weevils reproduced rapidly when
acorns were available and when necessary
moved to seed-rich trees, especially when

4.3. Larvae of acorn weevil (*Curculio* spp.)
inside an acorn with an existing weevil exit
hole. *(Shealyn A. Marino)* SEE COLOR PLATE

acorn production at the population level was lower. In other words, the starvation and satiation effects of masting did not occur. Even in red oaks, satiation only occurred when acorn crops were synchronized across individual trees in a population. When individual crops were less synchronized, weevils moved to higher producing trees and there was no greater increase in acorn survival (satiation) (Bogdziewicz et al. 2018b).

These analyses suggested in our system masting was not tuned well to insect populations. Perhaps, instead, it was better aligned with the rodents. As I suggested earlier, for predator satiation to be fully effective it should favor not only satiation and starvation of seed predators but also increased effectiveness of acorn dispersal and survival.

EFFECTS OF ACORN PRODUCTION ON MAMMAL POPULATIONS

Not long after we began monitoring acorn production, I was pulled in yet another but related direction—a direction I had considered in the past but it was a huge investment in money, energy, and time. In 2002, I was contacted by my former undergraduate research advisor, Dr. David Zegers, of Millersville University, who had learned that I was monitoring oak production at Hawk Mountain. He asked if I would be interested in also monitoring small mammal populations at the sites. I credit Dave as my first role model in science and the first to turn me on to mammalian ecology at the ripe old age of 18. His invitation seemed like a good idea at the time and, although I had plenty to do, I signed on.

We decided we would establish three small mammal trapping grids. Typically, a small mammal live-trapping grid consists of a network of live traps evenly placed across approximately 1 ha (10,000 m²) of habitat. Our grids each consisted of a 6 × 6 grid of trap stations with stations equally spaced 20 m. At each station, we placed two live traps within approximately one meter of the station marker. Thus, each trapping grid consisted of 72 live traps. The concept is rather simple. By effectively concentrating one area with enough traps, we could sample the small mammal populations (primarily rodents and shrews) and assess everything from individual characteristics (species ID, age, sex, body mass, and reproductive condition) to population densities of each species on the grid. Then, by conducting periodic samples through time we can monitor changes in individual characteristics, reproductive patterns, and population densities.

The live traps we used were the industry standard, although numerous variations on the standard trap are also available. We used standard-sized Sherman live traps (25 cm × 7 cm × 7 cm; Fig. 4.4). The animal is drawn to the trap by bait placed just inside the back door. In our studies, this bait was a mixture of oats and sunflower seeds. Inside the trap, the animal is also provisioned with

4.4. Sherman live trap used to capture small mammals such as mice, voles, shrews, and chipmunks. Approximate size is 9 cm × 8 cm × 24 cm. *(Shealyn A. Marino)*

nest material (e.g., cotton). Traps are set and then checked daily soon after sunrise. Upon capture, the back of the trap is covered with a plastic bag, the door opened, and the animal is gently shaken into the bag. The rodent or shrew is first weighed and then held by the nape of the neck to note and collect all the age and reproductive data. Researchers then record the data and endow the creature with a uniquely numbered ear tag in each ear. Generally, this procedure works well for a diversity of rodents. At our sites, this included mice, voles, chipmunks, and squirrels, although larger traps are often used for larger species, such as the eastern gray squirrel (Fig. 4.4).

Our primary goal was to survey the white-footed mouse (*Peromyscus leucopus*), the most common species on our sites, as well as the southern red-backed vole (*Myodes gapperi*), short-tailed shrew (*Blarina brevicauda*), eastern chipmunk (*Tamias striatus*), and southern flying squirrel (*Glaucomys volans*). As we describe later in this chapter, Richard S. Ostfeld and colleagues had already been monitoring small mammal populations in oak forests of upstate New York well before we had initiated our small mammal surveys. In fact, several years earlier, they had definitively documented how the energy provided by oak mast was the underlying driver of rodent populations in eastern deciduous forests (Jones et al. 1998). Then, why repeat this? Our reasons for conducting such surveys were motivated by several slightly different questions. I was interested in understanding how rodents influence acorn fates, either as seed predators or as agents of dispersal. This we were sure required knowledge of not just acorn abundance but also rodent densities as well. In other words, the per capita number of acorns per rodent seems to be the ultimate measure for assessing rodent-acorn interactions (Xiao et al. 2013b; see Chapter 9).

However, several other factors motivated these small mammal surveys, some of which were not at all related to our focus on oak ecology. Dave and I, both professors at predominantly undergraduate institutions, saw this as an opportunity to train student researchers and to integrate this small mammal research into our courses in ecology, animal behavior, and vertebrate biology. The surveys were thus designed to address several objectives and as they did, they grew in size and effort. For example, we had learned a great deal about long-term mammal surveys from previous studies by our colleague, Dr. Joseph F. Merritt (Merritt et al. 2001). We knew Joe would advise to survey more than one grid for replication. Joe conducted a small mammal survey every month for 20 years at Powdermill Biological Station in western Pennsylvania, where Joe served as the director. During that time, Joe was also the primary field team, so the labor involved in conducting his survey simply precluded site replication, which he emphasized repeatedly to both of us. So, we started with three grids in 2003 and then expanded the project to include six sites by 2005.

Joe is also a shrew guy, an expert, in particular, on the physiological ecology of shrews. We knew from both Joe's work and my previous small mammal community studies in the Southern Appalachian Mountains that the use of live traps alone would not account for at least four species of shrews (*Sorex* spp.). The shrews found on our site include the smoky shrew (*S. fumeus*), masked or *Cinereus* shrew (*S. cinereus*), American pigmy shrew (*S. hoyi*), and the larger short-tailed shrew. Although exclusively insectivorous, shrews may also indirectly depend on acorns as they feed on the weevil larvae and pupae when in the soil (Andersen and Folk 1993), although we know little of the extent to which this dietary item varies over time.

Shrews are less likely to be attracted to the traps and further unlikely, due to their small size (1–15 g), to trigger the trap even if they entered it. We therefore decided to use pitfall traps to survey for shrews. Our pitfall traps consisted of nothing more than a cottage cheese container (about 20 cm in diameter at the opening and 30 cm in depth). We placed 10 pitfalls along a diagonal transect from the top left corner of each grid to the bottom left side of the grid, with each pitfall placed approximately 10–15 m along the transect. As their name suggests, pitfall traps function by providing a hole in which the shrew falls as it is travelling along the edge of a fallen log or a large rock. Pitfalls therefore must be carefully placed along the edge of rocks and logs by digging a hole so the pitfall is buried in the ground with its top flush with ground level, and one side of the pitfall tight against the edge of the rock or log. Shrews, which sometimes travel in family units, remaining close to large objects on the forest floor simply fall into the container and are unable to escape. Shrews are also tagged and released, but because of their high metabolic rate and need to eat every few hours, shrews sometimes expire in the pitfalls. Based on previous studies and the literature, however, a few days of sampling every few months appear to have no effect on shrew numbers. Pitfall surveys appear the best approach for approximating shrew numbers.

With our trapping grids established, we began sampling small mammal populations on the original three grids in the summer of 2003, and, by 2005, expanded the number of grids to six. To ensure independence between grids, each was established 0.5 km to 1 km to the next closest grid, although several were located significantly farther from the nearest grid. In our early surveys, we conducted approximately four trapping surveys throughout the year: three from early spring to fall and another before snowfall. Once snow was on our sites it was too difficult to safely maneuver on the grids, so trapping was suspended until spring.

During each trap session, live traps and pitfalls were set one day and then checked early each morning shortly after sunrise every day of the trap session. Traps and pitfalls were then closed at the end of the session. From 2003 to 2012, trap sessions were conducted over five consecutive days. Thereafter, we shortened the session to four days, as it became evident that a four-day session was effective in trapping most of the small mammals on the site. In 2012, we also slightly modified grid size and trap number during a two-year experiment we conducted concurrently with colleagues Kurt Vandergrift and Peter Hudson at Pennsylvania State University (PSU) and Richard Ostfeld at The Cary Institute of Ecosystems Study in New York.

Finally, for a few years we also monitored litter sizes of the mice. My son Tyler, as part of an Eagle Scout project, had constructed more than 100 small mammal nest boxes, and in 2006 we placed 18 boxes on each of the six grids, with each box strapped about 0.75 m above the forest floor on the trunk of the tree. Boxes were then monitored 1–2 times annually to assess occupancy and litter size. This effort was only continued for six years due to time constraints and the loss of boxes due to rotting and bear activity.

The above methodological details are necessary to convey the time and human energy needed to secure information on small mammal populations. It illustrates not only the effort required to understand ecological processes but many of the unexpected interactions that might be uncovered. In our early surveys, Dave and I, along with a small army of students, completed most of this work initially, but by 2005 we had hired Shealyn Marino, a graduate of PSU's wildlife and fisheries science program, who has largely managed this survey since then. Shea's tremendous commitment and attention to detail cannot be over emphasized.

Figure 4.5 summarizes the relationship between acorn mast and small mammal populations. My goal here is to provide a general overview and some brief

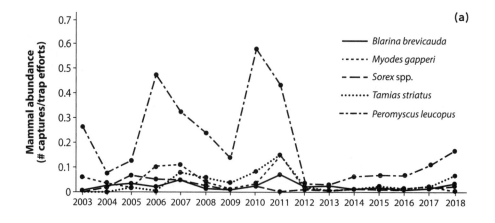

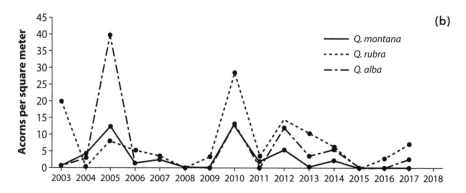

4.5. Rodent and shrew abundance and acorn production at the study areas at Hawk Mountain Sanctuary. Shown are **(a)** capture rates for short-tailed shrew (*Blarina brevicauda*), red-backed vole (*Myodes gapperi*), shrew species (*Sorex* spp.), eastern chipmunk (*Tamias striatus*), and white-footed mouse (*Peromyscus leucopus*). Also shown is **(b)** acorn production in chestnut oak (*Q. montana*), red oak (*Q. rubra*), and white oak (*Q.alba*). *(Michael A. Steele, unpublished data)* SEE COLOR PLATE

highlights of these data. At the time of this publication, colleagues and I were beginning more formal analyses for the scientific literature. The small mammal data in Figure 4.5 are summarized across all study areas. Most evident are the two oak masting events, both of which were associated with a peak in densities of white-footed mice. The first peak in mouse densities occurred in 2005, the year after a peak in white oak mast. The second peak occurred nearly concurrently with the peak in red oak acorn production. Both peaks in mouse density were then followed by a steady decline. This second decline in mouse density was followed by five years of exceptionally low density, despite the availability of some acorns during this time. Based on our observations we suggest that mouse densities in 2004–2006 were so low that there was little opportunity for populations to recover. Although chipmunk populations remained relatively low throughout the study, we observed their highest densities following the two masting events. And, although the same appeared true for gray squirrels, we do not include data for this species because the Sherman traps were not a reliable method for monitoring their densities.

The results here echo those of other studies (Ostfeld et al. 1996; Jones et al. 1998), which suggest that acorn crops serve to satiate rodent seed predators during masting years, after which rodent populations increase and then decline due to limited food availability (starvation) during the periods of low acorn abundance, most likely to follow mast years. Thus, it follows that many of the predictions of the predator satiation hypothesis are supported when considering the effects of oak mast on these rodents. However, for satiation to benefit the oaks it must also result in an increase in dispersal and caching as well. In an initial attempt to explore this, my lab conducted seed dispersal experiments during high and low mast years and found strong support for the PSH. We found that during a mast year, dispersal and caching of red oak and pin oak acorns (both red oak species; section *Lobatae*) increased significantly compared with acorns of white oak (section *Quercus*) (see Chapter 8). Dispersal and caching of white oak, in contrast, remained low regardless of acorn abundance. In our experiments, this was most likely because the more perishable white oak acorns were always presented to rodents in the presence of the more storable, dormant red oaks. With red oaks to store, white oaks are selectively eaten (Smallwood and Peters 1986; Steele et al. 2001a) even when mast is abundant.

OAK MASTING: THE PHENOMENON OF A PULSED RESOURCE

Our acorn and mammal surveys at Hawk Mountain Sanctuary provide some insight into how fluctuations in acorn crops are likely to influence rodent and weevil populations. In subsequent chapters, I discuss how the relative numbers of acorns and rodents likely drive patterns of acorn predation by rodents and insects (primarily weevils) and acorn dispersal by rodents. Before moving on, however, it is worth considering how masting in oaks often results in an energy pulse in the environment that can trigger a series of significant trophic interactions across the ecosystem well beyond factors that influence acorn fate. As reviewed by Ostfeld and Keesing (2000), pulsed resource events occur in both terrestrial and aquatic systems and, although the initial pulse of energy often creates a bottom-up flow of energy in the ecosystem, this, in turn, can result in a significant wave of trophic cascades up, down, and across the ecosystem.

Perhaps one of the most thoroughly studied resource pulses to occur in terrestrial systems is that which follows bumper acorn crops in the oaks. Failures in oak production can exert equally significant effects. Such cascading effects to follow from both oak masting and failures in acorn production have been eloquently documented by Dr. Richard S. Ostfeld of the Carey Institute of Ecosystems Studies of New York and his collaborators Drs. Clive R. Jones, Felicia Keesing, Kenneth A. Schmidt, and others.

By monitoring acorn production, small mammal populations, gypsy moth outbreaks, songbird nest survival and fledgling success, Lyme disease outbreaks, and several other events over the past 25 years, Ostfeld and his collaborators have carefully documented how all of these ecological processes are interrelated and ultimately tied directly or indirectly to acorn abundance. The research tells a fascinating story, one that illustrates how acorn production can, in many ways, define the ecology of an oak forest (Ostfeld et al. 1996; Ostfeld 1997; Ostfeld and Keesing 2000).

In eastern deciduous forests of the northeastern US, the story begins in a year of high acorn production. Small mammals, such as the white-footed mouse, do exceptionally well concentrating on a diet of acorns, a food source that is high in energy, that is efficiently metabolized, and one that stores well at least over a single winter. The result is an increase in overwinter survival, a higher reproductive rate, reproduction even in the winter, and an increase in population size. Over several years of monitoring acorns and mice, much the way I describe above, Ostfeld et al. (1996) demonstrated a highly significant positive association between acorn production in a given year and the size of the mouse populations the following summer. This relationship between white-footed mouse populations and acorn numbers, evident in my lab's work as well (Fig. 4.5), has also been reported throughout the Central Hardwoods Region of the United States (Wang et al. 2009). Consider also, that the mouse is not the only benefactor of a bumper acorn crop. Other rodents, especially the squirrels, including eastern chipmunks, the red squirrel (*Tamiasciurus hudsonicus*), and the tree squirrels, all highly adept at hoarding acorns, show similar population increases in response to oak masting (Steele and Koprowski 2001). However, the demographic evidence of this is less definitive than for mouse populations. Other acorn consumers are also drawn to forests with numerous oaks during mast years. White-tailed deer, for example, move into oak sites from surrounding forests dominated by non-oak species such as maples (*Acer* spp.), as do other species (Ostfeld et al. 1996).

Increases in mouse populations as a result of acorn crops causes a signature domino effect across several trophic levels in the forest. Here I briefly touch on a few of them. For more detailed treatments from which I have borrowed much of this brief discussion, consult the reviews of Ostfeld and his collaborators.

One obvious effect of an increase in mouse populations is the increasing focus of top predators. Increases in rodent abundances result in numerical increases in predator numbers and/or behavioral shifts in the diet of generalist predators, such as the raptors and the mustelids (e.g., long-tailed weasels [*Mustela frenata*] and the American mink [*Neovison vison*]). These predators appear to move where population densities of mice are high. During our mammal surveys, we observed weasels and an occasional mink on our sites during peak population levels. At other times they appeared absent from the site. Although we did not survey for weasels, they frequently enter closed Sherman traps to eat the mice, sometimes leaving only a headless mouse behind—clear evidence that a weasel was there. Occasionally though, we would find the weasel stuffed inside the small trap. These observations only occurred when our mouse densities were at peak levels. My colleague, Travis Knowles, and I previously witnessed the wrath of weasels during a small mammal survey in the Southern Appalachian Mountains at the peak of a mouse population. We had set a transect of 30 traps at a site along the Blue Ridge Parkway and on the first day, 27 of our traps contained white-footed mice. As we went down the line, we discovered one headless mouse after another. More than 20 of the mice had been decapitated by a weasel that was then unable to pull the mice from the traps.

Another significant outcome to follow from high acorn numbers and subsequent peaks in mouse production is the risk of Lyme disease (Ostfeld 1997). The deer and the white-footed mouse, in particular, are the most important hosts for the black-legged tick (*Ixodes scapularis*), which hosts the bacterium, *Borrelia burgdorferi*, which, in turn, causes Lyme disease (Ostfeld et al. 1996). During a year of high acorn abundance, large numbers of adult ticks, common on deer, are dispersed to oak forests. The deer often move in from other forest patches without oaks to eat acorns, and they bring with them their ticks. The ticks then overwinter in the soil or leaf litter. The following year just as mouse populations are increasing, ticks mate, lay eggs, and larvae hatch by early summer and, because of their availability, seek mice and ground nesting birds for a blood meal. Among all available hosts, the white-footed mouse is the most likely to successfully host the Lyme bacterium (LoGiudice et al. 2003). The white-footed mouse is considered a reservoir host, which means it carries the Lyme bacterium at high levels with no ill effect. Compared with other hosts, the white-footed mouse is the most competent of all potential hosts, meaning that it is most likely to acquire, retain, and transmit the bacterium to the tick host (LoGiudice et al. 2003). The larval tick takes a blood meal, drops off the host, molts into a nymph, and overwinters. The following spring the nymphs search for a host and again feed. If the nymph had previously acquired the bacterium and then feeds on a human, the disease can be transmitted. The acorns, in a way, bring the deer, mice, and ticks together and hence provide a pathway for a Lyme out-

4.6. Low acorn mast can result in many unexpected ecological effects. Shown is **(a)** red squirrel (*Tamiasciurus hudsonicus*) just moments before preying on a wood thrush chick in the spring following an oak mast failure, and **(b)** another nest where at least one wood thrush was eaten and where we later captured photos of this white-footed mouse storing acorns; this same nest was then subsequently visited by a southern flying squirrel and soon thereafter a screech owl (*Megascops asio*). *(Jeffrey Stratford)*

break. The deer introduce the ticks and the increase in mouse densities ensure a high transmission rate of the bacterium.

The result can be a perfect storm (Ostfeld et al. 1996). In the forest, soon after the mouse densities peak, they then crash. Because of the high energy invested by oaks during the mast year, acorn crops are usually low the following year or two after a mast crop (Ostfeld 1997; Koenig and Knops 2002). Results from our own research support this as well (Fig. 4.5). This means acorn crops are low just as small mammal populations increase. The result is an equal, if not more significant, impact on forest dynamics, as the rodent diet shifts away from acorns. With little else to eat, rodents often focus their diet, out of sheer necessity, on the eggs or young chicks of ground-nesting birds (Ostfeld 1997), such as the common thrush, the veery (*Catharus fuscescens*) (Schmidt and Ostfeld 2003). This effect is not limited to eastern deciduous forests. For example, Haarsma and Kaal (2016) reported high rates of predation on hibernating bats by the wood mouse across a dozen hibernacula in the Netherlands following an oak mast failure. Although only an anecdotal observation, following a low mast year in 2016, we caught a photo of a red squirrel just before preying on a nestling wood thrush (Fig. 4.6).

In addition to the rather compelling relationship between acorn crop size and mouse numbers are the same results yielded in experimental manipulation of both acorns and mouse densities (Jones et al. 1998). By simulating a year of mast production on some sites and comparing control sites where acorn production was low, Jones et al. (1998) convincingly showed this placement of acorns in the forest drives this same series of ecological events. The increase in acorn densities resulted in higher mouse populations, which brought mice, deer, and ticks together along with a higher risk of Lyme disease. Similarly, experimental reductions in mouse populations resulted in outbreaks of gypsy moths, whereas higher densities of mice resulted in higher rates of predation of moth pupae by the mice and thus lower probability of gypsy moth damage.

The extensive research of Ostfeld and associates reveals a number of cascading effects to follow from the masting of oaks. A year without acorns can be devastating for rodents, especially when populations are high when acorn crops crash. However, in eastern deciduous forests, as in many oak forests across the globe, other seed producing trees often yield seeds or nuts when the oaks fail. Depending on the availability of these alternative crops, other outcomes are possible, which follows from the composition of the forests. Although discussed in subsequent chapters, few studies to date have considered how forest composition influences oak dispersal.

THE GHOST IN THE OAK FOREST:
THE LOSS OF THE AMERICAN CHESTNUT

It is important to appreciate that our knowledge of oak masting in eastern deciduous forests follows almost exclusively from studies conducted after the loss of the American chestnut (*Castanea dentata*) (Freinkel 2007) to the chestnut blight fungus (*Cryphonectria parasitica*). This resulted in the near elimination of this species by the middle of the 20th century. Prior to this, oak masting occurred in the context of the chestnut, which tended to produce seed annually, so it was often likely available to rodents when oak crops failed. And, because of its consistently high output of seed, the American chestnut was abundant wherever it occurred. The chestnut seed had a dispersal advantage over white oak but not red oak because of its nutrient content, lower tannin levels, and especially, its short period of dormancy (Lichti et al. 2014; see Chapter 7). Acorn dispersal in the context of the American chestnut is thus quite different and more variable than when oaks mast in the absence of this species. Although few studies have explored the effect of forest composition on patterns of oak dispersal, it follows from these observations that the relative abundance of other tree species will likely influence the effects of oak masting.

The next two chapters focus on a range of factors that influence acorn dispersal by rodents and birds, which center on intrinsic factors, characteristics of the acorn, and the oak that drive the process of dispersal, followed by factors extrinsic to the oak that influence the process.

5

INTRODUCTION

With a background in the basics of seed dispersal and oak biology in hand (see Chapters 1–4), Chapters 5–8 now focus on the many factors that influence oak dispersal. I divide these many variables into two categories: intrinsic factors, which include primarily acorn characteristics, and extrinsic factors, which include those of the broader environment that also drive oak dispersal. I also discuss some of the interactions between both sets of factors. This discussion follows primarily from research in my lab and that of many of my colleagues and collaborators.

This chapter and Chapter 6 provide a review of those aspects of oak dispersal that follow directly from acorn traits—the physical and chemical characteristics of the fruit. This discussion is organized around seven themes: (1) the significance of germination schedules of red and white oak species on scatter-hoarding decisions and patterns of dispersal; (2) the similar effects of insect infestation on acorn perishability and dispersal; (3) the detection of seed dormancy by rodents, the behavior of embryo excision by tree squirrels, and the characteristic of multi-seeded acorns; (4) a broader, more global perspective from both Mexico and China on the response of rodents to acorn traits; (5) the adaptive significance of early germination in acorns of white oak species for escaping acorn pruning and predation by birds and rodents; (6) the adaptive significance of chemical gradients in acorns for facilitating tolerance to acorn predation by a diverse suite of insect and vertebrate seed predators; and (7) an alternative view of the acorn as not just a package of energy for the developing seedling but also as a fruit for promotion of dispersal and tolerance to seed predation.

ACORN CHARACTERISTICS AND THE FEEDING
AND STORING DECISIONS OF RODENTS

My lab's foray into the effects of acorn characteristics on oak dispersal began with an investigation on how a potential seed disperser, the eastern gray squirrel (*Sciurus carolinensis*), reacts to some of the fundamental differences in acorns of different oak species. Our initial work on this question was hatched from an important study carried out by my colleague and friend, Dr. Peter Smallwood (Smallwood and Peters 1986). At the time of this breakthrough study, considerable debate, especially in the wildlife literature, surrounded the question of which acorns animals preferred for food: those of red oak species (section *Lobatae*) or those of white oak species (section *Quercus*).

As reviewed by Smallwood and Peters (1986), many authors argued that red oak acorns with their higher energy content were selected over their white oak counterparts (Smith and Follmer 1972; Lewis 1980). In contrast, other studies showed similar evidence to support a preference for white oak acorns, arguably more palatable because of their lower tannin levels (Short 1976). Tannins, secondary phenolic compounds, are common in both acorns and the leaves of oak and are important in the defense against microbes, as well as the invertebrate and vertebrate herbivores and the seed predators that feed on these structures. Assumed to evolve first to defend against microbes, tannins are well known to bind to both salivary and digestive enzymes, thereby reducing both palatability and digestive efficiency. The result, so it would seem, is a major deterrent to herbivores and a vast array of seed consumers (Schultz and Baldwin 1982).

Among the acorns, however, tannin content is not the only characteristic to vary. Consider the differences in red oaks (section *Lobatae*) and white oaks (section *Quercus*)—the two major groups of oak in North America, including Mexico. The red oaks, while high in tannin (5–15% dry mass) are also usually high in lipid (8–31% but usually ~20%), which is the primary source of energy for the many species that feed on acorns (Vander Wall 2001; Smallwood et al. 2001; Table 5.1). Most species of white oaks, in contrast, exhibit lower levels of both tannin (<2%) and lipid (2.6–11.5% but usually ~10%). The result is lower palatability and digestibility, but higher energy for red oak acorns, and for those of white oak acorns, the result is improved palatability but less energy. These differences translate into confounding acorn traits and, in turn, most likely account for the many conflicting observations in the literature on wildlife preferences for the two groups of acorns.

Smallwood and Peters (1986) resolved much of this confusion by testing the feeding responses of eastern gray squirrels to artificial acorns made by the researchers. To experimentally evaluate the effects of tannin and lipid on acorn

Table 5.1 Characteristics of acorns of the red oak and white oak subgenera in eastern North America.

CHARACTERISTIC (REFERENCES)	RED OAKS	WHITE OAKS
Lipid concentrations	High (ca. 20% by dry wt.)	Low (ca. 10% by dry wt.)
Tannin concentrations	High (5–15% by dry wt.)	Low (usually <2% by dry wt.)
Germination schedule	Spring (remains dormant for months)	Autumn (germinates soon after falling)

Source: Modified from Smallwood et al. 2001.

Note: There are some exceptions to these comparative patterns. For example, the white oaks *Quercus montana* and *Quercus macrocarpa* have been found in some studies to have high tannin concentrations.

preferences, they shelled acorns of a white oak species (*Q. alba*) collected under two trees, ground the cotyledon, and made acorn dough to which they then added tannin and/or fat to produce acorn dough balls that varied in both tannin content (0–6% added tannic acid) and fat content (20–40% crude fat). They then presented free-ranging gray squirrels on the campus of Ohio State University with these artificial acorns early in the autumn when conditions were mild and again in the winter when energetic requirements were higher.

Tannins significantly reduced the time the animals fed on the dough balls, especially in the autumn. In both feeding experiments, higher fat levels attenuated the effects of the tannin, but this effect was more pronounced earlier in the season when the squirrels appeared to avoid higher tannin content. In the winter, the squirrels seemed to prefer the artificial acorns with higher fat levels regardless of tannin content, probably because of the higher energetic demands at that time of year. Based on their observations coupled with those from previous studies, Smallwood and Peters (1986) hypothesized that perhaps in the early autumn, squirrels and other rodents might prefer to eat acorns of white oak species and at the same time selectively cache those of red oaks. This hypothesis made particularly good sense when you also considered another important difference between the two groups of acorns.

The acorns of nearly all species of red oaks undergo a period of dormancy prior to germination. These species will not germinate until the acorns experience a period of cold stratification (or drought in lower latitudes) that allows the acorn to break dormancy and then germinate (Table 5.1). In contrast, white oaks germinate immediately, often soon after they are shed and deposited on the ground. In Florida and central Mexico, some species begin to germinate while still attached to the tree. And although the dormancy is shorter for many red oaks in lower latitudes, a short period of dormancy is still evident.

THE IMPORTANCE OF SEED PERISHABILITY AND
HANDLING TIME OF SEEDS FOR SCATTERHOARDERS

A few years after the publication of Smallwood and Peters and the completion of each of our dissertations on markedly different topics, Peter and I began to share ideas regarding interactions between acorns and seed consumers and the potential for testing this tantalizing hypothesis. Concurrent with our discussions, however, two other ideas were beginning to emerge regarding the proximate factors that influence food hoarding in animals. Jim Reichman at Kansas State University was studying caching behavior in wood rats (*Neotoma floridana*) and found that this species selectively ate more perishable foods and stored items with a longer shelf life (Reichman et al. 1986; Reichman 1988). For example, Reichman (1988) found that grapes were consistently eaten when presented with blocks of rodent chow; the rodent chow, in contrast, was larderhoarded for later use. Likewise, Reichman et al. (1986) found that when wood rats stored seeds that then later grew moldy, they would eat infected seeds, and manage the other seeds to reduce the mold and increase their storability. It followed from such studies that perishability of stored food was a critical factor influencing caching decisions.

Soon after Reichman's research came the important dissertation of Lucy Jacobs at Princeton University. One of her studies directly countered this perishability hypothesis with a fully reasonable alternative. Jacobs's alternative hypothesis was simple but made very good sense for any food-hoarding animal that depended on ephemeral pulses of food, such as an autumn acorn crop. She reasoned that when food is temporarily abundant, as it is when masting species such as oaks shed their acorns, time is at a premium, and food-hoarding animals must minimize their time spent feeding in order to maximize opportunity for storing food. Thus, she argued that handling time would function as a more critical proximate cue driving caching decisions, and that in some instances, perishability may be confused with handling time as a driver of food-hoarding decisions.

Working with captive gray squirrels, Jacobs tested her handling-time hypothesis against the competing perishability hypothesis by first presenting the animals with whole, intact hazelnuts and shelled hazelnuts. Whole nuts with shells required 29.9% more time to consume than those without shells. And, as predicted, the animals selectively ate the shelled nuts and stored those that were intact. However, whole, intact nuts are also far less perishable in a cache than just the nut kernel. Hence, Jacobs followed the experiment with a second experiment in which she presented the animals with shelled hazelnuts and blocks of rodent chow—two food items of similar nutritional value but markedly different handling times and perishability. Rodent chow required 66.0% more time to

eat than the hazelnut kernels but are also substantially more perishable (blocks of lab chow
quickly disintegrate when buried in moist
soil). Consistent with the handling-time hypothesis, blocks of rodent chow were stored
at a rate of four times that of the shelled hazelnuts. Jacobs thus concluded that in many
situations, squirrels may appear to respond to
perishability when in fact handling time may
be the more important proximate factor to
which they are responding.

This was the context in which Peter Smallwood and I decided to further explore the influence of perishability and handling-time hypotheses in the dispersal and scatterhoarding
of acorns by rodents. Our goal was to follow
up with Peter's earlier experiments and his
original hypothesis that acorns of red oaks
may be selectively stored over those of white
oaks because of the differences in germination
schedules and tannin levels of the two types of

5.1. Photograph of a tagged acorn with
painted brad nail to follow seed fate. Rodents
typically cache a whole intact acorn or eat the
acorn, discard the tag, and leave behind an
obvious record of food consumption. Both
tags and tagged acorns are readily recovered
with metal detectors and large magnets.
(Shealyn A. Marino)   SEE COLOR PLATE

acorns. Red oak acorns, we predicted, should be selectively dispersed and cached
because of their delayed germination (dormancy), whereas acorns of white oak
should be eaten immediately because of their habit of germinating in autumn.

To test this, we relied on a simple procedure in which we presented free-
ranging forest rodents with tagged acorns (see Chapter 7). Metal brad nails
(3 mm × 9 mm) could easily be inserted into the basal end of acorns until flush
with the outer pericarp of the nut (Fig. 5.1). Rarely do the tags interfere with
germination. And, the rodents simply eat around the nail when consuming the
kernel (cotyledon) of the nut. Following removal or consumption of the acorns
at the original site of presentation, we then could search with metal detectors
for tagged, cached acorns, and the lone brad nails, which provide an obvious record of food consumption. In experiments in which we presented two or more
acorns in the same enclosure, we could color code the brad nails in order to distinguish the seed type or the specific enclosure from which the acorn originated.

In these experiments, tagged acorns are presented inside a small wood-frame
enclosure (1 m × 1 m × 10 m) covered with hardware cloth on all sides with small
openings on each side at ground level (Fig. 5.2). These openings allow access by
small mammals (e.g., gray squirrels, white-footed mice [*Peromyscus leucopus*] or
deer mice [*P. maniculatus*], and eastern chipmunks [*Tamias striatus*]), but prevent access by other acorn consumers (e.g., eastern wild turkey [*Meleagris gal-*

5.2. Shown are **(a)** exclosure in which tagged acorns were selectively presented to only rodents that could access acorns from openings on each side of the box. Note the gray squirrel in the blue circle. *(Andrew Bartlow)* **(b)** Flagged locations where observers found cached acorns and brad nails, indicating where acorns were cached and eaten, respectively. *(Michael A. Steele)* SEE COLOR PLATE

lopavo], white-tailed deer [*Odocoileus virginianus*], and raccoons [*Procyon lotor*]) that generally act as acorn predators. Over the years, we conducted numerous experiments of this type with various minor modifications to test different hypotheses (see Chapter 7). Immediately following the placement of acorns, rodents readily access the seeds usually within 24 hours. They then either eat acorns at their origin or disperse them outside the enclosures where they eat or cache them. In years of high acorn abundance (mast years), it often took as many as 30 days for acorns to be eaten or dispersed. But, in low mast years, we have seen most acorns eaten and dispersed within 48 hours.

In our first experiment of this type, we were interested in the response of rodents to early germinating acorns of white oak (*Quercus alba*) and those of a red oak (*Q. rubra*), which exhibit a delayed germination. We presented acorns to animals on small piles of 200 tagged red oak or white oak acorns, with each pile placed in a single enclosure. We then revisited the enclosures every few days to record removal rates until nearly all of the acorns were eaten or dispersed and cached. We then systematically searched the area under the enclosure and the area within 30 or more meters on all sides of the enclosure with the metal detector and a large magnet. When a cached acorn or brad nail was located, we flagged the location, recorded the acorn fate, and the distance and compass direction from the enclosure (Fig. 5.3).

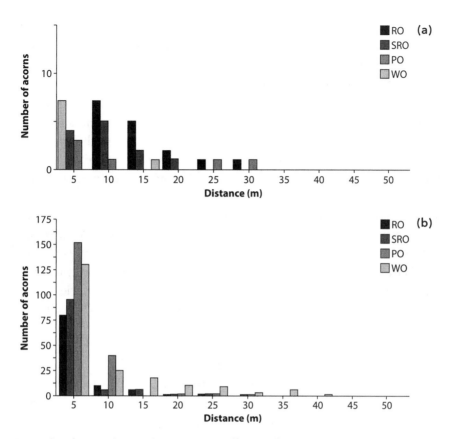

5.3. Results of acorn dispersal experiment in forests of northeastern Pennsylvania. Shown are patterns of acorns **(a)** dispersed and cached intact and acorns **(b)** dispersed and eaten. Acorn types are RO = red oak (*Q. rubra*), SRO = small-seeded red oaks, PO = pin oak (*Q. palustris*), and WO = white oak (*Q. alba*). *(Modified from Steele et al. 2007)*

Results of this first experiment were compelling. White oak acorns were consistently eaten at the origin, and, although moved short distances, rarely cached. Red oak acorns, in contrast, were dispersed farther and frequently cached. Our first attempt at testing the effect of perishability clearly supported our hypothesis that red oaks would be cached over those of white oak, but this experiment did nothing to control for the issue of handling time. It was, however, an exciting result that inspired further exploration.

We followed our first metal-tagging experiment with a more controlled experiment. Dr. Leila Hadj-Chikh, an undergraduate at the time, was determined to design and carry out the definitive experiment that tested both the perishability and handling-time hypotheses in one set of experimental trials, and she accomplished this (Hadj-Chikh et al. 1996). Leila and I spent nearly four months designing and redesigning what we considered to be the required combination of trials to test not only the perishability and handling-time hypotheses but also other factors that likely contributed to caching decisions. The experiment in fact took longer to design then to execute.

We decided to present individual free-ranging gray squirrels with pairs of acorns that varied in both acorn size (handling time) and germination schedule (perishability). Trials were carefully selected to vary either of these characteristics or other key characteristics (i.e., lipid or tannin content) that might also influence hoarding decisions. In the end, we decided on six experimental trials (Fig. 5.4). In trial one, for example, we presented acorns of pin oak (*Q. palustris*), a small-seeded red oak species with delayed germination, and larger white oak acorns, an early germinating species. Thus, the perishability hypothesis predicted that the pin oak acorns should be cached, and the white oak acorns should be eaten. In contrast, the two acorn species differ significantly in size and therefore handling time. The mass of white oak acorns typically varies between 4 and 6 g whereas that of pin oak acorns varies between 1 and 2 g. Feeding trials with gray squirrels indicated that these mass differences translated into approximately 200–250% difference in feeding time. Thus, the handling-time hypothesis predicted that the larger white oak acorns should be cached and those of pin oak should be eaten. As a second example, in trial four, the perishability hypothesis predicted that both small pin oak acorns and large red oak acorns should be cached. The handling-time hypothesis, however, predicted that the larger red oak acorns should be cached and the pin oak acorns should be eaten.

Lipid (fat) and tannin content were also included as secondary alternative hypotheses because these acorn characteristics also differ between acorns of the two oak groups. Hence, predictions from these two additional hypotheses often corresponded with either the handling time or the perishability hypothesis. In trial five, for example, we paired a white oak with the chestnut oak (*Q. montana*), an early germinating white oak species that has unusually high tannins. This allowed us to independently test the possibility that acorns with higher tannin are cached regardless of perishability. The perishability hypothesis predicted that both should be eaten, whereas the handling-time hypothesis predicted that both should be cached, and the tannin hypothesis predicted that the chestnut oak should be cached and the white oak should be eaten. Both were eaten, supporting the perishability hypothesis.

We conducted the study in a suburban park (Kirby Park, Kingston, PA), where gray squirrels were relatively habituated to human activity but still exhibited much of their normal scatter-hoarding and social behavior and were regularly subjected to other natural ecological challenges, such as predation risks (see Chapter 8).

This Kirby Park site, while somewhat simplistic compared with natural for-

		Acorn		Predictions of Alternative Hypotheses							
Experiment Number		A	B	Perishability		Handling Time		Fat Content		Tannin Content	
	1	PO	WO (L)	○	●	●	○	●	○	○	●
	2	NRO	WO (L)	○	●	○	○	○	●	○	●
	3	PO	WO (S)	○	●	●	●	○	●	○	●
	4	PO	NRO	○	○	●	○	●	○	○	○
	5	CO	WO (L)	●	●	○	○	●	●	○	●
	6	WO (S)	WO (L)	●	●	●	○	●	○	●	●

○ cache   ● eat

5.4. Experimental design and overview of results of a study testing the perishability and handling-time hypotheses. Shown are the acorn pairs presented to free-ranging animals in each experimental trial and the predictions (cache or eat) for four competing hypotheses (the perishability, handling-time, and two additional hypotheses based on tannin and lipid content). Shaded blocks indicate where results of the experimental trial statistically support the predictions. Acorn species are pin oak (PO, *Q. palustris*), white oak (WO, *Q. alba*), NRO (norther red oak, *Q. rubra*), and chestnut oak (CO, *Q. montana*). S and L refer to small and large acorns, respectively. *(Figure modified from Hadj-Chikh et al. 1996)*

est stands, had many advantages for experimental work. For example, the site afforded us the opportunity to conduct experiments with large numbers of free-ranging squirrels and to make direct behavioral observations on individual animals in a seminatural setting. As a result, we could assess responses of individual animals in daily experimental trials. This allowed us to generate statistically meaningful sample sizes and address important questions that simply could not be attempted in a forest. Despite the occasional reminders of my closest friends and colleagues that my research now involved "feeding squirrels at the park" or my own father's witty quip that I was becoming "too soft for field biology," I was convinced that this site provided the perfect setting for some of my experiments. This decision was bolstered by the sound advice of one of my former graduate professors, Dr. William Connors: "Never let anyone tell you that you can't do meaningful research at a small university; you just need to be creative about how you do it." And, so it was, that several of my lab's studies on scatterhoarding and oak dispersal by eastern gray squirrels were based on research at this site (Steele and Smallwood 1994; Hadj-Chikh et al. 1996; Steele et al. 1996, 2001ab, 2011, 2014, 2015).

Leila Hadj-Chikh's experiment on perishability and handling time was one of the first to be conducted at the Kirby Park site. For each trial, we sequentially presented individual gray squirrels with one pair of acorns, randomly determining the order of acorn type for each of approximately 20 squirrels per trial. For each acorn, we determined whether it was eaten or cached, the time to eat or cache the acorn, and the distance the acorn was moved (dispersed) prior to consumption or caching. The primary results were based on the feeding and caching decisions. When the results were in for all six trials and statistical analyses was complete, the results were clear: there was overwhelming support for the perishability hypothesis. Regardless of acorn size and thus handling time, squirrels selectively cached the acorns that showed delayed germination (i.e., those of the red oak group [*Lobatae*]). Results in all six trials supported the perishability hypothesis, whereas none of the six experimental trials supported the alternative handling-time hypothesis. Because predictions for both the lipid and the tannin hypotheses sometimes coincided with that of the perishability hypothesis, results of some trials supported these hypotheses. However, acorn pairs in each experimental trial were specifically chosen so that in at least some of the trials opposite predictions followed for the perishability and lipid hypotheses (three trials), and the perishability and tannin hypotheses (one trial). In all trials, results supported the perishability hypothesis.

This study clearly demonstrated for the oaks we studied that perishability was an overwhelming determinant of the caching decisions of gray squirrels, at least in the context in which we approached this problem. However, Jacobs's handling-time hypothesis is compelling, nonetheless. It makes abundant sense for any food-hoarding animal trying to maximize its caching efficiency during autumn seed fall. I therefore return to this issue later in this chapter when exploring the caching decisions of the Mexican gray squirrel (*Sciurus aureogaster*)—a species for which seed perishability, although relevant, seems to be less of a driver in their caching decisions.

THE EFFECTS OF INSECT INFESTATION ON ACORN PERISHABILITY AND CACHING DECISIONS

I continue with the issue of perishability of acorns and its impact on scatter-hoarding decisions. It followed that if the shelf life of acorns influenced these caching decisions, then other factors that influence perishability other than germination schedules may also drive caching decisions. Thus, concurrent with Hadj-Chikh's experiment, we also sought to determine how insect larvae might influence the storability of acorns. Indeed, other studies have suggested that infested seeds are often recognized as such and should be rejected.

The larvae of several insects frequently infest acorns and feed on the cotyledon of the fruit before moving on to the next stage in their development (see Chapters 4 and 11). Larvae of the genus *Curculio* are the most common insect predator of acorns. Following metamorphosis and emergence from below ground, adult beetles ascend nearby oak trees and mate. Within a few weeks the females then oviposit in the acorns. This usually takes place late in the development of the acorns a few weeks before acorns are shed from the tree. Following seed fall, larvae excavate an exit hole in the pericarp of the acorn, through which they emerge and then burrow into the soil. The larvae then undergo a diapause of 1–3 years before metamorphosing into adults and emerging from the soil.

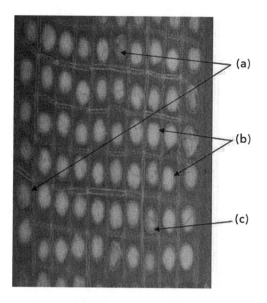

5.5. Shown are **(a)** an X-ray image of white oak (*Quercus alba*) acorns, showing acorns infested with weevil larvae, **(b)** sound acorns, and **(c)** a fungal-damaged or rotted acorn. *(Michael A. Steele)*

Although weevil infestation of acorns can on occasion exceed 90% of the annual crop of an individual tree, infestation rates typically average 30% or less. Even at this rate, however, infested acorns can significantly influence the quality and storability of the acorn. If cached with larvae inside the acorn, the larvae may continue to feed on the cotyledon. Whereas most acorns when infested house a single weevil, we have observed as many as 15 larvae in a single red oak acorn. And whereas one larva may inflict only partial damage to the cotyledon, the potential for complete destruction of the cotyledon prior to recovery of the cached acorn is high.

We therefore suspected that squirrels might be equally sensitive to the perishability of acorns caused by weevil infestation. Two students decided that this was an ideal follow-up to the perishability-handling time experiment. But how were we to test the animal's response? If we simply presented infested acorns with exit holes, the obviously damaged acorns were likely to be rejected. We decided to try to catch infestation early enough in the season so that mature acorns still housed weevils but damage to the cotyledon was minimal.

One of the students, also a hospital technician, suggested X-raying the seeds. We tested it and it was very effective. X-rays allowed us to distinguish sound acorns from both those infested with weevil larvae and those that were damaged by fungi or mold, the latter of which we discarded (Fig. 5.5). Recognizing the opportunity, the student then befriended the lead director of the hospital's radiology lab and one day returned from the hospital with a stack of X-rays of thousands of red oak and white oak acorns. I avoided any additional questions about how she secured these X-rays, and we sorted the acorns and ran another experiment with gray squirrels at Kirby Park in much the same way we had conducted the perishability and handling-time study.

This time we compared sound red oak acorns with those infested with larvae, and both infested and noninfested white oak acorns as well. As a control, we shelled acorns of both species to determine if squirrels also recognized these intact, but exposed cotyledons as perishable (Steele et al. 1996).

As predicted, sound red oak acorns were selectively cached compared with infested acorns of the same species. The infested red oak acorns were eaten. In contrast, both infested and noninfested white oak acorns were consistently eaten. And, shelled acorns of both species were eaten. In all cases the acorns that were perishable—due either to insect infestation or germination schedule—were selectively consumed. The only nonperishable acorns, the sound acorns of *Q. rubra*, were stored.

Even more striking was the response to the weevils. When the squirrels encountered a weevil, they consistently ate it. The infested acorns were not rejected but consumed. The sound acorn was quickly sequestered in the cache. Perhaps the squirrels sought the weevils? Consumption of animal material by squirrels is widely reported in the literature (Koprowski 1994). From a variety of arthropods to bird eggs, young nestlings, and even rabbits, the diet of squirrels is far more eclectic than suggested by their simple designation as granivores—those that specialize on seeds and nuts (Koprowski et al. 1994; Steele and Koprowski 2001). However, using rough calculations of the costs and benefits of opening acorns to only consume weevils, we concluded that the caloric value of a weevil simply does not support such a strategy (Steele et al. 1996). And, regardless, it was clear the squirrels were consuming both the acorn cotyledon and the weevils.

We concluded instead that the weevils were likely an important nutrient supplement. Acorns represent a complicated food item. While an important source of energy (from the lipid in the cotyledon), the tannins they contain bind to both salivary and digestive enzymes and, as a result of the latter, reduce digestibility of the cotyledon. While eastern gray squirrels have been reported to acclimate to high tannin foods with changes in their enzymes (Chung-MacCoubrey et al. 1997), it appears that this response is still not significant enough for gray squirrels to rely only on high tannin acorns for any extended period of time. In fact, when raised on only acorns of red oak, gray squirrels will lose approximately 2 g of body mass per day, which, over a period of three weeks, puts a gray squirrel at significant risk of mortality (Havera and Smith 1979; Steele personal observation).

Perhaps then, the weevils represent a critical protein supplement that counters the negative effects of tannin? One of the surprises from this experiment

Table 5.2. Response of rodent species of eastern deciduous forests to acorns of red oak (*Q. rubra*) and white oak (*Q. alba*). The "X" indicate where data from published studies and unpublished observations by the author support either the selective dispersal or selective caching of red oak acorns over those of white oak.

SPECIES	DIFFERENTIAL DISPERSAL	DIFFERENTIAL CACHING
Gray squirrel (*Sciurus carolinensis*)	X	X
Western fox squirrel (*S. niger rufiventer*)	X	X
Eastern chipmunk (*Tamias striatus*)	X	X
White-footed mouse (*Peromyscus leucopus*)		X
Deer mouse (*P. maniculatus*)	X	X
Southern flying squirrel (*Glaucomys volans*)		X

was the consistency with which the squirrels consumed the weevils. But even more compelling was the accuracy with which the acorns were detected. With X-rays, we were about 85% accurate at differentiating infested acorns from sound acorns. But after the experiments were completed and the data were analyzed, to our amazement, the squirrels were successful at detecting infested acorns in about 93% of the trials. Only discovered years later by Jacobs's team, the squirrels were likely relying on a rapid head shake to detect the soundness of the acorns (Preston and Jacobs 2009; see Chapter 9).

In China, my colleagues and I have reported that the Siberian chipmunk (*Tamias sibiricus*) frequently shells acorns of *Q. mongolica* before caching them. When presented with insect-infested acorns and sound acorns, chipmunks consistently shelled the acorns and cached those that were sound (Yi et al. 2012b). More recently Yang et al. (2018), working in the same system, showed that this chipmunk eats the weevils and, by caching shelled acorns, reduces losses to pilfering as well as insect and fungal damage. And, as a result, these cached shelled acorns are more likely to establish as seedlings. Similar preferences of rodents for sound acorns of *Q. aliena* over those infested with insects has also been demonstrated by my colleagues in other areas of China (Zhang et al. 2018).

The keen sensitivity of rodents (caviomorphs, murids, sciurids) to the perishability of seeds extends across the globe. As we review in Lichti et al. (2017), many of these rodents use emergent shoots to detect germinating acorns and other seeds prior to pilfering caches, or they prune germinating seeds before caching them. In the Central Hardwoods Region of the United States, we have found that nearly all rodents that handle acorns are sensitive to their perishability. Moreover, in both captive and field studies with different rodent species, my colleagues and I have found that gray squirrels (*Sciurus carolinensis*), the western fox squirrel (*S. niger rufiventer*), mice (*Peromyscus* spp.), eastern chipmunk (*Tamias striatus*), and southern flying squirrel (*Glaucomys volans*) cache acorns of red oak species over those of white oaks (Table 5.2).

HOW DO SQUIRRELS DETECT SEED DORMANCY?

The ability of many rodents to detect that dormancy has broken when acorns are germinating should not come as a surprise. However, that some rodents distinguish dormancy when acorns have no emerging radicle raises an interesting question: How is dormancy detected? In the experiments of Hadj-Chikh et al. (1996), we used acorns without any visible signs of germination, but gray squirrels consistently cached acorns of red oak species over those of white oaks regardless of size (Hadj-Chikh et al. 1996). As discussed in Chapter 3, we tested this question by presenting acorns that we altered with respect to either the outer pericarp (shell), the internal cotyledon, or a shell made odorless by first soaking it in acetone (Steele et al. 2001a). Squirrels cached all artificial acorns constructed with a red oak shell regardless of its internal contents. Only when red oak shells were first soaked in acetone, potentially removing any volatile compounds, were these acorns eaten. Sundaram (2016) took these initial experiments much further to demonstrate that dormant acorns are coated with a plant wax that likely prevents squirrels from detecting odors, thereby causing them to store these "dormant" acorns (see Chapter 3).

EMBRYO EXCISION BY SQUIRRELS

Despite the above similarities in the response of many acorn consumers to the perishability of acorns for caching, there is one fascinating behavioral adaptation nearly exclusive to the family Sciuridae (squirrels), which seems to set them apart from almost all other acorn scatterhoarders—the behavior of embryo excision. As described in Chapter 3, this response to early germinating acorns (such as those of white oak species, section *Quercus*) involves first the detection of nondormancy. Once detected, the nondormant acorn is flipped apical-end upward and the squirrel inserts its incisors at the apical tip, and with a few quick scrapes of these front teeth, removes the embryo. The result is an acorn that will not germinate (Fox 1982) but will store well for up to six months with its cotyledon intact (Steele et al. 2001b; see Fig. 5.6).

As far as we know, this behavior is common for gray squirrels across their native range. In subsequent studies, and in observations by other researchers, it became evident that in eastern deciduous forests, gray squirrels were the only species performing this behavior. Although other rodents in these forests seemed to show the same preference for storing acorns of red oak over those of white oak species, my colleagues and I observed none of these species to perform embryo excision. And, although one colleague noted embryo-excised acorns in the nest boxes of white-footed mice, it was later reasoned based on observations of captive individuals that the acorns deposited in the nest boxes were likely collected after they were excised by squirrels. In the autumn during rapid seed fall, gray squirrels will often excise hundreds of acorns under maternal trees to arrest germination, and then return to recover, disperse, and cache them. Thus, it seems logical that other rodents may occasionally recover excised acorns. However, when maintained in captivity, both southern flying squirrels and white-footed mice showed no evidence of the behavior, leading us to suspect that embryo excision may be limited to the gray squirrel.

That is, until 1997 when I found myself in central Mexico working on oaks and acorn dispersal by rodents in the epicenter of oak diversity—the southern portion of the Sierre Madre Oriental and the surrounding environs. I soon learned that the behavior of embryo excision of nondormant acorns was also frequently performed by Mexican gray squirrels (*Sciurus aureogaster*), a close congener of the eastern gray squirrel. And, soon thereafter, I heard anecdotal reports of embryo excision by the western fox squirrel in the Midwestern US, although I am yet to fully confirm the behavior in this species.

THE INNATE BASIS OF THE BEHAVIOR OF EMBRYO EXCISION

The growing evidence that the behavior of embryo excision was limited to squirrels led us to speculate that perhaps there was some inborn (genetic) basis for this behavior. I carried this nagging thought with me for a few years, until a few colleagues, students, and I decided to tackle the question. Dr. Michael Pereira, then at a neighboring institution, was studying physiological characteristics of gray squirrels and was housing a group of young squirrels born in captivity at the university's primate facility. I met with Michael and soon learned that the young squirrels had never eaten acorns. It is rare to meet a gray squirrel that can make such a claim and rarer yet to be sure that it was true. When Michael assured me of the controlled diet on which they were maintained, a plan was hatched. Michael, a primatologist by training, was completing some of his research on the animals and agreed to turn them over to me for their ultimate experience—a date with their first acorn.

To do this we needed to plan carefully and ensure that we tested each individual's response to acorns. With the help of a few students and my son, Tyler,

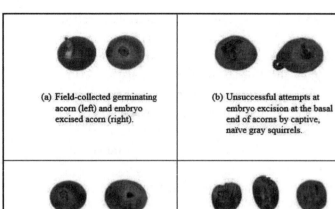

(a) Field-collected germinating acorn (left) and embryo excised acorn (right).

(b) Unsuccessful attempts at embryo excision at the basal end of acorns by captive, naïve gray squirrels.

(c) Unsuccessful (shallow) attempts at embryo excision (left) and successful embryo removal (right) by captive, naïve gray squirrels.

(d) Unsuccessful attempts at embryo excision at both apical and basal ends of acorns by captive, naïve gray squirrels.

5.6. Northern red oak acorns (*Q. rubra*) with evidence of **(a)** embryo excision by wild gray squirrels (*Sciurus carolinensis*) and **(b, c, d)** both successful embryo excision and unsuccessful attempts at embryo excision by captive, naïve squirrels. *(Michael A. Steele)*

we first constructed a large outdoor enclosure (8 m × 2.5 m × 2.5 m) on my personal property. Then we transferred 10 eastern gray squirrels—four wild-caught squirrels and six squirrels reared in captivity. We also acquired two additional squirrels from Princeton University, where my former student, Leila Hadj-Chikh, was completing her PhD. She too had raised several squirrels in captivity and had not presented them with acorns. We then maintained these 12 squirrels in captivity by providing adequate nest structures and a balanced, acorn-free diet for another 11 months before conducting our experiments. In preparation of our experiments, we constructed an individual cage (1.0 m × 2.2 m × 1.0 m) for each squirrel. We provided each squirrel with two small nest boxes attached inside the top of the small enclosure, and a water bowl, and a sand box on the floor. We also covered each cage with opaque, black plastic on three sides, leaving the face and small door uncovered and accessible by observers. We then positioned all cages so that squirrels were unable to observe any other squirrel during the experimental trials.

We conducted the experiments between early October and mid-December 2001, so they coincided with the same period of autumn seed fall and the peak period in which free-ranging squirrels were storing acorns. We conducted five experiments. In each experiment we presented each squirrel with eight dormant red oak (*Q. rubra*) and eight nondormant acorns, each day for three consecutive days. In each of the five experiments, dormant red oak acorns were paired with non-germinating but nondormant white oak acorns (experiment 1); germinating white oak acorns (experiment 2); germinating chestnut oak acorns, a white oak species that also exhibits no dormancy (experiment 3); red oak acorns that had broken dormancy and were germinating (experiment 4); and red oaks that had broken dormancy but were not germinating (experiment 5). The acquisition of non-germinating red oak acorns meant that red oaks had to be collected during the autumn of 2000 and stored in the cold until they emerged from dormancy.

During the experiments, squirrels were presented acorns in the morning hours. Then, each evening, we carefully inspected the entire cage and two nest boxes and sifted the sand for any acorns that were stored since the morning's acorns were presented. At the end of each day, we collected all the acorns and made sure each squirrel had ample access to additional food (e.g., laboratory chow, seeds, and fruits). Originally, we hypothesized that squirrels might cache the dormant red oak acorns over the non-germinating acorns and indeed that is what we observed. Across all trials

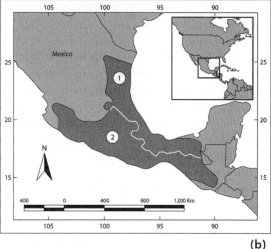

5.7. Shown are **(a)** the Mexican gray squirrel (*Sciurus aureogaster*). *(Michael A. Steele)* **(b)** Mexican gray squirrels' distribution in Mexico and Central America. *(From Koprowski et al. 2016)*

combined, more dormant acorns of red oak were stored than the nondormant acorns. However, this was most compelling (statistically significant) in the first two trials and then dropped off in the subsequent three trials. As we suggest in the final published work (Steele et al. 2006), it is likely that acorn caching dropped in later trials because we were repeatedly robbing the animals' caches and/or because the differences between the paired acorns became increasingly more subtle with each subsequent experiment.

Regardless, captive-reared squirrels—those without any previous experience with acorns—could distinguish dormant red oak acorns. But even more exciting was the evidence of embryo excision. Embryo excision was performed or attempted on eight acorns by four of the captive-reared squirrels and, as expected, by all four wild-caught squirrels. Interestingly, however, when attempted by the naïve animals, embryo excision was often not fully successful at removing the embryo. In fact, on a few occasions, these inexperienced animals attempted embryo excision on the wrong part of the acorn (Fig. 5.6). Hence, there is indeed a genetic basis for this behavior, although it may require further refinement through learning, either by observation of adults or trial-and-error learning. Fox (1982) too reported a few young squirrels that attempted the behavior but were unsuccessful at removing the entire embryo.

SOUTH OF THE BORDER: BROADENING THE PERSPECTIVE ON CACHING DECISIONS

Long before we conducted the captive experiments demonstrating an innate basis of this behavior in gray squirrels, I was afforded an ideal opportunity to extend my research to central Mexico with the generous support of both Wilkes University and the US Fulbright Foundation. From 1997 to 1998, I had the life-altering opportunity to accompany my entire family to the enchanting, historic city of Puebla, Mexico, where my wife Margaret, our daughter Emily, sons Tyler and Michael, and I logged the most memorable experience a family could ever imagine. And, in the midst of this family adventure, I also had the wonderful opportunity to conduct research in the epicenter of the world's oak diversity. With as many as 150 species of oaks (depending on the taxonomist you consult), Mexico boasts the highest diversity of oaks in the world.

Following six months of intense planning, countless discussions with the Mexican consulate in Philadelphia on how best to prepare and execute our journey, and the long, arduous but truly memorable trip from NE Pennsylvania to Puebla, we arrived by jeep, with a modified utility trailer and an impressive load of various field and lab gear—everything I thought I would need to conduct a year of study on oak ecology in Mexico.

Among the several field and lab studies I hoped to accomplish in Mexico was a parallel study to that of Leila Hadj-Chikh's comparing the Mexican gray squirrel's (*Sciurus aureogaster*) feeding and caching responses (Fig. 5.7). However, before such a study could be attempted, I needed to introduce myself to the oaks. Fortunately, one of the hosts for my Fulbright Fellowship, Maricela Rodriguez, knew the oaks of Mexico well, perhaps as well as anyone in the world. At the time of my visit, Maricela, Director of the Jardin Botanico "Louise Wardle de Camacho" in Puebla and the Herbarium at the Universidad de Benemerita in Puebla was working closely with colleagues at the Nacional Autonoma de Mexico (UNAM) in the Federal District of Mexico (Mexico City).

The year prior to our stay in Mexico, my eldest son, Michael, and I spent two weeks in and around the massive metropolis of Mexico City, exploring the sites, introducing ourselves to the enchanting culture and people of Mexico, and attempting oak seedling surveys in the forested volcanoes surrounding the valley that is the Federal District of Mexico, Mexico City. One of the highlights of our trip was a visit with Dr. Gerardo Cellabos, a mammalian ecologist at UNAM and a leading international authority on the conservation of mammals. Gerardo hosted us for a wonderful barbeque at his home in Toluco, and then accompanied by his wife and young children, Gerardo introduced us to a field site where my son and I spent a week or more measuring and mapping oak seedlings. One of my fondest memories of this trip was measuring seedling densities on our hands and knees in the pouring rain when my son looked up to pronounce this day, his 16th birthday, "the best birthday ever!" This was either a clear indictment of my past parenting skills or just simply one of those rare moments between a father and son. I took it as the latter.

During our brief stay in Mexico that year we also had the opportunity to visit the Instituto de Ecología at UNAM where we were introduced to Gerardo's colleagues and some of the graduate students working out of the center. That afternoon we also wandered next door to the Instituto de Biología to the National Herbarium to see if we could peruse the oak collection. There, the director of the herbarium introduced himself and, as luck would have it, his visiting colleague—Maricela Rodriguez Acosta. It was this fortuitous encounter that set the stage for a year with the oaks of Mexico.

Fast forward 15 months and I am back in Mexico working out of Benemerita Universidad de Puebla. Just a few weeks in and I was beginning to grasp some of the many challenges before me, not the least of which was identifying the oaks. Among the red oaks in particular, many species shared the same leaf structure, leaf arrangement, and even acorn morphology. To all but the taxonomists they were often indistinguishable. And, on a number of occasions I would return to the lab to proudly pronounce the acorn species I had collected in the field only to have an undergraduate or technician humbly shake their head in disappointment. Overcoming these challenges again depended on some good fortune as well as perseverance. The good fortune came in the form of Mr. Juan Radillo, a Mexican native who was pursuing a master's degree in the wildlife ecology program at the University of Maryland, Frostburg. Introduced to me by a former student, Juan elected to join me in Mexico and conduct some of his research on oak-animal interactions for his thesis. Juan quickly became my right arm. He first flew down several weeks ahead of me and worked his way through the immense red tape that characterizes the paperwork, time, and frustration involved in securing permits from the Mexican government to not just live-trap small rodents, but to collect acorns. Yes, collect acorns. Maricela warned me about this hurdle by citing an example of a federal conservation officer who was stopped, questioned, and jailed for a few days in a small village for collecting acorns without a permit, and then only released after he was able to verify his identity. Juan secured the permit before my arrival, learned how to identify many of the oaks, and even helped my family and me navigate the contract for an apartment.

Juan also taught me some simple tricks of how to interact with the local citizens and, on occasion, unexpected encounters with wildlife and even wild dogs. On one early occasion, Juan and I were in the field, just south of Chulula, Mexico collecting acorns on a hillside in an open grove of mature white oaks. At the top of the hill I heard what I soon gathered was a pack of howling wild dogs, headed in our direction. Juan immediately shouted, "Mike, you might think about picking up a rock." My first reaction was to find a tree to climb. And, in my panic to do so I must have looked a bit comical when Juan shouted again about the rock. I was not having anything to do with fighting off a pack of wild dogs. But then he made it clear, "Mike—just pick up a rock!" And, I did, just as these hounds were within 20 meters. The dogs saw me with the rock and off they went. Juan knew from his undergraduate days in veterinarian medicine at UNAM, in which he was required to "collect" a feral dog for autopsy, that wild dogs were accustomed to being pelted if they caused trouble. And so, it was within a few weeks of working and interacting with Juan, that he was thereafter dubbed "Juan in a Million."

It was Juan that I charged with conducting experiments that paralleled those of Hadj-Chikh et al. (1996). After several weeks of collecting acorns from more than a dozen oak species, from several sites in and around Puebla, including the forested mountainsides surrounding the volcano Popocatepetl, we carefully selected acorn pairs of both red and white oak species for the experimental trials. In total, we were able to use 10 species of acorns and create 17 different acorn pairs (for a total of 18 experimental trials, 13 of which predicted opposite predictions for the perishability hypothesis) (Fig. 5.8). This contrasted considerably with the simple six experimental paired treatments used by Hadj-Chikh et al. (1996).

The greater number of acorn species available to us, coupled with additional variation in seed size and even variation in germination schedules for the same species (e.g., *Q. crassifolia*), allowed us to create a greater range of paired treatments to test the perishability versus handling-time hypotheses. Acorn sizes varied from less than 1 g to a nearly 5 g across species, while some acorn species also varied significantly with respect to acorn size (the white oak, *Q. obtusata*, and the red oak, *Q. crassipes*).

The study site we chose was in the central region of Mexico City, close to where Juan's family lived. The site, Vivero Coyoacán, DF, is a 1 km² forested urban park, with a high density of Mexican gray squirrels (*S. aureogaster*). The site was ideal for our study for several reasons: the squirrels were habituated to human activity, they regularly cached or ate acorns when presented to them, and

Acorn Pair		Alternative Hypotheses			
		Perishability		Handling Time	
Q. microphylla (SWO)	Q. mexicana (SRO)	●	○	●	●
Q. microfila (SWO)	Q. acutifolia (SRO)	●	○	●	●
Q. conspersa (LRO)	Q. acutifolia (SRO)	○	○	○	●
Q. conspersa (LRO)	Q. obtusata (LWO)	○	●	○	○
Q. mexicana (SRO)	Q. obtusata (LWO)	○	●	●	○
Q. acutifolia (SRO)	Q. obtusata (LWO)	○	●	●	○
Q. obtusata (SWO)	Q. obtusata (LWO)	●	●	●	○
Q. obtusata (SWO)	Q. mexicana (SRO)	●	○	●	●
Q. obtusata (SWO)	Q. acutifolia (SRO)	●	○	●	●
Q. crassifolia (SRO, germ)	Q. acutifolia (SRO)	●	○	●	●
Q. crassifolia (SRO, germ)	Q. crassifolia (SRO)	●	○	●	●
Q. crassifolia (SRO, germ)	Q. mexicana (SRO)	●	○	●	●
Q. glaucoides (SWO)	Q. acutifolia (SRO)	●	○	●	●
Q. obtusata (SWO)	Q. obtusata (LWO)	●	●	●	○
Q. crassipes (LRO)	Q. crassipes (SRO)	○	○	○	●
Q. rugosa (LWO)	Q. laeta (LWO)	○	●	○	○
Q. laeta (LWO)	Q. obtusata (LWO)	●	●	○	○
Q. laeta (LWO)	Q. conspersa (LRO)	●	○	○	○

● cache ○ eat

5.8. Effects of acorn perishability (red versus white oak species) and size (handling time) on caching decisions by Mexican gray squirrels (*Sciurus aureogaster*) in central Mexico. Shown are predictions of whether to cache (black circles) or eat (white circles). Predictions were supported where boxes are shaded. LRO, SRO, LWO, and SWO represent large red oak, small red oak, large white oak, and small white oak species, respectively. (*Michael A. Steele, unpublished data*)

each animal possessed unique pelage characteristics. Squirrels varied in markings from solid gray to a mixture of orange-reddish patches, mixed with gray and even completely melanistic individuals. This made identification of individuals possible during each daily experimental trial.

During late autumn of 1997, we conducted each experimental trial on a single day. During each of the 18 trials, we presented the same pair of acorns to each of approximately 20 squirrels. The order of acorns in each pair was altered randomly with each squirrel. For each acorn, we recorded whether the acorn was eaten or cached, the caching and eating times, and, if cached, the dispersal distance.

The results were opposite of what we expected. As I tell my students, this is usually a sign of something exciting to come and certainly a clear indication that

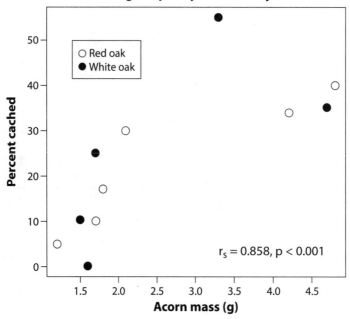

Caching Frequency of Acorns by Mass

○ Red oak
● White oak

$r_s = 0.858, p < 0.001$

Percent cached

Acorn mass (g)

5.9. Caching frequency of acorns of five species of red oak (section *Lobatae*) and six species of white oak (section *Quercus*) by Mexican gray squirrels (*Sciurus aureogaster*). As shown, caching frequency increases with acorn mass (handling time). (*Michael A. Steele, unpublished data*)

we did not just erroneously produce results we expected to see. Perishability, it appeared, drove none of the initial caching decisions of these squirrels. In fact, perishability was not supported in any of the 18 trials. Moreover, the handling-time hypothesis was supported in 13 of the 18 trials (Fig. 5.8). In other words, the larger acorns (those with substantially higher handling costs) were selectively cached over smaller acorns. The probability of caching was directly related to acorn size regardless of whether the acorn was a red oak or a white oak (Fig. 5.9).

This was an interesting outcome but also one for which I did not have an immediate explanation. But when Juan handed me one of the crumpled paper cups in which he stored the cached acorns he recovered, a more complete picture emerged. I opened one paper cup and inside I found two relatively large white oak acorns, both with their embryos excised. And, when all the data were collected, among the cached acorns that were white oak species, many had their embryos excised (Steele et al. 2001). In other words, the Mexican gray squirrel was indeed sensitive to the early germination in white oaks. The perishability due to early germination did not just drive their immediate decision of whether to cache or eat an acorn.

My best explanation for this followed from the environment in which these animals were living. Yes, the squirrels stored acorns. And yes, they could detect acorn perishability (acorn dormancy). However, perishability was just not a top priority when making a decision on whether or not to cache a seed. The milder environments and the conditions for stored acorns may explain the result. The reduced seasonality in the temperate forests of central Mexico mean that squirrels are not required to store food for extended periods of time as compared to an eastern gray squirrel in the Central Hardwoods region of the United States. In addition, in Mexico, storage time may also be limited by competition with other acorn predators such as ants, which I have observed frequently attacking acorns soon after seed fall in Mexico. Thus, the Mexican gray squirrel is only caching for a short period of time and, if given the choice, it chooses the larger, more profitable acorn in a pair. If this larger acorn is not dormant, then it also pays to quickly remove the embryo, so the cotyledon is not lost to germination during storage. These results suggest that squirrels may have a hierarchy of criteria on which they base caching decisions, and, given varying conditions, the priority of each criterion in the hierarchy may shift.

EXTENDING THE GLOBAL PERSPECTIVE ON EMBRYO EXCISION

Following our publication in 2008 (Steele et al.) in which we found evidence for an innate basis of the behavior of embryo excision in eastern gray squirrels, I soon learned that a colleague was turning up evidence of embryo excision by a squirrel in Southeast Asia. Dr. Zhishu Xiao, of the Chinese Academy of Sciences, who I had met earlier at a conference in Brisbane, Australia, published a paper with coauthors in 2009 demonstrating that the Pallas's squirrel (*Callosciurus erythraeus*) performs embryo excision on acorns, predominantly those of *Q. variabilis*, a white oak species (Xiao et al. 2009). Soon thereafter, Zhishu and another colleague, Zhibin Zhang, asked me to coauthor a study (Xiao et al. 2010) on how acorn abundance influences the frequency of embryo removal in another squirrel species, Pére David's rock squirrel (*Sciurotamias davidianus*). And by 2012, Zhishu and Zhibin reported yet a third species, the Asian red-cheeked squirrel (*Dremomys rufigenis*), that also performed the behavior.

What was particularly compelling about the third observation was that it was made in a National Nature Reserve (in Yunnan Province) where two white oak species (*Q. variabilis* and *Q. franchetii*) were extirpated from the site several hundred years earlier and the closest extant populations of white oaks were found greater than 5 km from the study area. Thus, the red-cheeked squirrel appeared to possess an innate tendency for the behavior (Xiao et al. 2010) as we had reported for the eastern gray squirrel (Steele et al. 2008).

5.10. The species of squirrels now reported to perform embryo excision on non-dormant acorns. Shown are **(a)** the eastern gray squirrel (*Sciurus carolinensis*). *(Barbara Evans)* **(b)** The Mexican gray squirrel (*S. aureogaster*). *(Herson Guevara)* **(c)** The western fox squirrel (*S. niger rufiventer*) *(Steven Mlodinow)* **(d)** Pallas's squirrel (*Callosciurus erythraeus*). *(Bryan J. Smith)* **(e)** Pére David's rock squirrel (*Sciurotamias davidianus*). *(Devotram Thirunavakkarasu)* **(f)** The Asian red-cheeked squirrel (*Dremomys rufigenis*). *(Sheau Torng Lim)* SEE COLOR PLATE

Overall, at the time that this book was written, the behavior of embryo excision by rodents was reported in the literature for four genera of squirrels occurring across three subfamilies within the family Sciuridae—the subfamily Sciurinae (which includes the Holarctic tree squirrels and flying squirrels), the subfamily Callosciurinae (which includes the Southern Asian tree squirrels), and the subfamily Xerinae (which includes all three tribes of ground squirrels). Therefore, the taxonomic breadth over which this behavior was reported, coupled with further evidence that some species of flying squirrels (*Glaucomys volans*), ground squirrels (*Tamias striatus*), and Holarctic tree squirrels (*Sciurus vulgaris* and *Tamiasciurus hudsonicus*) appear not to perform the behavior, suggests that behavior may have evolved independently in several species and locations across the globe (Figs. 5.10 and 5.11).

Such convergent evolution of the behavior in different lineages appears to be the best explanation for the evolutionary history of this adaptation in the squirrels. However, it is also worth noting that the behavior has been observed on a limited basis in a few individuals of the Korean field mouse (*Apodemus peninsulae*). Xianfeng Yi, a former postdoctoral researcher from my lab, directed a captive study on individual mice of this species in China (Yi et al. in prep.)

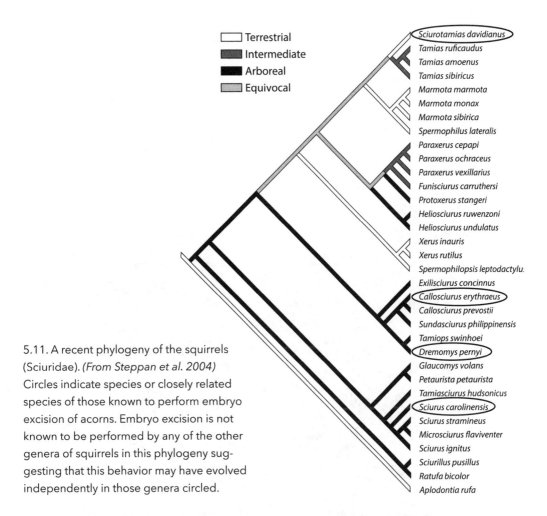

5.11. A recent phylogeny of the squirrels (Sciuridae). *(From Steppan et al. 2004)* Circles indicate species or closely related species of those known to perform embryo excision of acorns. Embryo excision is not known to be performed by any of the other genera of squirrels in this phylogeny suggesting that this behavior may have evolved independently in those genera circled.

when consuming and caching acorns of *Q. variabilis*, an early germinating white oak, and *Q. acutissima*, a species with delayed germination. In these experiments, however, mice only performed the behavior on a small percentage of only the germinating acorns of *Q. variabilis* and often failed at removing the embryos.

This lower rate of evolutionary embryo removal, performed only on acorns with emerging radicles by a limited number of individual mice, coupled with the fact that the behavior is often unsuccessful at arresting germination, strongly suggests that this behavior may be in the early stages of evolutionary convergence in a species other than a squirrel (Yi et al. in prep.). The behavior appears absent in other populations of the wood mice in China as well as in other rodent species of several genera that are known for scatterhoarding acorns (*Apodemus* spp., *Niviventer* spp., *Maxomys* spp., and *Rattus* spp.) (Xiao et al. 2009; Cao et al. 2011).

In the next chapter, I continue the discussion on how acorn characteristics influence patterns of acorn predation and dispersal, with a particular emphasis on traits that are best interpreted as adaptations of the oak for escaping predation.

THE OAK DISPERSAL PROCESS: INTRINSIC FACTORS II

6

INTRODUCTION

Chapter 5 focused on rodent responses to basic characteristics of the oaks. I now shift the discussion to the adaptive significance of some of the acorn's characteristics with respect to their seed predators. This review centers on (1) the adaptive significance of early germination in acorns of white oak species for escaping acorn pruning and predation by birds and rodents, (2) the adaptive significance of chemical gradients in acorns for facilitating tolerance to acorn predation by a diverse suite of insect and vertebrate seed predators, and (3) an alternative view of the acorn as not just a package of energy for the developing seedling but also as a fruit for the promotion of dispersal and tolerance to seed predation. I close this chapter by briefly revisiting the topic of multiple seeded acorns and the potential for this characteristic to thwart embryo excision by squirrels.

THE ADAPTIVE SIGNIFICANCE OF EARLY GERMINATION IN WHITE OAKS

Early germination in acorns of white oak species is widely held in the literature as an adaptation in the oaks that allows the acorn to escape predation by rodents as well as insect larvae (Lewis 1911; Barnett 1977; Fox 1982; Vander Wall 1990; Steele et al. 2005; Steele 2008). The rapid transfer of energy in the cotyledon into a fibrous radicle, heavily fortified with lignin and cellulose, results in a taproot that is universally avoided by rodents. Thus, the early germination in white oaks is, in part, a race against time.

During seed fall, especially during years of heavy mast production, rodents can consume and sequester only so many acorns before many begin to germinate in the moist, dark conditions just below the fresh leaf litter.

In northern latitudes, it is not uncommon for acorns of some white oak species (e.g., Q. alba and Q. montana), prior to the onset of winter, to produce radicles in excess of 20–30 cm. However, this early germination is not restricted to the north. In Central Mexico, for example, I observed acorns of *Quercus microphylla* with obvious radicles of 0.5–1 cm while still attached to the shrubby oak. Yes, these acorns were already germinating before they hit the ground. These and similar observations of rapid germination in white oaks across their range suggest another, not mutually exclusive, advantage of this strategy of rapid germination. In the north, germinating white oak acorns, in addition to escaping rodent predation, may be racing the onslaught of winter. And while winter is not an issue for many of the oaks in Mexico and Central America, impending conditions of drought during a seasonal dry period represent a comparable challenge.

Regardless of the origins of early germination in the white oaks, the adaptation of embryo removal by squirrels provides a tremendous advantage at dealing with this strategy in the oaks. But it also would seem to provide a significant advantage for squirrels in their ability to compete with other rodents that do not, or cannot, perform the behavior. This would place strong selective pressure on other rodents to either adopt the behavior or master another means of manipulating early germination in acorns. The latter certainly appears to be the case.

THE COTYLEDONARY PETIOLE AND ACORN PRUNING

Although only the squirrels perform the behavior of embryo excision, nearly all rodents are vulnerable to early germination and show a sensitivity to seed perishability that follows from germination schedules. At a minimum, this sensitivity involves a hoarding preference for the acorns that show delayed germination (e.g., those of red oak species). However, upon radicle emergence in white oak acorns, many rodents are likely to prune the radicle before handling it. Removal of the radicle of white oak acorns allows rodents to access acorns even after the radicle is well rooted (at 20 cm or more). My colleague Xianfeng Yi observed this in China and was thus well prepared when he joined my lab in the early autumn of 2010—a year of heavy mast production of chestnut oak (*Quercus montana*). In fact, our records indicated that it was the highest production for white oak species that we had seen in 12 years (Yi et al. 2013a).

After his first trip to the field, with Andrew Bartlow and Rachel Curtis, to a study site in nearby Mountaintop, PA, Yi (as he preferred to be called) was excited at what they observed: clear evidence of acorn pruning by rodents (Fig. 6.1). They developed a plan. Yi immediately returned to the field, and

6.1. Rodent feeding station in northeastern Pennsylvania where, most likely, an eastern gray squirrel (*Sciurus carolinensis*) pruned and ate numerous chestnut oak (*Q. montana*) acorns soon after autumn germination. (*Xianfeng Yi*)

with Rachel's and Andrew's help ran several extensive transects on which they mapped and tagged germinating acorns of both chestnut and white oak. The plan was then to return and determine which of these acorns were pruned following the establishment of the radicle in the soil, and to further follow their fate into the spring. This experiment was inspired by previous field observations in China by Yi in which he observed rodents pruning acorns from the established radicles of two white oaks: the oriental cork oak (*Q. variabilis*) and the Mongolian oak (*Q. mongolica*). His observations, although new to Asia, were preceded by earlier reports in Europe (Bossema 1979; Gómez et al. 2003) documenting that acorns of holm oak (*Q. ilex*) are often removed from radicles by jays and rodents in the autumn of their germination. In my time in central Mexico, I also observed rock squirrels (*Otospermophilus variegatus*) pruning acorns from those of several white oak species with well-anchored radicles both in the field and in an open greenhouse.

Before his observations in the United States, however, Yi had also documented that after the removal of the acorn, the undisturbed radicle will sometimes develop an epicotyl (leaf shoot) the following spring and survive seedling establishment even without the benefit of the acorn cotyledon. This raised further questions about the process of oak seedling establishment that required a closer look at the morphology of germinating acorns.

The acorn itself is comprised mostly of cotyledon, which contains the energy and nutrients for the developing seedling—or alternatively—the seed predator that eats it. At the apical end of the fruit, set just inside the remaining style, is the root tip (radicle) to which is attached the plumule (Fig. 6.2; see Fig. 3.4). The entire acorn of the oak is comprised of the cotyledon, radicle, and plumule. The first stage of germination is characterized by the emergence of the radicle,

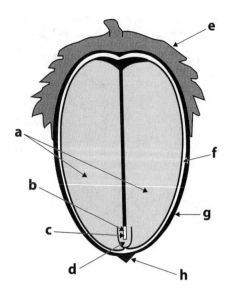

6.2. Anatomy of an acorn showing **(a)** cotyledons, **(b)** epicotyl, **(c)** hypocotyl, and **(d)** radicle. Also shown are the **(e)** cupule, **(f)** seed coat (testa), **(g)** pericarp (fruit wall), and **(h)** remains of the style. *(Shealyn A. Marino)*

which, under normal circumstances, pushes out at the distal end of the fruit, and then into the soil. As the root tip pushes into the soil, the radicle grows in length and carries with it the plumule, which is the embryonic stage that develops into the epicotyl (the young shoot of the seedling). As first reported by Lewis (1911), when the white oak (*Q. alba*) germinates, the plumule, still attached to the radicle, emerges from the body of the acorn and is eventually positioned some distance (1–3 cm) from the distal end of the acorn (Fig. 6.3a). This extension of the radicle, between the distal end of the acorn and the plumule, is the cotyledonary petiole. This separation of the radicle and the plumule away from the acorn allows the acorn to be pruned, long after the acorn has germinated. As a result, the germination process can continue after the acorn is removed.

The morphology of the germinating acorns of white oak species in North America, however, differs in some important ways to that of the oaks from China. In the germinating oriental cork oak and Mongolian oak, the cotyledonary petiole and the plumule are retained in the acorn, and the hypocotyl, the distal edge of the cotyledonary petiole, extends approximately 1 cm from the apical end of the acorn (Fig. 6.3b).

We hypothesized that the cotyledonary petiole was an adaptation that allowed the white oak, and possibly other early germinating oaks in North America, to sustain acorn pruning and still survive seedling establishment. In contrast, we predicted the oaks from China would not successfully sustain acorn removal. Yi devised a simple experiment to test it. During heavy mast production of the oriental cork oak (2009), Yi and an assistant mapped and tagged germinating acorns so they could revisit and document patterns of acorn pruning

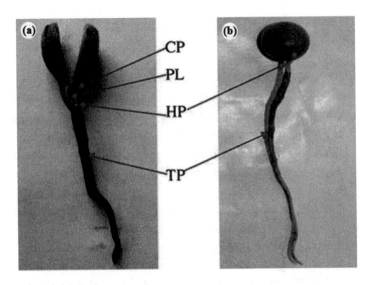

6.3. Contrasting morphological differences between **(a)** the white oak (*Q. alba*) native to North America, and **(b)** the oriental cork oak (*Q. variabilis*) native to China. In the white oak, the cotyledonary petiole (CP) extends well beyond the apical end of the cotyledon to the hypocotyl (HP), beyond which is the taproot (TP). The plumule (PL) will give rise to the developing epicotyl (stem). In the cork oak, the hypocotyl extends to just outside the apical end of the acorn, while the cotyledonary petiole and the plumule are retained inside the pericarp. The morphology of the oriental cork oak allows for two seedlings to develop, one from the pruned acorn and another from the pruned taproot (see text). *(From Yi et al. 2012b)*

and seedling survival. Likewise, after arriving in my lab, Yi carried out a similar experiment for chestnut oak in 2010 and arranged for an assistant to conduct a similar experiment with Mongolian oak in 2010 in China (Yi et al. 2012b, 2013a).

The results were compelling. For the *Q. alba* from North America, the frequency of pruning was low but for 11 of the acorns pruned above the plumule, seven survived to seedling establishment the next spring. As expected, those pruned below the plumule did not survive. In China, pruning of the oriental cork oak and the Mongolian oak occurred frequently—38% and 27% for these two species, respectively. These acorns were all pruned below the plumule. However, contrary to our predictions, 42% and 24% of these pruned acorns survived to seedling establishment the following spring (Yi et al. 2012b).

Similarly, in laboratory experiments in which we simulated pruning, we verified that acorns of North American white oaks (white oak and chestnut oak) survived pruning above the plumule. In fact, approximately 50% of pruned radicles of both species survived when only 6 cm in length; 75% survived when the radicle was 16 cm. And, as observed in the field, the radicles of both oriental cork and Mongolian oak again survived pruning but only significantly so (>50%) when radicle length exceeded 12 cm.

This led us to several important conclusions regarding the adaptive strategy of early germination in white oaks. First, the rapid transfer of energy from the cotyledon to the taproot is indeed a race against time for white oaks. However, it is not just about transferring energy to an inedible taproot. It is also critical for the taproot to achieve a critical size before the attached acorn is removed by a seed consumer. We also learned that in the two Asian oaks, enough of the developing embryo is moved along with the hypocotyl to allow a pruned radicle to develop into a full seedling, provided enough energy reserves are first transferred to the taproot. However, this also means that the pruned acorns of oriental cork and Mongolian oak, which still contain the plumule, will also germinate. In lab experiments, pruned acorns, with either attached or internal plumules, consistently germinated at rates >85% (Yi et al. 2012b). In case the math is not yet clear, this means that one pruned acorn, along with its pruned radicle (of either oriental cork or Mongolian oak) can produce two oak seedlings (Yi et al. 2012b). And the battle between the rodents and the oaks continues.

Yi and my lab team extended this work by examining the development of the cotyledonary petiole in germinating acorns and the impact of acorn pruning at different stages of development of this structure (Yi et al. 2013a). In the oaks from North America, development of the seedling from either the acorn or the radicle depended entirely on whether the plumule was attached to either the acorn or the radicle following the pruning event, and this potentially would depend on the size of the developing cotyledonary petiole, which increases in size as the radicle develops.

This further underscored the possibility that the cotyledonary petiole evolved specifically for this function in white oaks. This was my hunch when I first measured the length of the cotyledonary petioles of several germinating white oak and red oak species several years earlier in Mexico. The difference was dramatic—the length of the cotyledonary petiole in white oaks often exceeded that of red oaks by threefold (Table 6.1). In fact, for some of the red oak species there was virtually no separation of the epicotyl and the acorn—the epicotyl appeared to emerge directly from the cotyledon (Yi et al. 2013a). Interestingly, however, this marked difference in acorn morphology between the early germinating white oaks and the red oaks did not hold up for the two groups of oaks in North America, north of Mexico (Table 6.1).

Why the difference? The answer to this question remains open; however, at the time of this publication, we are exploring changes in a germinating acorn morphology, and specifically the cotyledonary petiole in the northern red oaks across a latitudinal gradient. One of the ancillary hypotheses we are testing is that the cotyledonary petiole is more pronounced in red oaks in more northern

Table 6.1. Length of cotyledonary petioles (distance between the apical end of the acorn and the point at which radicle and epicotyl diverge) of several white oak (subgenus *Quercus*) and red oak (subgenus *Lobatae*) species. Sample sizes are number of seedlings measured. Seedlings of *Q. montana* are from germinated acorns without developing epicotyls, randomly collected in the field. All other seedlings had produced epicotyls. Seedlings of *Q. alba*, *Q. rubra*, *Q. velutina*, *Q. palustris*, and *Q. macrocarpa* are from germinating acorns in laboratory experiments at Wilkes University. All other results are for seedlings from greenhouse experiments conducted in Central Mexico. (From Yi et al. 2013a)

LOCATION	SUBGENUS	SPECIES	(n)	LENGTH (mm) MEAN ± SE
Northeastern deciduous forests	*Quercus*	*Q. alba*	39	12.77 ± 0.53
		Q. montana	25	12.68 ± 0.47
		Q. macrocarpa	39	9.84 ± 0.33
	Lobatae	*Q. rubra*	31	14.25 ± 0.78
		Q. velutina	20	11.57 ± 0.73
		Q. palustris	30	7.07 ± 0.50
Mexico	*Quercus*	*Q. glaucoides*	8	18.3 ± 2.4
		Q. laeta	6	27.0 ± 2.3
		Q. microphylla	51	21.6 ± 0.2
		Q. peduncularis	6	17.0 ± 2.4
	Lobatae	*Q. crassipes*	8	6.2 ± 0.9
		Q. crassifolia	9	7.0 ± 0.6
		Q. dysophylla	9	2.3 ± 0.6
		Q. mexicana	17	3.5 ± 0.4
		Q. scytophylla	5	2.6 ± 1.1
		Q. conspersa	19	2.9 ± 0.7

Source: From Yi et al. 2013a.

latitudes. We knew from previous observations that red oaks in more northern latitudes, which only germinate in the spring, are also sometimes pruned by rodents when their epicotyls emerge from the soil.

TOLERANCE OF ACORN REMOVAL: THE WHITE OAK ADVANTAGE

The previous discussion strongly suggests that the cotyledonary petiole facilitates acorn pruning, which further raises the important question of how well various oaks tolerate early acorn removal as the germinating acorn presents itself above the ground surface. To test this in the laboratory, we simulated cotyledon (acorn) removal at different stages of epicotyl (stem) development in seven species of oaks: three red oak species (red, black, and pin oak), three white oaks (white, chestnut, and bur oak [*Q. macrocarpa*]), and the sawtooth oak (*Q. acutissima*) (Yi et al. 2019). We removed acorns at six regular intervals between stem heights of 2 cm and 16 cm. We then measured growth rates, leaf numbers, and shoot-root ratios following acorn pruning.

As expected, growth and survival were negatively affected by acorn removal; however, this effect decreased with stem length. Most important, we found that there was a clear stem length at which the three white oak species saw no decrease in survival or growth—a pattern not evident in the other four oak species. In a separate experiment, we also measured nitrogen uptake in the soil in developing seedlings of three red oak and two white oak species. Rates of uptake were highest in both white oak species (Yi et al. 2019). Thus, there appears to be a higher tolerance to acorn removal in white oaks than in the other species that exhibit a delayed germination. These results strongly support the idea that white oaks possess a suite of characteristics (adaptations) that enable these species to tolerate acorn pruning by rodents in the face of early germination.

THE CONUNDRUM OF PARTIAL ACORN CONSUMPTION

My first entrée into oak ecology came somewhat fortuitously as a graduate student long before I was ever thinking much about many of the subjects covered in this book. At the time, I was completing a doctorate on the foraging ecology of southeastern fox squirrels (*S. niger niger*). Much of this work involved strict seed predation by fox squirrels on cones of the longleaf pine (*Pinus palustris*) and patterns of tree and cone selection during the squirrel's late summer decimation of the longleaf pine cone crop in the southeastern coastal plain of North Carolina. However, among the many field and lab studies I conducted, I was also part of a team that was monitoring the population ecology of fox squirrels under the direction of Dr. Peter Weigl, my graduate advisor. Part of that responsibility included monitoring several hundred nest boxes distributed across 15 or more study areas. Nest box checks involved periodic surveys to determine recent use or occupancy of fox squirrels in these nest structures, and the live capture and tagging of any squirrels recovered. Because of the diurnal nature of the squirrels, it was necessary to conduct nest box surveys at night, after the animals were in their nests.

It was during one of these nest box checks that a simple natural history observation would change the direction of much of my future research. We had arrived at one of our study areas a bit before sundown and, just as we were beginning our box checks, my colleague and close friend, Travis Knowles, and I headed through a grove of densely populated scrub turkey oaks (*Q. laevis*). Below our feet were thousands, if not tens of thousands, of half-eaten acorns. All were consumed from the top or proximate (basal) end of the acorn. In most cases, only 30–50% of the cotyledon was consumed. From our knowledge of the literature on foraging behavior, this was an interesting observation. Why would any animal pick up an acorn, eat a small portion of it only to pick up another and repeat the process, over and over again? This seemed quite inefficient unless there was a good reason to avoid the bottom half of each acorn. Even more interesting was what this potentially meant for the fate of each acorn. We were familiar enough with the anatomy of the acorn to speculate that the small embryo at the distal (apical) end of the fruit meant that these partially eaten acorns could still germinate.

We first scoured the literature for similar observations. Finding none, we kept our eyes open in the field. And, in the months and years that followed, this simple natural history observation grew to include several additional species of red oaks (e.g., the northern red oak [*Q. rubra*], willow oak [*Q. phellos*], pin oak [*Q. palustris*], water oak [*Q. nigra*], scarlet oak [*Q. coccinea*], and black oak [*Q. velutina*]).

Observations of the seed consumers responsible for the behavior also grew. We concluded that our initial observation on turkey oaks was entirely the work of eastern gray squirrels, which we determined by matching the incisors of museum specimens to the incisor marks left behind on the oak cotyledon. Many subsequent observations were direct behavioral observations on gray squirrels while they were feeding on individual acorns. However, we soon learned that this occurred in other oaks, and other seed consumers were also involved (Fig. 6.4). In one particularly heavy mast year, we made numerous systematic observations on the towering willow oaks on the main campus of Wake Forest University in Winston Salem, NC. Here we observed blue jays (*Cyanocitta cristata*), flocks of common grackles (*Quiscalus quiscula*), and again gray squirrels consistently feeding on only the top of each acorn, sometimes consuming less than 30% of the cotyledon (Steele and Koprowski 2001). In subsequent observations, we also observed that both deer mice (*Peromyscus maniculatus*) and white-footed mice (*P. leucopus*) selectively consumed cotyledon from the basal half of the acorn (Steele et al. 1993).

Each of these acorn consumers, however, accomplished this task in a different way. Rodents such as gray squirrels rolled the acorn in their front paws until the basal end was upright, then they began gnawing at the basal scar and the surrounding pericarp until enough of the cotyledon was exposed for consumption (Fig. 6.5). Blue jays held the acorn against a branch with their feet and effectively stabbed or pounded the basal end of the acorn with their bill. Grackles, in contrast, relied on another approach. Holding the acorn in the bill, they

6.4. Pin oak (*Q. palustris*) acorns partially predated by Eastern gray squirrels (*Sciurus carolinensis*) and blue jays (*Cyanocitta cristata*). Note that all acorns are consumed from the basal end of the acorn and most have far less than 40% of the cotyledon removed. (*Shealyn A. Marino*)

scored around the equator of the acorn until it was possible to split the acorn in half. They then dropped the bottom (apical) half and used their tongue to flick out the cotyledon from the top (basal) half of the remaining pericarp. However, this technique was on occasion unsuccessful, resulting in either the whole acorn dropping to the ground with a well scored pericarp, or both halves simultaneously slipping from the birds' bill. More often than not, though, the birds were successful, and the bottom half dropped to the ground along with an empty pericarp from the top half of the acorn.

The grackles feeding technique meant that when a flock of these birds (of 20 or more individuals) descended on a single willow oak, it rained acorn fragments—sometimes whole acorns with their shells scored, or both basal and apical halves, but usually many more apical halves along with the empty pericarps of top halves. This was mixed with the apical halves of acorns dropped by squirrels and jays and the various fragments associated with their feeding. Only after countless hours of observing individual animals could we conclude that each acorn consumer was working the acorns for the top half of the fruit. Although now extinct, the passenger pigeon (*Ectopistes migratorius*), hypothesized to contribute to long-distance dispersal of the oaks and other nut species (Webb 1986), likely fed on many red oak species in the same manner, quite possibly after first dispersing them in their crops. Large flocks of this nomadic species were expected to converge on oak groves and then move considerable distances before feeding on the acorns. Although these acorns would not have been scatterhoarded, it is quite possible that given the large numbers moved, some would have successfully germinated when dropped to the ground.

 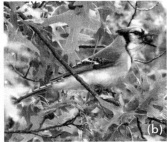 

6.5. Three vertebrate species that consistently consume acorns from only the basal half of the fruit early in the season and when acorns are particularly abundant. Shown are **(a)** common grackle (*Quiscalus quiscula*). *(Lynette Spence)* **(b)** Eastern blue jay (*Cyanocitta cristata*). *(Jeffrey Stratford)* **(c)** Eastern gray squirrel (*Sciurus carolinensis*). *(Richard Stead)*   SEE COLOR PLATE

Our observations on partial acorn consumption soon led to other hypotheses, additional experiments, and the constant reminder from my graduate advisor Peter Weigl, "Don't forget, you have a dissertation to complete." He was right and this was pulling me away from the work at hand. I did as he suggested and returned to my writing. However, a few months later I was searching in a closet at home for some papers I had collected and came across several bags of the acorn fragments as well as whole acorns we had originally collected from the pin oak stands. I had placed wet paper towel in each of the bags and stored them in the dark hoping to determine if they could germinate. They did—the whole acorns and nearly all the fragments as well. In fact, even fragments with less than 30% of the cotyledon had produced radicles (Fig. 6.6). Of course, this was just an interesting observation with the potential to suggest that oaks may be able to sustain partial seed damage and still germinate. It was a stretch to assume that just because the partially damaged acorns can germinate, they also survive long enough to produce a viable seedling.

My appetite for these questions grew and before I completed my degree, several colleagues and I had answered a few more pieces of this fascinating puzzle. First, we reasoned that if the vertebrates preferred the basal half of the acorn, then perhaps the weevil (*Curculio* spp.) larvae do as well (see Chapter 12). This was easily determined by having a class of undergraduates determine the location of the weevil larvae in the acorn and the amount of damage they inflicted on each acorn. Consistently we found that, as predicted, weevil damage was significantly higher in the top halves of the acorns of white oaks than in the bottom halves (Steele et al. 1993). More precisely, we found that weevil occurrence was almost three times as high in the basal half of the acorn. Weevil damage was two times higher in the basal half than the apical half. This likely meant adult female beetles selectively oviposit in the top half of the acorn. More re-

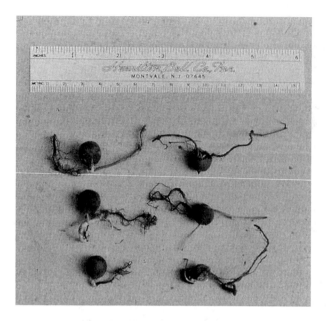

6.6. Germination of partially eaten acorns (left) and whole acorns (right) of willow oak (*Q. phellos*). Acorn damage was due to feeding by eastern gray squirrels (*Sciurus carolinensis*). *(Michael A. Steele)*

cent estimates in both red oak and white oak acorns regularly confirmed this pattern (Fig. 6.7 a b). And, as discussed further in Chapter 12, in three of our long-term studies, weevil larvae occurred significantly more often in the top half of the acorn than in the bottom (or apical) half where the embryo (seed) is located.

A FIRST LOOK AT ACORN CHEMISTRY: THE COMPLEXITY OF TANNINS

The second piece of this fascinating puzzle, which we discussed as early as that evening following our first observations of these half-eaten acorns, focused on the question of why would any animal consume only half the acorn? It seemed like a tremendous waste of potential food and thus contrary to one of the fundamental premises of foraging theory. Competing suggestions seemed to flow from others—perhaps the shell at the basal end of the acorn was thinner and easier for seed predators (both vertebrates and adult weevils) to penetrate or, alternately, maybe the shape or geometry of the acorn contributed to the behavior? Or, perhaps the suggestion that bugged me the most—who cares? Why does it matter if an animal eats half of an acorn? However, given the frequency with which it occurred and the likelihood that partially eaten acorns could germinate, several colleagues and I were convinced it did matter. Thus, our first stab at this question focused on a third specific hypothesis to follow from the

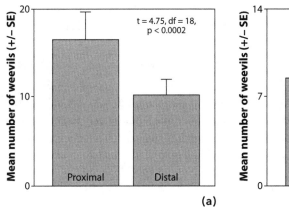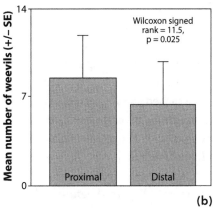

6.7. Distribution of weevil larvae in top (proximal) and bottom (distal) halves of white oak and red oak acorns during a single year. Comparison is for weevil distribution in **(a)** 18 white oak and **(b)** 12 red oak trees (error bars = +/- 1 SE). *(Michael A. Steele, unpublished data)*

biology of acorns—maybe the chemical deterrents in acorns are higher in the apical half of the acorns?

Tannins immediately came to mind. Tannins or tannic acid—highly variable, water-soluble, phenolic compounds—are well-known secondary metabolites in many plants that help to defend against microbes, as well as herbivorous insects and vertebrates. Moreover, tannins are common in oak leaves and acorns, so they seemed like a reasonable starting point. Although early assessment of the direct function and adaptive significance of tannins was in many cases debated (Bernays 1981; Schultz 1988, 1989), tannins were widely known to deter herbivory by both insects and vertebrates. In insects, for example, they impede development (Howe and Westley 1988), and in vertebrates, tannins reduce digestibility and palatability by binding to proteins and salivary enzymes (Robbins et al. 1987).

TANNINS AND RODENT SCATTERHOARDING

In a more recent review (Lichti et al. 2017, supplemental Table S3), two of my colleagues and I summarize all of the papers to date that provided evidence for the effect of secondary metabolites on seed use by rodents. This global review revealed that the vast majority of studies reported that the removal (11 of 16 studies), consumption (15 of 16 studies), and caching of seeds (10 of 12 studies) was influenced by polyphenols (tannins) over any other secondary plant metabolite (primarily glycosides and alkaloids). In the majority of these studies, seeds with higher tannins were removed less often (14 of 16 studies) and eaten less often (13 of 16 studies). Approximately half of the studies found that seeds with higher tannins were also cached more often (Lichti et al. 2017), although most of the other studies pursuing this question found only equivocal evidence for an effect of tannins on caching. Two studies reported that seeds with higher tannin levels were cached significantly less often (Xiao et al. 2008; Zhang et al. 2013b). In this latter study, several of us under Dr. Hongmao Zhang's leadership found that in both captive and field studies with two rodent species (the Siberian chipmunk [*Tamias sibiricus*] and the Korean field mouse [*Apodemus peninsulae*]), tannin levels were negatively correlated with both caching frequency and the distance seeds were dispersed before caching.

TANNINS AND PARTIAL ACORN CONSUMPTION

The overall take-home message, however, is clear—seed use, especially by rodents, is influenced by tannins. While we were not armed with all of the above information at the time of our first investigation on partial acorn consumption, even at this point, we had a strong hunch that tannins could play a role. So, we recruited a colleague, Dr. Kenneth Bridle, a plant biologist with a knack for biochemistry, to conduct some simple assays to determine tannin levels in the acorns. We followed the original techniques of Bate-Smith et al. (1973) modified slightly by Schultz et al. (1981). We prepared acorn tissue much the way we still do today; that is, we cut the acorns transversely, and shelled the top and bottom halves to create two separate composite samples of top and bottom halves of a dozen or so acorns for each tree. We dried them to remove moisture and then ground the samples with a ball grinder to a fine powder. At that time, the tannin assay was a simple colorimetric test, in which a known sample of the tissue was mixed and centrifuged in a buffer solution to which a small amount of bovine serum was added. The solution was then incubated and then its light absorbance measured with a spectrophotometer. The binding of the tannin with the blood protein provided a measure of the relative amount of tannin. Each experimental measure was then standardized with a curve produced by running the test with tannic acid, which allowed us to generate a tannic acid equivalence (TAE) curve. Although this test did not provide a complete picture of all the phenolics (condensed and hydrolysable tannins) in the acorns, I have over the years preferred this test because it measured what I believed to be the most ecologically relevant tannins—those that bind with proteins and thus influence palatability and digestion efficiency. I must, however, warn against this rather simplistic and naïve perspective as I have been reminded on more than one occasion by Dr. Jack Schultz, one of the world's leading experts on the ecological significance of tannins. And, while there is considerable discussion regarding this issue, I still today use a similar technique for measuring the protein-precipitable phenolics—the radial diffusion assay (Hagerman et al. 1987). My students, research associates, and I have over the years regularly consulted Dr. Hagerman, another renowned expert on plant tannins. Dr. Hagerman's technique, still widely used, is a straight-forward assay that is easily replicated and quickly mastered by undergraduates.

Our first set of results on acorn chemistry (tannins) was exciting (Steele et al. 1993; Steele and Koprowski 2001; Table 6.2). For acorns of both turkey oaks (Q. laevis) and willow oaks (Q. phellos), protein-precipitable phenolics (tannins) were significantly higher in the apical end of these acorns. Just as we expected, tannin levels were higher near the embryo. These results, coupled with observations of higher seed predation by a suite of acorn consumers (birds, rodents, and insect larvae) as well as evidence for seedling survival following acorn damage, all fit together nicely. However, our first inclination that these tannin gradients may represent an adaptive strategy in the oaks was short-lived. We were quickly reminded by reviewers and editors that the more plausible explanation for these tannin gradients is that they first evolved to protect the embryo from microbes and that seed consumers are just responding to this already present characteristic of the acorn. We thus settled for the safer explanation that the tannin "gradients" represented a preadaptation that helped the oaks to thwart seed predation. Although the complete explanation had to wait for additional research, this first foray into oak-animal interactions certainly whet my appetite for more. However, it was this simple side project during my years of graduate work that sparked a lifetime commitment to understanding the interrelated complexities of acorn characteristics and oak dispersal.

Table 6.2. Comparison of tannin activity (tannic acid equivalence, TAE) in the basal and apical halves of Q. phellos and Q. laevis acorns.

	WILLOW OAK Q. phellos	TURKEY OAK Q. laevis
Basal	5.07 ± 0.41	9.74 ± 0.44
Apical	9.54 ± 0.33	10.96 ± 0.23
Difference	4.27 ± 0.18	1.22 ± 0.29
Paired Comparison	15.32*	4.25*
N	20	28

Source: Modified from Steele et al. 1993.

Note: Statistical comparison of TAE in basal and apical halves of acorns was based on paired t tests. Both samples were significantly higher in the apical halves. *$P < 0.0005$.

OTHER FACTORS UNDERLYING PARTIAL ACORN CONSUMPTION

Although our first stab at the underlying causes of partial seed consumption seemed fruitful, several of the other alternative hypotheses also appeared worthy of investigation, if for no other reason than to rule them out. Among these alternatives, variation in seed abundance continued to rear its head. Partial acorn consumption, so it seemed, was more common in years of high acorn mast and it certainly made sense that this method of foraging would be more likely to occur when acorns were abundant. To document this, however, could take years. Instead, we decided to measure this across the autumn season in individual pin oak trees. Small-seeded pin oaks in urban settings frequently produce prodigiously and their acorns, once ripe about the end of September in Northeastern Pennsylvania, rapidly decline in abundance due to the feeding activity of eastern gray squirrels and blue jays.

Once acorns were ripe and feeding had begun, two of my students and I selected five trees, and then every 7–10 days we raked all acorns below each tree. We counted all the acorns, measured whole acorn size, and, for a sample of 50 partially consumed acorns per sampling period, weighed and estimated the percentage of each acorn consumed. We continued this through the season until seed fall had ceased. We then compared estimates of seed abundance to the amount of each acorn consumed from the basal end of the fruit. Early in the season, squirrels consumed little more than 10% of the cotyledon. This amount steadily increased until the end of the feeding season when acorns had 50–60% of the cotyledon removed from the basal end. Although this series of observations supported the seed abundance hypothesis as an explanation for partial acorn consumption, we were not able to control for other factors that may have occurred during the period of seed fall, such as seasonal changes in seed chemistry and increasing energetic costs for the squirrels. Nonetheless, it seemed seed abundance was in part related to this behavior.

Another hypothesis we encountered frequently was that the shape of the acorn may direct the behavior. Whereas this explanation may account for rodent damage, it was far less likely to explain avian consumption of the basal acorn halves, and it certainly seemed unlikely to influence weevil behavior, such as the location on the acorn where female weevils oviposited. However, because we were frequently studying food hoarding in gray squirrels, we knew enough about the squirrels' behavior to see some merit in this hypothesis. Gray squirrels engage in a rather involved "rolling behavior" of food items, such as seeds and acorns, before eating or caching them. This behavior, lasting a minute or more, often involves licking behavior, which is now recognized as a rather involved decision-making process. By tasting the surface of an acorn, the squirrel is gathering knowledge about whether the acorn had broken dormancy or not, and thus whether the seed should be eaten or cached, respectively. As my colleague, Dr. Mekala Sundaram, has now demonstrated, dormant acorns can be detected by the presence of a wax on the surface of the acorn (Sundaram 2016; see Chapter 3). During the rolling behavior, squirrels also engage in the rapid head shake (Preston and Jacobs 2009), which provides information on whether the acorn is sound or damaged by insects or other pathogens. Clearly, this short period of information gathering provides a wealth of information to the squirrel.

We designed a simple experiment to test the effects of acorn shape on partial consumption of acorns by free-ranging squirrels. We systematically presented individual squirrels with whole acorns, shelled acorns in which the cotyledon of the acorn retained a similar shape to whole intact acorns, and shelled acorns that were carefully carved so the true top of the acorn (basal end) appeared as the bottom (apical end) and the true apical end appeared rounded and flat like the true basal end. We simply reversed the geometry of the cotyledon (Steele et al. 1998; Steele and Koprowski 2001). The results were clear—squirrels fed from the basal end of whole intact and shelled acorns. But when we carved the cotyledon, thereby flipping the geometry of the acorn, squirrels almost always ate from the apical end of the acorn. However, what was most telling was the squirrels' finer responses. The rolling behavior, for example, was significantly extended, indicating that we were indeed messing with their heads. When they began to eat from the true apical end of the acorn, consumption time was truncated, as the higher levels of tannins—or even more caustic quinones that form when tannins oxidize—likely influenced the squirrels' feeding behavior. This is best interpreted as just one seed characteristic among perhaps a suite of traits that contribute to partial acorn damage. And, geometry is unlikely to influence seed predation by insect larvae and the birds.

Yet another characteristic frequently suggested to us was that the shell of the acorn (pericarp) may influence why seed predators prefer the top of the acorn. Our initial measurements of the pericarp thickness of several red oak species produced no evidence of this. As a result, we moved on to other questions. However, more recent observations by Yi and Yang (2010a) suggest otherwise. They reported significantly thicker pericarps at the apical end of Q. variabilis (cork oak) acorns (0.5–7.5 mm) compared with that of the middle or basal end (0.35–0.55 mm). And, as hypothesized, the frequency of weevil larvae also corresponded with the thinner pericarps, suggesting that sites selected for ovipositing by female beetles may be based on pericarp thickness. Yet, whereas these thicker pericarps would certainly direct weevils to other parts of the acorn for ovipositing, they are unlikely to influence predation by birds and even the smallest rodent seed predators.

Overall, our attempts to rule out most of the alternative hypotheses outlined above were rather unsuccessful; they were in fact quite the opposite. It appeared, instead, that there may be a suite of characteristics—both physical and chemical traits—that collectively guide seed predators toward the basal end of the acorn.

PARTIAL ACORN CONSUMPTION AND THE ADAPTIVE SIGNIFICANCE OF CHEMICAL GRADIENTS IN ACORNS

Our initial reports of this behavior of partial seed consumption in a popular article in *Natural History* (Steele and Smallwood 1994) caught the attention of Dr. Ted Stiles, a leading authority on fruit chemistry and frugivory in birds (Stiles 1980). At his urging, I decided to look more closely for evidence of chemical gradients in acorns that may correspond with this behavior in seed predators. Ted graciously offered his help in the analysis. We collected a large sample of acorns from 20 individual trees of three oak species (red, pin, and white oak). For each tree, we selected a random sample of 10–20 sound acorns per tree, cut each acorn in half laterally to produce "top" and "bottom" samples per tree, shelled the top and bottom halves, and then freeze-dried and ground the samples to a fine powder. Ted then ran specific chemical tests that he and his graduate students regularly used to evaluate seed and fruit quality (Choo 2005). In contrast to many studies in which seeds are sent off for analyses to commercial companies, Ted and his graduate students had adopted techniques that measured total utilizable lipid, protein, and carbohydrate (Choo 2005). They argued that these measures were more meaningful from the frugivore's perspective because they assessed what was directly available for their use. To extend the analyses, we also sent a subsample of each top and bottom composite sample for nutrient/mineral analyses to the Agricultural Analytical Laboratory at Pennsylvania State University (PSU). These analyses allowed us to compare the relative amounts of several nutrients and minerals (i.e., P, K, Ca, Mg, Mn, Fe, Cu, B, Al, Zn, Na). Tannin analyses were completed in-house using the radial diffusion assay.

The results were overwhelming. Upon receipt of the raw data, we jumped on the statistical analyses and immediately saw patterns we simply had not expected (Fig. 6.8). Univariate comparisons of top and bottom halves revealed significantly higher lipid levels in all three oak species: red, pin, and white oak. Likewise, tannins were higher in the bottom halves of all three oak species, but not significantly so. In red oak acorns, Boron (B) and Sodium (Na) were also higher in the top of the acorns. And, as expected, some nutrients (Calcium [Ca], Potassium [K], and Magnesium [Mg]) were higher in the bottom end of the acorn, closer to the embryo and developing seed. Multivariate statistical analyses (discriminate function analysis and jackknifed classification) consistently

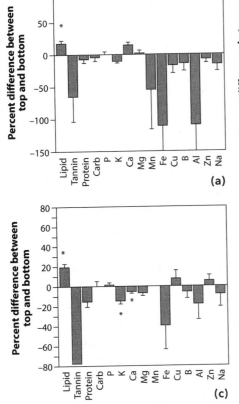
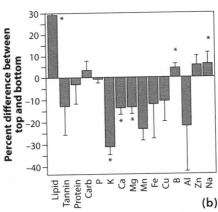

6.8. The relative difference in percent concentrations of lipids, tannins, carbohydrates, and 11 minerals in the top and bottom of acorns. Measurements were based on composite samples of acorns per tree for 19, 22, and 21 trees of white oak **(a)**, northern red oak **(b)**, and pin oak **(c)**, respectively (*indicates significant difference [$p < 0.05$]). *(Michael A. Steele, unpublished data)*

distinguished 85–90% of the top and bottom halves of the acorns of all three species based on these chemical differences.

Yi and Yang (2010a) had found no differences in nutrient and metabolites in their study on weevil distributions within acorns. However, their results may be attributed to the fact that they calculated total chemical composition and not the utilizable measures of various nutrients; likewise, total hydrolysable and condensed tannins were calculated rather than just the tannin that bound to and precipitated protein.

These results raised an interesting question and quickly refocused our earlier attention on the possibility that such chemical gradients may represent an adaptive trait in the oaks. In red oaks, for example, higher lipids (and sodium) in the top of the acorn render the top more nutrient rich. After all, lipid (fat) is the reason seed consumers eat acorns and sodium is often a limiting nutrient for many herbivores (Weeks and Kirkpatrick 1976), such as squirrels (Weeks and Kirkpatrick 1978). The most plausible explanation for selective consumption of the top of acorns was this apparent suite of chemical gradients that make the basal end of the acorn more palatable, digestible, and more energy-rich. In fact, these estimates of energy content in the basal and the apical half of the acorn showed that just based on differences in energy content, squirrels could maintain a higher feeding efficiency when feeding only on the basal end of acorns.

GERMINATION AND ESTABLISHMENT OF PARTIALLY EATEN ACORNS

Our results suggested that perhaps this story of the half-eaten acorn may represent an important reproductive strategy for the oaks. I guess we were already suspicious of that possibility before we received the data from the Stiles's lab and PSU. For the 20 trees of the three species from which we had collected acorns for chemical analyses, we also collected another sample of approximately 50 whole acorns and 50 partially eaten acorns. Or, when damaged acorns were not available, we cut whole acorns in half, simulating squirrel damage. We then planted these whole and partially damaged acorns in an outdoor plot (Fig. 6.9) early in the spring and followed their germination, establishment, and growth through an entire growing season. We planted acorns of each species (red, pin, and white oak) and acorn treatment (whole acorn and half-damaged acorns), randomizing the location of acorn type. In addition, we separated each treatment of 50 acorns into two groups of 25 acorns each, randomly located in the research plot. This controlled for any position-effect in the plot that might influence acorn performance. Then, at the end of the growing season, we carefully harvested each seedling and measured the root and stem length and the wet mass of the stem (including leaves) and root system. We dried each of these three components in a drying oven and recorded their total dry mass.

6.9. Flagged plots where germination and establishment of whole and damaged acorns of red oak (*Q. rubra*), white oak (*Q. alba*), and pin oak (*Q. palustris*) were tracked for one growing season. One hundred (50 whole and 50 damaged) acorns were planted for each of 20 trees of each species. A random subsample of one treatment of 25 acorns is planted between flags. (*Michael A. Steele*)

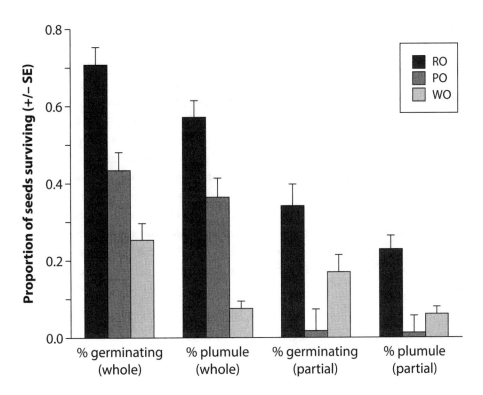

6.10. The proportions of whole acorns (whole) and damaged acorns (partial) germinating or establishing (plumule emergence) over the course of one growing season. RO = northern red oak (*Q. rubra*), PO = pin oak (*Q. paulstris*), and WO = white oak (*Q. alba*). (*Michael A. Steele, unpublished data*)

The results indicate the potential for partially consumed acorns to both germinate and establish. Yes, germination rates and plumule emergence (i.e., establishment of the seedling) were generally lower in half-eaten or comparably damaged acorns, but the rates of survival of these damaged acorns was instructive (Fig. 6.10). In all three species, some of the partially damaged seeds germinated and/or established. For red oaks, for example, 70% and 58% of whole acorns germinated and established, respectively, whereas 38% and 26%—approximately half as many of the damaged acorns—germinated and established. In contrast, germination and establishment of whole white oak was considerably lower (<30% and 8%, respectively) and establishment of both whole and damaged white oak acorns were nearly the same (Fig. 6.10).

These results, coupled with measures of seedling mass of these partially damaged acorns, indicated that these acorns may be able to sustain partial damage, germinate, establish, and survive through an entire growing season. Survival beyond the first year seemed quite plausible. Yet, the critics of such a claim were plentiful. Most plant population biologists argue that significant loss of cotyledon from an acorn will drastically reduce the ability of the seed and/or seedling to compete, establish, and ultimately survive. And, our simple germination and growth experiments do not prove otherwise. While my colleagues and I do not deny this, we also maintain a broader perspective that follows from a number of observations. In high mast years, large numbers of acorns—in fact, a significant proportion of a tree's crop—will sustain partial acorn consumption. Depending on the relative abundance of seed predators, in some years this may account for the majority of predation events. Moreover, some of these damaged acorns will be dispersed and cached (Steele et al. 1998). Whereas many partially consumed acorns will not survive, a small percentage may germinate and establish. And, in the life of an oak—150 years or more for some species—long-term survival of partially damaged seeds must only occur under some limited circumstances for this strategy to make a significant contribution to the tree's reproductive success.

ACORN SIZE AND ITS MULTIPLE ROLES IN SEED DISPERSAL, SEEDLING GROWTH, AND TOLERANCE TO SEED PREDATORS

The above discussion on partially damaged acorns and their ability to germinate and sometimes establish, raises yet another question regarding acorn size. Perhaps acorn size affords multiple benefits to the developing oak? Widespread studies in ecology and evolution have demonstrated how offspring size influences both offspring and parent fitness. Studies in plants, for example, have demonstrated how larger seeds aid both dispersal and growth rates as well as tolerance to seed damage (see review in Bartlow et al. 2018a). Across oak species, acorn size varies considerably from 1 g to >10 g. As we discussed earlier in this chapter and again in the following chapter, acorn size directly influences dispersal patterns. Typically, rodents disperse larger acorns farther and are more likely to cache these acorns than smaller acorns of the same species (Steele and Smallwood 2002; Moore et al. 2007; Steele et al. 2014; but see Bartlow et al. 2011). Acorn size would also be expected to influence growth rates. When all else is equal, larger acorns would be expected to benefit seedling growth by maintaining a higher rate of growth for a potentially longer period during seedling development, thus perhaps increasing the probability of seedling establishment (see review in Bartlow et al. 2018a).

Acorn size, however, may also afford another benefit—tolerance to partial acorn damage. To test this, several of my lab members and I conducted an experiment in which we simulated both insect and vertebrate damage to acorns

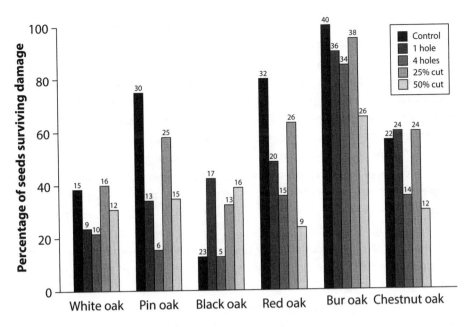

6.11. The survival of each species in response to four experimental treatments and the control acorns, not subject to damage. The species are ordered smallest to largest by mass. Values indicate numbers of individuals surviving. *(From Bartlow et al. 2018a)*

of six oak species that ranged in mean size from 1.5 g to about 5.9 g. Species, in order of acorn size, were white oak, pin oak, black oak, red oak, and bur oak. We then simulated weevil damage by drilling either one hole (light damage) or two holes (heavy damage) in the cotyledon. And, we simulated vertebrate damage by severing approximately 25% (light damage) or 50% (heavy damage) of the cotyledon from the basal (top) half of the acorn (Bartlow et al. 2018a).

We predicted that if the cotyledon is most critical as an energy source for the seedling, we would expect to see a significant negative effect of damage on seedling survival and growth rates. In contrast, if seed size is more critical for other functions, such as dispersal and tolerance to damage, we expected limited effects on growth rates. The latter result was indeed the outcome we observed. Although fewer damaged acorns survived to seedling establishment than control acorns, some acorns in all damage treatments survived (Fig. 6.11). Overall, various statistical analyses revealed that seed mass explained only a limited amount of variation in data on survival and seedling growth. Smaller acorn species only showed a reduction in performance in the simulations of heavy weevil damage.

Studies on the specific relationship between weevils and acorn size also suggest that size facilitates tolerance to damage by these seed predators (Xiao et al. 2007; Bogdziewicz 2018b) although a possible downside to this is that larger acorns result in larger larvae, higher survival, and possibly higher weevil abundance (Bonal et al. 2012). However, analyses of acorn size of holm oak across a latitudinal gradient in Spain revealed significant evidence of tolerance to partial damage by weevils in more southern latitudes where acorn size ranges as much as four times the size of acorns in the north (Bogdziewicz 2018b). Acorns in the north sustained greater damage due to smaller acorn size. Evidence of this latitudinal gradient of tolerance occurred despite higher frequency of the small-bodied weevil, *Curculio glandium*, in the north and an inverse pattern for the larger weevil, *C. elephas*, which was more common in the south (Bogdziewicz 2018b).

Results of these studies showed that not all of the energy in the acorn is essential for seedling survival and growth. Instead, much of the excess energy stored in the fruit promotes dispersal as well as an extra energy source to aid in tolerance to partial damage by seed predators (Bonal et al. 2012; Bartlow et al. 2018a; Bogdziewicz 2018b). Given that acorn size can vary from 1 g to >10 g across oak species, it is likely that size influences oak seedling survival and establishment in numerous ways. Additional research is needed to more fully understand the adaptive suite of characteristics that collectively influence tolerance of partial seed damage, seed dispersal, and scatterhoarding of the acorn and the energy necessary to ensure seedling survival and establishment.

MULTI-SEEDED ACORNS

I close this section with a brief reminder of the characteristic of multi-seeded acorns in several oak species reviewed in Chapter 3. Although this was previously discussed as it relates to embryo development, I consider it here to emphasize its potential as an adaptation of the oaks to thwart embryo excision by squirrels (see Chapter 5). The ability of squirrels of several genera to detect nondormancy in acorns and excise the embryos of these perishable foods, even without previous experience, suggests that many squirrels may have an evolutionary advantage over acorns that have broken dormancy. However, as discussed in Chapter 3, numerous oaks of both the red and white oak sections occasionally produce acorns with more than one embryo (McEuen and Steele 2005). Is this characteristic (Fig. 6.12) a specific counter-adaptation to the behavior of embryo removal? Perhaps. But it is equally likely a developmental anomaly given the relatively low frequency with which it occurs in each oak species. Although acorns with more than one embryo can clearly germinate after embryo excision, more research is needed to determine the origins and potential significance of this trait.

6.12. Multi-seeded white oak (*Quercus alba*) acorns. Note multiple radicles in some acorns and atypical radicle development on the side of others. (*Michael A. Steele*)

In the following chapter, I move to factors external to the oak and the acorn that influence dispersal. This discussion centers on factors such as the effects of the availability of other seed types, acorn abundance, forest composition, forest fragmentation, and the influence of caching on oak dispersal.

(a)

(b)

(c)

Plate 1.1 (a)–(c)

Plate 1.2 (a)–(b)

Plate 1.2 (c)

Plate 1.4 (a) & (b)

Plate 1.4 (c) & (d)

Plate 1.5 (a) & (b)

Plate 2.2

(a)

(b)

Plate 2.4 (a) & (b)

Plate 2.5 (a) & (b)

Plate 3.1

Plate 3.2

Plate 4.1 (a) & (b)

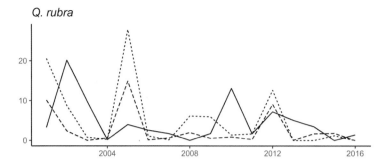

Q. rubra

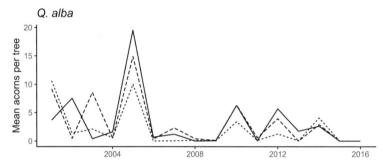

Q. alba

Mean acorns per tree

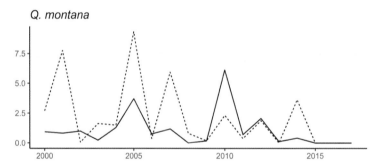

Q. montana

Plate 4.2

Plate 4.3

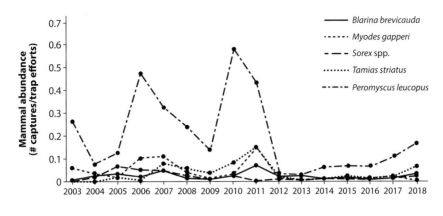

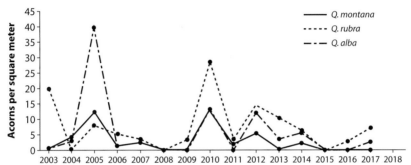

Plate 4.5

Plate 5.1

Plate 5.2 (a) & (b)

Plate 5.10 (a)–(f)

Plate 6.5 (a)-(c)

Plate 7.2

Plate 7.4 (a) & (b)

Plate 7.6

Plate 9.3

Plate 9.4 (a) & (b)

Plate 10.1

Plate 10.2

Plate 10.3 (a) - (c)

Plate 10.3 (e) – (g)

Plate 10.11 (a) & (b)

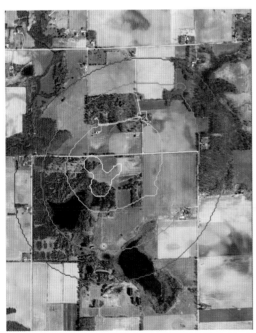

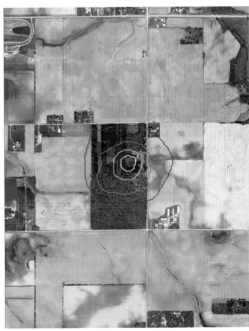

0 125 250   500 Meters

Plate 10.13

(a)

(b)

(c)

(d)

Plate 12.1 (a) - (d)

Plate 12.3 (a)–(e)

Plate 13.1 (a)-(e)

Plate 13.5 (a) & (b)

Plate 14.1 (a)-(c)

Plate 14.2 (a) & (b)

Plate 14.2 (c)

Plate 14.3

Plate 14.4 (a) & (b)

Plate 14.6

Plate 14.7 (a) & (b)

THE OAK DISPERSAL PROCESS: EXTRINSIC FACTORS I

7

INTRODUCTION

Chapters 5 and 6 focused on the influence of acorn characteristics on the oak dispersal process. Therein, I demonstrated how numerous factors such as seed size, germination schedules, seed chemistry, and even chemical gradients within the acorn directly influence acorn dispersal, survival, and seedling establishment. But as I emphasize here and in Chapter 8, the dispersal process and the mutualism between oaks and the animals that disperse their fruits is a highly conditional relationship—one that can shift quickly from mutualism and thus successful dispersal and establishment, to one in which these animals become exclusive seed predators, often consuming entire acorn crops.

Extrinsic (environmental) factors that influence oak dispersal are numerous, but my coverage here is not intended to be exhaustive. Instead, I focus only on those factors for which we have experimental evidence to support such claims—those environmental and ecological factors that we now know, in concert with acorn characteristics, drive oak dispersal. These include the effects of (1) re-caching behavior, (2) the cache environment on seed characteristics, (3) acorn abundance (masting) and ambient food availability, (4) other seed types and the relative frequency of seed types, (5) forest structure, and (6) forest composition on the oak dispersal process. In Chapter 8, I focus on the potential effects of competitors and predators on the dispersal process.

To clearly decipher the influence of extrinsic factors on oak dispersal, it is necessary to conduct such experiments under natural conditions (e.g., in a forested environment). This means it is often not possible to directly observe the behavior of the dispersal agents. Instead, indirect methods are often required to follow seed fates (patterns of seed consumption and hoarding behavior). I therefore begin this chapter with a necessary diversion on how to tag and follow seed fates under natural field conditions (Table 7.1).

Table 7.1. Summary of most-commonly used methods for tracking seed and acorn fates. Excluded here are molecular techniques discussed in further detail later in this book.

NAME OF TECHNIQUE	RELATIVE COST	ADVANTAGES	DISADVANTAGES	REFERENCES
Metal-tagging	Moderate expense (<$400)	Enables long-term tracking of seed fate	Labor intensive	Sork 1984; Steele et al. 2001b; Moore et al. 2007; Lichti 2012; Lichti et al. 2014
Tin-tagging and thread-tagging methods	Inexpensive (<$50)	Allows rapid assessment of initial seed fate; limited effort involved	Only allows determination of initial seed fate; rodents may gnaw threads; hole in cotyledon may destroy acorn	Xiao et al. 2006, and references therein; Yi et al. 2008; Steele pers. obs.
Magnetic-tagging	Moderate expense ($600–$700) for magnet locator	Rapid location; can be used to follow artificial cache fate by placing beneath acorn	Rodents often remove magnets attached to acorns	Steele pers. obs.
PIT (Passive Integrated Transponders)-tagging	High expense for both tags ($500 per tag) and for PIT-tag readers (>$1000)	Ability to follow re-caching events	Relocation of acorns difficult; useful in limited experimental situations	Steele et al. 2011; Suselbeek et al. 2013
Radioactive labelling (Scandium)	Limited expense	Rapid location of seeds or debris from consumed seeds in relatively small area	Difficult acquiring permission in some regions	Vander Wall 2000, and references therein; Waitman et al. 2012
Radiotelemetry	Highest expense (>$75) for radio-tags and (>$1000) for radio receivers and antenna	Ability to track seeds over long distances and across multiple caching events for 2–3 months or longer	Costs; data collection is often labor intensive; if data collection is automated, data management is intensive	Hirsch et al. 2012; Lichti 2012; Bartlow et al. 2018b

7.1. Shown are **(a)** exclosure in which tagged acorns were selectively presented to only rodents that could access acorns from openings on each side of the box. Note the gray squirrel in the blue circle. *(Michael A. Steele)* **(b)** Flagged locations where observers found cached acorns and brad nails, indicating where acorns were cached and eaten, respectively. *(Michael A. Steele)*

TRACKING ACORN FATES IN THE FOREST:
THE METAL-TAGGING METHOD

Our first attempt at tagging and tracking acorns borrowed a novel, relatively simple but effective, technique of metal-tagging (see brief description in Chapter 5) developed by Dr. Victoria Sork of UCLA (Sork 1984). This procedure involved placing a small brad nail (3 mm × 9 mm) in each acorn and then using metal detectors to relocate the tagged acorns when cached or the metal tags after the animal has eaten the cotyledon and dropped the metal tags, leaving a distinct record of acorn consumption (Fig. 7.1). The technique was effective for several reasons. Most important, the tag could be inserted flush with the pericarp of the acorn (with a brad punch), thus not significantly altering acorn perishability. Prior to insertion, the nail could also be spray-painted to mark the source and acorn or other seed type in which it was inserted. Upon recovery, researchers could then easily record these data.

Our first attempts at following seed fate in the forest were encouraging. However, almost as soon as we began, we discovered some shortcomings in our experimental design. In the first study, we placed individual piles of either tagged red (*Quercus rubra*) or white oak (*Q. alba*) acorns in a predominantly white oak forest. We covered each pile of tagged acorns with chicken wire staked about 10 cm above the ground level, allowing access by only rodents. We returned a few

7.2. Oaks are often dispersed to sites suitable for storage and germination but not establishment when they escape recovery. Shown here are acorns of *Q. petrea* deposited most likely by an avian (corvid) disperser in the bracts of a palm in Lisbon, Portugal. *(Michael A. Steele)* SEE COLOR PLATE

days after placement and, sure enough, many of the brad nails were found at the seed source, whereas other tagged acorns were dispersed and either eaten or cached some distance from the source. However, we soon realized we had introduced some unexpected variables. Each patch of acorns, for example, were located in a slightly different habitat thus we were unable to know for sure if the differences were due to habitat structure or acorn type. We saw evidence that the white oaks were eaten more and their embryos excised more than those of red oaks but the range of variables that could have contributed to the behavior seemed worrisome. We also soon realized that the forest had very few red oaks. That meant that the animals were acclimated to dealing with white oaks but likely had little experience with red oak acorns. We decided that too could influence their feeding and caching decisions.

Other notable limitations with the technique were also apparent. Whereas rodents systematically gnaw around the acorn, other seed consumers (e.g., bear and deer) masticate the acorn in the mouth and still others may even swallow the acorn whole (e.g., turkey). We controlled for these concerns by selectively provisioning the rodents with the acorns and excluding access to other acorn consumers. Early in our experiments we did this by staking a 1.5 m × 1.5 m-piece of chicken wire about 7–10 cm above the ground surface, which allowed rodents access under the chicken wire. We soon then modified this technique by presenting acorns in 1 m × 1 m × 10 cm-boxes constructed of a wood 2″ × 2″-frame covered with mesh hardware cloth (Fig. 7.1a). Small holes in the hardware cloth on each vertical side of the box allowed exclusive access by rodents that were then provisioned with nuts from a small hinged opening on top. In many of these experiments, seed source boxes were also equipped with a camera trap to determine the rodent species visiting the box (Fig. 7.1a).

The other major limitation with the technique was more difficult to address. Although relatively inexpensive, metal detectors (<200) were all that was needed to relocate tagged nuts or discarded tags. The recovery process, however, was labor intensive and never more effective than accounting for about 25% of the tagged seeds. Yet, overtime we learned that the majority of the tagged acorns that were not eaten or scatterhoarded within 30 m of the source box were moved to larders in the tree cavities or chipmunk burrows. These sites of course are dead-ends for acorns, where, even if they were to germinate, they would not produce viable seedlings (Fig. 7.2).

In two metal-tag studies (Moore et al. 2007; Lichti et al. 2014) coordinated by doctoral students at Purdue University under the direction of Dr. Robert Swihart, the process of metal-tag retrieval was significantly improved. In these studies, my colleagues addressed this issue of search costs by creating a grid of 5 m × 5 m-search blocks around each seed source. Then, with the exception of blocks directly adjacent to the seed sources, searches were conducted in alternate blocks in a checkerboard arrangement. And, prior to official field inventories, research assistants were also tested on their rate of recovery of lone tags, tagged nuts, and different tag types, which were necessary for some seed types. Correction factors could then be applied to each measure and used to estimate rates in uncensored blocks (Moore et al. 2007). This meant approximately half the search effort was required to estimate patterns of seed consumption and scatterhoarding for each seed source.

Over the years, my colleagues have perfected the technique, and we, along with a number of other colleagues and students, conducted numerous acorn-tagging studies to test an array of questions concerning oak dispersal. These include studies on the effects of seed characteristics, alternative seed types such as an invasive oak and an American chestnut (*Castanea dentata*), and the effects of seed abundance on the oak dispersal process. Over the past 25 or more years, I estimate that over 500,000 seeds have been tagged in my lab alone. The many undergraduates who have worked on such seed-tagging studies number well over 50, all of whom I am sure would have choice words for the experience. I recall one student research assistant now a PhD who, upon discovery of several new students in the lab, immediately held a brief induction ceremony handing them each a metal detector and informing them of their new responsibilities and the badge of honor they would someday wear for their commitment to metal-tag recovery. As he saw it, he was free at last, never to have to recover another metal-tagged seed.

Before highlighting some of the insights from seed-tracking experiments, I digress a bit further to highlight other seed-tracking techniques, many of which we have used in my lab at various times depending on the specific question before us (Table 7.1).

The tin-tag method, developed and made popular by two of my colleagues in China (Xiao et al. 2006; Yi et al. 2008), overcomes many of the challenges inherent in the metal-tag method. The tin-tag method involves drilling a small hole in the basal end of the acorn, inserting a fine tin wire (32 gauge) through the hole and attaching it at one end to the acorn and attaching a light, plastic tag at the other end of the tin wire, approximately 10 cm from the acorn (Fig. 7.3). The advantages of this method are many, including it is inexpensive, it does little to influence the perishability and the likelihood of germination, and each seed can be individually marked with a number on the plastic tag. Most appealing, however, is the recovery time.

I learned this firsthand when some of my students and I used a modified version of this technique in a remote field site in northwest Costa Rica to assess acorn dispersal by the Central American agouti (*Dasyprocta punctata*). There, we used heavy monofilament fishing line and fluorescent flagging instead of the tin wire and plastic tag, respectively, a method sometimes referred to as the thread-tagging method (Table 7.1). We arranged the acorns (*Q. oleoides*) in a neat circle, as is necessary to prevent each seed from becoming entangled with another. Upon return the very next morning I was struck with an immediate feeling of awe. Distributed around the seed source were nearly all the fluorescent tags, some attached to a cached acorn, whereas many others, especially those at the source, had obviously been eaten (Fig. 7.4). In less than an hour, we were able to record the fate and location of >85% of the tagged acorns that were either eaten or cached in this case by an agouti (Fig. 7.4).

My immediate gratification with this technique, however, was short-lived. I soon realized that the primary disadvantage of the tin-tag method is not insignificant. Whereas the metal-tagging method allowed us to follow seed fate well into the spring and thus determine acorn survival in the cache and even seedling establishment after germination, the tin-tag method was an immediate glance of initial seed fate that provided no reliable information thereafter. Although seeds and nuts tagged with a tin tag can be followed after they are initially cached to determine future fates, any information gained after this initial caching event seemed meaningless. While the obvious tag above the ground surface of a cached seed is quite helpful to the researcher, it is also visible to any potential cache pilferer. Upon discovery that such tags are associated with a stored seed or nut, any naïve rodent or corvid would quickly recover these cached items.

7.3. Photograph of eastern gray squirrel (*Sciurus carolinensis*) handling a tin-tagged acorn. *(Devon Kriebel)*

Nevertheless, the tin-tag method still stands as a reliable technique for assessing initial responses to a seed source, and it is especially effective when one wishes to determine the early seed fates of a large number of seeds.

Another method that we have used in a limited number of circumstances is the PIT-tag method. PIT tags are passive integrated transponders. More expensive than the previous two methods, PIT-tagging offers some important advantages and some important recovery limitations. As we described in Suselbeek et al. (2013), a PIT tag is an electrical microchip contained in a glass cylinder

7.4. Shown are **(a)** Central American agouti (*Dasyprocta punctata*). *(Ashley Hockenberry)*
(b) Recovery of thread-tagged acorns of *Q. oleoides* dispersed by agouti within
24 hours of seed placement from near the box at the far right. Orange flagging indicates
tagged acorns that were either eaten or cached. Pink flags denote fate of each acorn.
Study site was located in northwestern Costa Rica. *(Michael A. Steele)* SEE COLOR PLATE

that is programmed with an alphanumeric code, unique to each chip. A handheld reader (Radio Frequency Identification Device, RFID) is then used to read
the chip and report the unique code. With an antenna, the code can be detected
within about 2 m and, without the antenna, detection range is about 0.5 m (Fig.
7.5). A more detailed summary of how the tag and reader function is reported by
Suselbeek (2013). The technical description aside, PIT tags are the same uniquely
numbered tags that our veterinarians and many animal researchers insert under the skin of animals for easy identification of individuals when animals are
lost and recovered, re-trapped in mark-recapture studies, or monitored in lab
studies. Among the advantages is their small size (9–12 mm × 2 mm) and easy
detectability, but only at short distances. Limitations include the high costs for

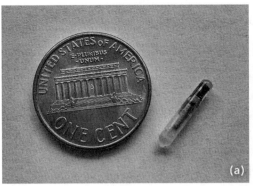 

7.5. Shown is **(a)** A passive integrated transponder tag (PIT-tag) that can be inserted into an acorn for tracking. (*Shealyn A. Marino*) **(b)** A PIT-tag reader used to detect tags and display their unique ID numbers. (*Shealyn A. Marino*)

both the tags and the detectors (Table 7.1). As we describe in Chapter 8, PIT tags are inserted into acorns and the hole is covered with a small amount of adhesive and ground pericarp without influencing seed perishability (Steele et al. 2011). They then can be detected when the nut/acorn is in the cache without uncovering or disturbing the cache.

Yet, another technique that we have relied on for tracking acorn dispersal is radiotelemetry. Most commonly used for real-time monitoring of wildlife, telemetry is easily modified for other similar uses. Typically, radio units, mounted to collars or backpacks, are placed on a live animal, which is released and then tracked with a receiver and antenna (Fig. 7.6; see Chapter 10). The transmitter consists of electronics that produce a high frequency, electromagnetic signal unique for each transmitter, which also requires an energy source (battery) and an antenna to magnify and detect the signal. A receiver with an antenna is then tuned to the unique signal produced by each transmitter to detect the pulsed signal produced by the transmitter. The upside of telemetry is that it allows for real-time monitoring of the tagged subject by direct relocation of the subject or remote monitor with GPS collars. And, in recent years, the technology has advanced far enough to conduct some rather sophisticated tracking procedures. For example, today it is possible to use telemetry to track insects (Kissling et al. 2014), even the long-distance movements in bees (Wikelski et al. 2010), or migrating dragonflies (Wikelski et al. 2006). Telemetry has also been used to follow dispersal of seeds of the palm tree, *Astrocaryum standleyanum*, by the red-rumped agouti (*D. leporina*). This was accomplished by placing radio-tags inside the seeds (Jansen et al. 2012; Hirsch et al. 2012). The researchers used thread tags that were monitored by automated telemetry towers (see below). The signal produced by these thread tags were only switched on when the seed was moved.

7.6. Radio-collar on anesthetized southern fox squirrel (*Sciurus niger niger*). (*Michael A. Steele*)   SEE COLOR PLATE

This allowed the seed to be tracked over extended periods with minimal cache disturbance, and, as a result, the investigators were able to document numerous re-caching events, including 36 secondary caching events of a single seed moved across numerous agouti territories (Jansen et al. 2012).

The limitations of telemetry center on the costs of equipment. One individual transmitter, handheld receiver, and antenna can often cost $100, $1200, $150, respectively. Add multiple transmitters, automated receivers, and multiple antennas and the costs can skyrocket. Nevertheless, telemetry can enable one to address questions that may not otherwise be possible. If the funds are available, by all means, consider it. To date, relatively few studies have relied on telemetry for deciphering patterns of seed dispersal, but those that have, often have uncovered exciting, unexpected findings (Jansen et al. 2012). Telemetry is especially effective for documenting relatively rare, but significant, long-distance seed dispersal events (Anderson et al. 2011).

Here I briefly review a study in which telemetry seemed the ideal option for tracking oak dispersal. Two of my colleagues from Purdue University and I received an NSF grant to explore oak dispersal by blue jays (*Cyanocitta cristata*) in Indiana, where forests are severely fragmented due to agriculture and, in northeastern Pennsylvania where, in contrast, fragmentation is far less severe. Our goal was to examine the effects of forest fragmentation and forest structure on patterns of acorn dispersal by blue jays. Among other goals for the study, we sought to radio-tag both the jays and a sample of the acorns that we presented to the same jays for dispersal (Fig. 7.7). This all looked great on paper but the challenges to this endeavor were numerous and revealed themselves with each stage of the study. We first used mist nets and vocal play back to lure jays in and when captured, fit them with a telemetry backpack. The backpack had nylon straps that were designed to decay and allow the jay to drop the pack 2–3 months later just before the batteries died. As with most transmitters, these backpacks were kept to <5% of the study animal's body mass. This size restriction in turn limited battery size and therefore longevity of the transmitter. For us, this meant 2–3 months of battery life. We had it all worked out. In the first year of the study, our ability to capture jays and successfully fit them with a transmitter was very successful. We then waited a week or so for the jays to acclimate, and then we set out to track them. In this first year, most of the tagged jays were gone. They had migrated. Lesson one—we learned from this experience that we had to wait longer for all migrants to move on before netting only the local residents.

Lesson two came when attempts at handheld telemetry proved unsuccessful. Research technicians were positioned at three locations (road intersections easily detected on aerial photographs) surrounding the estimated home range of the radio-tagged jay. Each assistant was equipped with a radio receiver, antenna, and a communication device. Assistants were instructed to take a bearing at exactly the same time. The handheld radios allowed for precise timing of when bearings were taken. Yet, no matter how hard we tried, triangulation from three locations rarely resulted in an accurate location. Quite often, two bearings suggested one location and the third suggested the jay had already flown to another part of its range. Accurately tracking the birds during their peak of acorn dispersal was simply not possible with handheld equipment.

My ingenious colleagues at Purdue first solved this problem by acquiring automated telemetry units, mounting them in sealed chambers on a flatbed trailer, and wiring them to a collapsible tower (10 m) equipped with six antennas at the top of the tower. We replicated the design for Pennsylvania. And with three of these systems strategically positioned around a tagged jay, we could monitor a jay "effortlessly" by doing little more than replacing the memory card in the receiving unit once per week. The automated system, identical to that first used in studies of agouti at the Smithsonian Tropical Research Institute on Barra Colorado Island in Panama, automatically acquired a signal from one or more tagged birds every minute, 24 hours per day. The satisfaction of knowing that we were collecting these data at the same time we were showering, sleeping, or teaching was overcome a bit with occasional equipment failure and when we realized the amount of data generated and the computer power that would be needed to process it. In Chapter 10 we describe several of the insights we gained from these procedures, although several papers are yet to come.

Additional challenges with the jays came when we presented the birds with radio-tagged acorns. Our goal was to systematically present jays with radio-tagged acorns of different species and then follow the jays as they cached the acorns and then the acorn fates after they were cached. Inserting small radio-tags in acorns required some practice. By drilling a small hole in the basal end of the acorn through the basal scar we could then wrap the whip antennae tightly around the transmitter, insert it into the acorn, and then fill the acorn with ground cotyledon, woody material from the basal scar, and odorless wood putty. With some light sanding, the radio-tagged acorn looked indistinguishable from a sound nut and with some additional practice, we could estimate the precise amount of cotyledon, so the acorn mass of a tagged acorn matched that of a sound acorn.

We then presented the tagged acorns on a platform to which a camera was attached allowing us to record the precise acorn that was taken (Fig. 7.7). Jays readily retrieved tagged acorns until they discovered they were fitted with transmitters. We soon realized that we were required to present plenty of sound acorns (>75%) with a smaller percentage of tagged acorns. Otherwise, the jays simply stopped feeding or dispersing acorns. The challenges with jays were plenty, and although we systematically overcame many, my colleagues and I eventually returned to our work with rodents.

Magnets also have been used to track seed fates (Iida 2006). However, their utility is generally limited, and a magnetic locator cost is approximately $600–$700. Magnets are inexpensive and can be purchased in thin sheets and then cut to sizes specific for their use. We have found that when magnets are attached to the top or side of acorns, many rodents first remove them before eating or dispersing the fruit, often making this tracking technique ineffective. Thus, magnets were most effective when inserted inside acorns. We have also had some success in placing magnets underneath artificially cached acorns to determine patterns of cache pilferage, in which case it was necessary to employ an additional tracking method to determine final acorn fate.

Finally, I mention briefly two techniques that, to date, have not been used to track acorn fates that may offer some opportunity in specific situations as well as a third technique that has been specifically modified for the study of oak dispersal. The first is high resolution GPS, which may be effective for documenting rare, long-distance dispersal events in the oaks. The second is the use of radioactive labels—in particular, scandium-46, a gamma-emitting radionuclide that has been used to track seeds of pine (*Pinus* spp.) (Abbott and Quink 1970; Vander Wall 1992; Vander Wall 2000) and the fruits of the Joshua tree (*Yucca brevifolia*) (Waitman et al. 2012). Although encouraged to consider this technique by my colleague, Dr. Stephen Vander Wall, I have not done so because of concerns of property owners.

The third approach is the use of molecular tools, which my lab has employed on occasion, under the guidance of plant molecular biologists, Drs. William Terzaghi and John Carlson (Steele et al. 2007; see Chapter 15). In this approach, seedling-parent spatial relationships are determined with the use of microsatellite DNA markers from seedlings and adult trees (Dow and Ashley 1998, and ref-

7.7. Feeding tray on which acorns of several species were randomly positioned for presentation to jays. A sample (< 25%) of each acorn type was equipped with transmitters. Trays were positioned 2–3 m above the ground surface and regularly monitored with camera traps (Reconyx) to determine removal by jays. *(Shealyn A. Marino)*

erences therein). Although this approach identifies definitive parent-offspring matches, it does not distinguish between maternal and paternal parents. As a result, we could distinguish the closest potential maternal sources and thus the minimal dispersal distance but not the specific source tree for acorns (Steele et al. 2007). This limitation was, in part, overcome by Grivet et al. (2005) that adapted procedures of Godoy and Jordano (2001) that involve isolating maternal DNA from a specific part of the plant propagule. Grivet et al. (2005) used maternally inherited tissue (from the pericarp of acorns) to measure genetic variability of seed pools moved by acorn woodpeckers (*Melanerpes formicivorus*). However, similar procedures could be used to identify the specific maternal sources of acorns as well (see Chapter 15).

Despite the diversity of techniques available for tracking seeds, I urge researchers to concentrate first on the question asked and then decide on the method that best suits their goals. Sometimes the easiest methods most directly address the question at hand.

7.8. Shown are **(a)** A small radio transmitter wrapped in its antennae and prepared for insertion into a red oak (*Q. rubra*) acorn. *(Shealyn A. Marino)* **(b)** Five radio-tagged acorns positioned over magnets to disable transmitters until removed from the enclosure. *(Michael A. Steele)* **(c)** Two student researchers tracking radio-tagged acorns. *(Shealyn A. Marino)* **(d)** Relocating one of the tagged acorns in a cache. *(Michael A. Steele)*

RE-CACHING BEHAVIOR AND ACORN DISPERSAL

The second dispersal study in which we employed radiotelemetry followed from the persistent urging of my students that we were missing important information when conducting metal-tagging studies. Metal-tag recovery was conducted after tagged acorns and other nuts were either eaten or dispersed, anywhere from a few days to 30 days after placement. At many sites, the process of metal-tag recovery then began around the middle of December, continued until snow fall, and then resumed in the spring as soon as the snow had melted. In other years, recovery of tags and tagged acorns could not be initiated until the spring. This meant that we generally had a reasonable picture of the final seed fate, but we knew little of what happened between the first dispersal event and the final seed fate. Were scatter-hoarding rodents recovering and re-caching acorns or were acorns dispersed to their final destination immediately in the autumn?

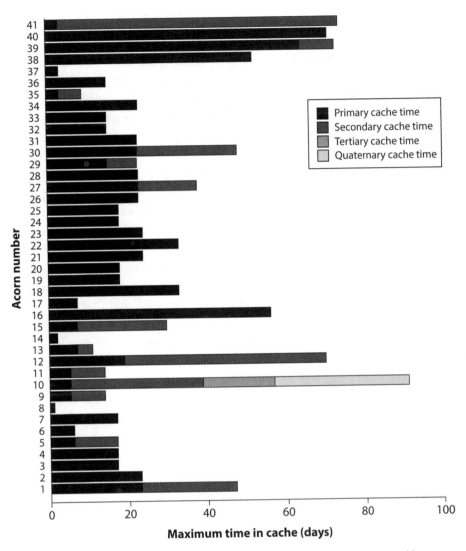

7.9. Time in the cache for radio-tagged red oak acorns likely dispersed by eastern gray squirrels (*Sciurus carolinensis*). Shown are frequency of multiple caching events. *(Based on data from Bartlow et al. 2018b)*

To address this question, northern red oaks were radio-tagged as described above and then placed in differential exclosures that allowed access by only rodents (squirrels, mice, and chipmunks). We presented 107 radio-tagged acorns at 17 seed sources and in each exclosure we presented approximately five tagged acorns each positioned over a magnet to keep the transmitter batteries off until dispersed by a rodent (Bartlow et al. 2018b; Fig. 7.8). We also presented 100 red oak and 100 white oak acorns (all untagged) to reduce the probability that rodents would open and subsequently reject the radio-tagged acorns. The white oak acorns were presented to increase the probability that red oaks would be selectively dispersed over the highly perishable white oak acorns. Over the course of three seasons, 102 of the 107 radio-tagged acorns were dispersed from their source. The five that were not dispersed were eaten at the source. We systematically relocated caches every 2–3 weeks through the caching season to determine acorn locations and fates. Among the 39 radio-tagged acorns initially cached, 19 (49%) were re-cached on one, two, or three additional occasions. Cached acorns were recovered and eaten anywhere from a few days to 93 days after initial dispersal, although these estimates of time were in part limited by battery life (Fig. 7.9). The majority of the caching events were the work of eastern gray squirrels.

Overall, the use of radiotelemetry allowed us to establish that cache recovery is common soon after initial scatterhoarding of acorns, and that an acorn may be recovered and re-cached several times before reaching its final destination and fate. However, contrary to our hypothesis that re-caching would result in acorns being dispersed farther from their sources, we found that acorns were typically re-cached close to their original cache site, likely in the same animal's home range (Fig. 7.10). These results, coupled with those of other studies, leave us to believe that these re-caching events at each seed source are largely the work of the same individual gray squirrel. It appears that gray squirrels revisit their caches throughout the food storing season, examine the cached acorn, consume the acorn perhaps if it is damaged or beginning to rot, or re-cache the acorn if it is still sound and storable.

THE EFFECTS OF THE CACHING ENVIRONMENT ON THE ACORN

Once an acorn is cached, the new environment can alter its composition or fate. This may seem obvious in that the primary reason that rodents scatterhoard seeds is to reduce their availability to competitors. But, in addition, the cache environment—where the acorn may be stored for six months before recovery or germination—has the potential to change the acorn. Although we now know that red oak acorns are selectively cached over those of white oak because of their delayed germination and reduced perishability in the cache (Hadj-Chikh et al. 1996), over the years, several authors tested the alternative hypothesis that acorn chemistry (tannin level) is modified while stored in the cache. Perhaps tannins are leached from the acorn by ground water, or they decline over time due to physiological processes?

Several studies tested this hypothesis by burying acorns, preventing their access by seed consumers and then periodically recovering a sample and quantifying their tannin levels. In all such studies published to date, however, data

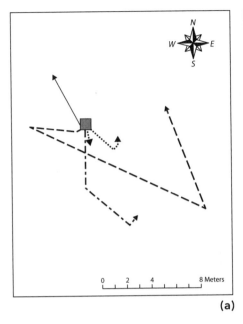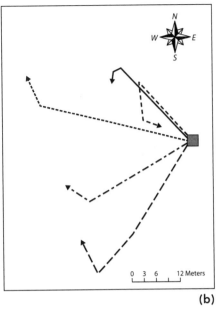

7.10. Dispersal trajectories for radio-tagged acorns from two seed sources. Each line indicates the distance and direction each radio-tagged acorn was moved through the food-hoarding season until its final fate. Triangles indicate location where the tagged acorn was eaten (opened). Note the differences in scale between **(a)** and **(b)**. *(Based on data from Bartlow et al. 2018b)*

did not support this hypothesis (Dixon et al. 1997a; Koenig and Faeth 1998; Smallwood et al. 2001). In some cases, tannin increased, which is what we observed for northern red oaks (Smallwood et al. 2001). However, all studies measured total tannin levels (both hydrolysable and condensed tannins), rather than just protein-precipitating phenolics, which may show greater relevance to rodent and insect behavior. As shown for foliage of six species of oaks, total phenolics (hydrolysable and condensed tannins) are not correlated with protein-precipitating phenolics, which are more likely a better measure of their ecological significance (Martin and Martin 1982).

Although not published, we attempted a similar experiment in which we buried acorns of both the white oak (the white oak and chestnut oak [*Q. montana*]) and the red oak group (pin oak and red oak). To simulate caches, we placed acorns in 10 enclosures (1.0 m × 1.0 m × 0.25 m) at a single hardwood forest in northeastern Pennsylvania. During the last week of November, approximately 140 acorns of each species were placed in each box and covered with a few centimeters of soil and a layer of leaf litter. Red oak and pin oak acorns were numbered to identify the tree from which they were collected, and equal samples

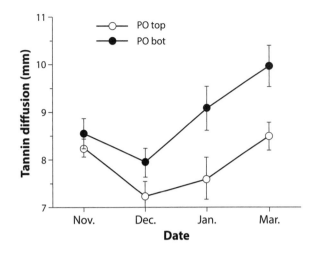

7.11. Summary of tannin activity in pin oak (*Q. palustris*) acorns during the four months of the experiment. Results of repeated measures ANOVA were significant (F = 6.56; DF = 3, 76; P < 0.002). In addition, tannin levels were significantly higher in the bottom halves of acorns in all but the control samples (P < 0.014). *(Michael A. Steele, unpublished data)*

of each of these two acorn species were placed in each box. Because the white oak acorns were unavailable, they were purchased from a seed company. Boxes were constructed of hardware cloth to prevent access by any seed predators. We collected random samples of each of the four acorn species on four dates during the study: late November (prior to placement in boxes), December 24, January 31, and March 14. For each collection period, we sampled 1–2 acorns of each species from each of the 10 boxes and then shelled and cut the acorns into top and bottom halves so we could also compare changes in tannin gradients. We quick froze and dried the cotyledon and then ground the cotyledon into a fine powder. In contrast to earlier studies, we chose to measure protein-precipitating phenolics rather than total levels of hydrolysable and condensed tannins. Our results, like those of previous studies, showed no evidence of a decrease in tannin activity. However, across all four species we observed a significant increase in tannin activity as illustrated for pin oaks in Figure 7.11. In all four species, tannin levels increased with burial time and was highest in three of the four species when acorns were about to germinate in the spring.

From the oak's and acorn's perspective, this seemed to make ample sense. Higher phenolic activity just at the time of germination may better ensure successful seedling establishment. However, for each acorn analyzed, we also carefully assessed any damage due to insects (primarily *Curculio* larvae) and other pathogens (e.g., mold and fungi). All acorns were first inspected for signs of insect damage and discarded, so levels of insect infestation were, as expected, low for all four species. However, during the time of burial, evidence of other pathogen damage steadily increased (Fig. 7.12), hence, suggesting an alternative benefit for increased tannin activity with time in the cache. Collectively, these studies indicate, whether it is precipitating-phenolics or measures of total

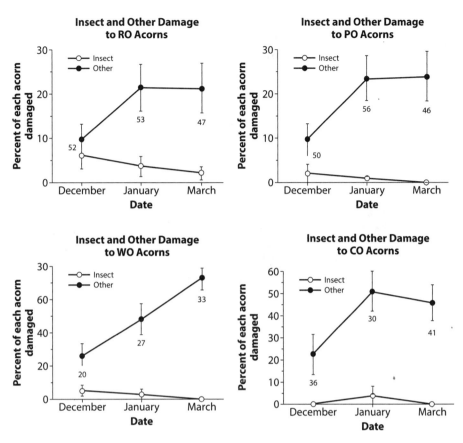

7.12. Summary of acorn damage resulting from insects (primarily *Curculio* larvae), fungi, and molds during time in simulated cache. Numbers indicate sample sizes. Note the increase in damage throughout the study. *(Michael A. Steele, unpublished data)*

tannin content, there is no evidence that phenolics decline during burial. In a related study, we modified tannin levels in artificial acorns presented to squirrels and found no evidence that such changes directed caching decisions (Steele et al. 2001a).

A second set of factors related to the cache site is whether the cache environment increases the probability of germination and establishment. In a model developed by Zwolak and Crone (2012), it was suggested that caching by scatterhoarders is likely to shift the relationship between the scatterhoarder and the seed along a continuum from one of seed predation to mutualism by enhancing the probability of seed survival in the cache. They conclude that many plant-rodent interactions tend toward a weak mutualism but can easily shift away from mutualism for some species and under some environmental conditions (Fig. 7.13). In a recent test of this model and these hypotheses, several of

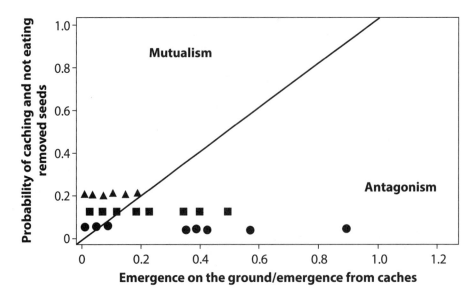

7.13. Classification of plant-granivore interactions based on the relationship between the probability of caching and not recovering seeds, and the ratio of seedling emergence from the ground to emergence from caches. Triangles = red oaks (*Quercus rubra*), Squares = Hybrid chestnut (*Castanea dentata* x *Castanea mollissima*), and Circles = white oak (*Q. alba*). The net effect of granivores is beneficial at any point above the line and antagonistic at any point below. *(From Zwolak and Crone 2012 and Sawaya et al. 2018)*

my colleagues and I (Sawaya et al. 2018) tested how the cache site influenced the probability of seedling establishment in red and white oak, American chestnut, and hybrid chestnut (*C. dentata* × *C. mollissima*). Fruits of each species and the hybrid chestnut were randomly buried or left exposed on the ground surface just below leaf litter, and all nuts were then protected from rodent predation with reinforced enclosures that were designed to prevent access by all rodents. Studies were conducted at 15 sites at each of five locations (Maine, Massachusetts, Pennsylvania, Indiana, and Virginia) over a three-year period. All sites were revisited the following summer and nut fates were determined (dead or seedling emergence). All three species and the hybrid experienced greater probability of emergence when buried. And there was no effect of soil temperature, site, and seed source on survival. However, when we used the model developed by Zwolak and Crone (2012) and accounted for the probability of caching (derived from other studies), results indicated a consistent mutualism for red oaks, most often an antagonistic relationship for white oaks in which establishment was rare, and an intermediate relationship for the American chestnut (Fig. 7.13).

EXPERIMENTAL EFFECTS OF ACORN
ABUNDANCE ON OAK DISPERSAL

A few years after my first metal-tagging experiments on acorn dispersal, some careful planning, and a small seed grant from the National Science Foundation (NSF), I decided to attempt an experiment that tested the effects of acorn abundance on patterns of dispersal. This time I was more ambitious (Steele et al. 2001b), perhaps a bit too ambitious. My goal was to conduct a large-scale experiment that tested the effects of acorn abundance on dispersal of red oak and white oak acorns under experimental conditions. I also varied the types of acorns presented to determine if the abundance of acorns of red oak, white oak, or both species influenced dispersal differently. We selected a 21-ha stand of middle-age to mature red oaks and white oaks mixed with other hardwoods during a year of low acorn production. The site was also surrounded by additional hardwood forests, so we were unlikely to introduce any edge effects along open fields or other forest types. We then established eight, 4 × 4-grids (100 m × 100 m) of seed sources. Each of the sixteen seed sources on a grid were spaced about 20 m apart and each of the eight grids were separated by 100–175 m from each of the other seven grids.

In mid-November, we presented 100 red oak and 100 white oak acorns at each seed source covered with chicken wire (1 m × 1.25 m) staked above the acorns. On each grid of 16 patches, we placed 100 red and 100 white oak acorns, which were subsequently monitored to determine removal rates. On each grid, we randomly identified five of these patches at which we presented 100 red oak and 100 white oak acorns that were metal-tagged for subsequent recovery. Finally, we randomly selected two grids in which we simulated high acorn crops of either red oak, white oak, or both red oak and white oak, and two control sites where no additional acorns were added. At the 16 patches on each grid, we placed either an additional 1000 red oak acorns, 1000 white oak acorns, 500 of each of the two species, or no additional acorns. The two grids to which no additional acorns were added were the control sites.

Following acorn placement in mid-November, we then determined the removal rates of acorns on each of the 128 patches of tagged seeds by counting the numbers of each acorn species remaining every other day for several weeks or until most of the acorns were removed. This in itself took a significant effort, with students doing much of the work. However, I recall several evenings long past sunset in which I trekked through the forest with a headlamp strapped to my head and my 2-year-old daughter, Emily, strapped to my back, her eyes wide open and her attention on the night sounds, and mine focused entirely on counting acorns.

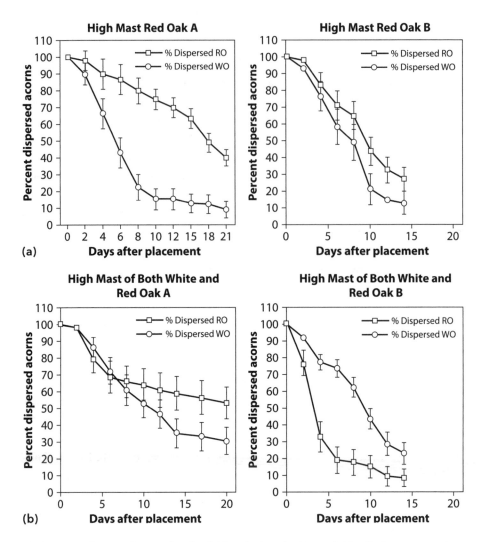

High Mast Red Oak A

High Mast Red Oak B

High Mast of Both White and Red Oak A

High Mast of Both White and Red Oak B

(a)

(b)

7.14. Removal rates of tagged red oak (*Q. rubra*) and white oak (*Q. alba*) acorns on sites in which we simulated **(a)** high red oak mast, **(b)** high red and white oak mast, **(c)** high

Removal rates were 3–4 times faster on control sites with limited acorn abundance (Fig. 7.14). On the two control sites, nearly all acorns were removed within five days. In contrast, at all other grids where we supplemented with additional acorns, acorn removal by rodents took 15 days or longer. And at four of the six sites, white oak acorns were removed faster than the red oak acorns, a pattern that we typically see when seed abundance is high. Consumption and removal rates were instructive, but only a small piece of the full dispersal picture. However, as I have often commented in my review of others' papers, removal rates of seeds, by themselves, provide limited information on seed fate. The ultimate

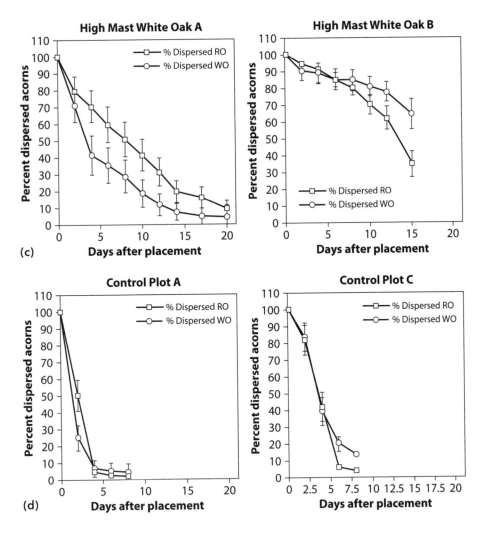

white oak mast, or **(d)** control sites in which additional supplements of acorns were not added to the sites. All treatments were replicated, and their location randomized. Note the rapid removal of acorns on the control sites. *(Michael A. Steele, unpublished data)*

fate of a seed type can often be the opposite of what is suggested by removal rates alone (Steele et al. 2001b).

After four weeks of following seed removal, we then conducted metal-tag recovery from mid-December to late spring. As a result of our effort, we recovered 21.6% (1725) of the 8000 tags or tagged acorns after a labor-intensive search of the 30 m surrounding each seed source. In the absence of any evidence of other seed disperser activity, we concluded that the acorn fates we observed were entirely the work of gray squirrels, white-footed mice, and eastern chipmunks. Moreover, given our search effort—both that within the 30-m and spot-checks

Table 7.2. Initial acorn fates following dispersal at all control and experimental sites. Notched acorns are those in which embryos were excised. Note high rates of caching of whole red oak (*Q. rubra*) and high rates of consumption and embryo excision of white oak (*Q. alba*) acorns.

ACORN DISPERSAL

Red Oak Acorns

TREATMENT	SITE	WHOLE	EATEN	NOTCHED
Control	A	30	7	0
Control	B	40	19	1
RO supplement	A	98	29	3
RO supplement	B	18	50	1
WO supplement	A	118	90	0
WO supplement	B	18	72	8
WO & RO supplement	A	71	22	0
WO & RO supplement	B	2	32	1

White Oak Acorns

TREATMENT	SITE	WHOLE	EATEN	NOTCHED
Control	A	14	24	7
Control	B	10	31	57
RO supplement	A	44	30	46
RO supplement	B	5	66	26
WO supplement	A	45	124	128
WO supplement	B	0	149	4
WO & RO supplement	A	25	28	58
WO & RO supplement	B	1	57	1

throughout the study area—led us to believe that the remaining tagged acorns were dispersed below ground to sites not suitable for germination (Steele et al. 2001b).

For those we recovered, the patterns were striking. As described briefly in Chapter 5 based on data from another experiment, red oak acorns were dispersed and cached significantly more often and farther from their source than those of white oak acorns. And, although white oaks were occasionally cached, when they were, they often had their embryos excised—the obvious work of a gray squirrel. Moreover, a comparison of data across the six months following acorn removal revealed yet another anecdote regarding embryo excision. In the winter, approximately 60% of the cached white oak acorns had excised embryos,

Table 7.3. Frequency of dispersed red oak (*Q. rubra*) and white oak (*Q. alba*) acorns surviving to establishment as seedlings. Results are from control sites and all other sites in which we simulated high acorn mast.

FINAL ACORN FATES

(No. of metal-tagged seeds germinating from an original sample of 8000 seeds [500 at each site])

TREATMENT	SITE	RO	WO
Control	A	3	0
Control	B	0	2
RO supplement	A	19	1
RO supplement	B	2	0
WO supplement	A	20	0
WO supplement	B	17	0
RO & WO supplement	A	2	0
RO & WO supplement	B	1	0

and just a few of the red oak acorns were processed the same way (<1%) (Steele et al. 2001b). In the spring, however, >80% of the white oak acorns had embryos excised. But, interestingly, on average across seed sources, 17% of the red oak acorns also had their embryos excised (Table 7.2) (Steele et al. 2001b). Based on these data and anecdotal observations in which we revisited cached red oak acorns, we concluded that closer to spring germination, after these acorns had emerged from dormancy, squirrels were likely revisiting caches and manipulating these red oak acorns as well (Steele et al. 2001b).

By early summer, we revisited all stored acorns and determined final acorn fates (i.e., seedling establishment). Despite our efforts to simulate acorn abundance, of the 1725 acorns for which we could track final seed fates (i.e., seedling establishment), only 67 survived: 3 white oak acorns and 64 red oak acorns (Table 7.3). All but five of these acorns that grew to the young seedling stage were dispersed from sites at which acorn supplements were provided.

The above demonstrates how careful tagging experiments in the forests can reveal patterns of acorn dispersal that simply could not be revealed in less controlled situations. Over the years, we conducted numerous such acorn-tagging experiments and learned quite a bit regarding how specific environmental factors can influence dispersal patterns of oak. Although we learned a great deal in this experiment, in retrospect, I was haunted by some design flaws. Most notably, I concluded after the experiment that our different sites were not isolated enough to ensure site independence and that supplementation of acorn crops to test the effects of different oak mast on dispersal were not likely effective. Nevertheless, we were able to determine how acorn mast in general affected dispersal and caching of red and white oak acorns.

THE EFFECTS OF NATURAL MAST ABUNDANCE ON ACORN AND OTHER NUT DISPERSAL

Once we secured a general understanding of red and white oak dispersal in response to simulated mast abundance, we sought to determine how acorn abundance due to natural masting patterns might influence variation in dispersal. With the assistance of two postdoctorates, Dr. Amy McEuen and Dr. Thomas Contreras, some generous support from NSF, and another band of highly motivated undergraduate researchers, we conducted another metal-tagging study. This time the study was conducted in three forests of similar tree composition across three consecutive years (Moore et al. 2007). At each of the three sites, we again selectively presented metal-tagged acorns to rodent dispersal agents via the differential exclosure boxes described earlier.

In this experiment, we provided five seed sources spread across each of the three study areas. This time, in each of the three years at each seed source, we presented rodents with four types of acorns: both large and small red oak acorns of the same species, as well as acorns of white oak and pin oak (*Q. palustris*). In addition to determining the effects of seed abundance, we also sought to test a second hypothesis nested within the larger experiment. We designed the experiment to also determine how acorn size might influence dispersal patterns. Pin oak is also a red oak species (section *Lobatae*) that has the same pattern of dormancy as northern red oak. Acorns of pin oak, however, are often <40% the size of a northern red oak acorn. To control for some of the subtle seed characteristics (e.g., tannin and lipid levels) that may vary between pin and red oak acorns, we also presented smaller acorns of the northern red oak species. At each of the three sites we also monitored acorn production on 45 trees: 15 each of the three most common species of oaks at each site. We did this by placing two collectors under each of these 135 trees and then monitored seeds trapped in each of these seed traps every two weeks from late August to late November (see Chapter 4). This allowed us to directly assess mast abundance of the oaks at each site across all three years.

In this field experiment, we determined the fates of 2670 acorns in eastern deciduous forests. Consistently, acorns were dispersed farthest during years of lowest mast production (Moore et al. 2007). When mast was abundant, dispersal distances were notably shorter, as low as 20–25% of what was observed during low mast production. Thus, when seeds were less abundant, rodents appeared to invest more energy per seed in dispersing and caching the seeds. The effects of seed size on dispersal varied as we expected. Northern red oak acorns were moved significantly farther than the other smaller acorns of red oak, as well as the acorns of pin oak and white oak, although this too varied with mast production. Differences in dispersal distance between acorn types were most pronounced during years of low mast production. However, this may not ultimately translate into comparable patterns of seedling establishment when mast is low.

INTERACTION OF MAST ABUNDANCE AND OTHER SEED AND RODENT CHARACTERISTICS

As reviewed by Lichti et al. (2017), when mast is low, few if any seeds dispersed by scatter-hoarding rodents survive to establishment (Crawley and Long 1995). Acorn survival and oak establishment is only likely when mast is abundant (Koenig and Knops 2002; Sone et al. 2002; Moore and Swihart 2006; Liu et al. 2013) but can be limited even under these circumstances when the relative abundance of seed predators is also high (Sun et al. 2004). Partial acorn consumption due to chemical gradients and subsequent survival is also more likely when mast is plentiful (Steele et al. 1998; Yang and Yi 2012). But, as demonstrated by Moore et al. (2007) and numerous other studies on scatter-hoarding rodents when mast is abundant, competition is reduced and less energy is invested per seed (Lichti et al. 2017). Consequently, dispersal distances are reduced under these conditions. However, as we describe in Lichti et al. (2017), the relationship between the relative seed abundance and seed fate depends on numerous other factors such as the relative abundance of rodents (Theimer 2005), the diversity and body size of other rodent species, and the availability of other seed types. All of these factors define the per capita seed availability for each potential dispersal agent and thus the circumstances in which the relationship between the rodent and the seed is one of "mutualism" or "seed predation" (Theimer 2005; Zwolak and Crone 2012; Lichti et al. 2017). This means that seed fate depends on other conditions in addition to mast abundance. Consequently, high seed abundance alone does not guarantee successful seed establishment.

THE EFFECTS OF SEED SIZE AND NUTRIENT CONTENT ON ACORN DISPERSAL

In addition to illustrating the effects of acorn abundance on seed dispersal, the above study also demonstrates how seed size and/or energy content influences the movement and caching of oaks (see Chapter 5). This selective caching of larger acorns over smaller acorns of the same or different species has been observed by numerous other authors for the oaks. Lichti et al. (2017) carefully evaluated 151 publications to determine the effects of seed size (within and among species), energy content, and other seed characteristics on the decision to cache and the investment in energy to handle (eat or cache) by scatter-hoarding rodents. Consistently, these studies show that larger seeds are dispersed and cached farther from their source in about 70% of such studies. Controlling for size, especially in studies that compare responses to two or more species of seeds, is quite difficult because of the many other confounding characteristics (e.g., plant secondary metabolites; nutrients such as protein, lipid, and carbohydrates; and germination schedules). Nevertheless, size and total energy content is a major determinant of dispersal and caching by scatter-hoarding rodents, especially those that move the oaks.

THE EFFECTS OF OTHER SEED TYPES ON ACORN DISPERSAL

It is likely that scatterhoarders not only respond to varying acorn characteristics when dispersing these fruits, but also to the ambient availability of other seed types. In one of the first studies to experimentally test this question, Nate Lichti, Robert Swihart, and I documented the response of rodents (primarily eastern gray squirrels) to pairwise presentations of red oak acorns, white oak acorns, and the American chestnut. Our goal was to track the animals' responses to dense patches of two seed types in each patch using red oak and white oak patches as a control. We specifically chose American chestnuts because they exhibit a short period of dormancy intermediate in length between that of white oak (no dormancy) and red oak (long dormancy) (Lichti et al. 2014).

As in so many studies before, we metal-tagged the nuts and presented them to rodents in 15 exclosures (described above) at two sites over two years. One site was a mature hardwood forest in Pennsylvania where the American chestnut had been lost to the chestnut blight caused by the parasitic fungus *Cryphonectria parasitica* nearly a century earlier. The second site was a 100-year-old stand of oak and chestnut in Wisconsin where the blight had only established a few years earlier and the chestnut trees were still regularly producing nuts. Our goal was to ensure that rodents had experience with chestnuts at least at one of our sites. At each site and in each year of the study, we presented three pairs of acorns: red oak and white oak, red oak and chestnut, and white oak and chestnut. Each treatment was replicated five times and in each exclosure we presented 200 tagged nuts of each of the two species. Our goal was to evaluate the rodents' responses to pairs of dense patches of nuts in order to evaluate how ambient availability of another nut type influenced dispersal patterns. We tracked nut fates and determined patterns of survival and germination in the spring following each year in which nuts were presented.

As predicted, red oaks were dispersed, cached, and survived when presented with either chestnuts or white oaks. And, chestnuts showed a similar dispersal advantage in the context of white oaks (Lichti et al. 2014). In a comparable experiment in which we paired acorns of an invasive oak (the sawtooth oak [*Q. acutissima*]) with that of either red or white oak, my lab found quite similar results (see Chapter 14). However, characteristics of this introduced oak (i.e., large acorn and short dormancy) appear to endow it with a dispersal advantage over both white oak and red oak, which likely contributes to its potential as an invasive, especially in forests of the eastern US.

EFFECTS OF MASTING PATTERNS OF OTHER SEED TYPES AND FOREST COMPOSITION ON OAK DISPERSAL

The studies reviewed above illustrate how seed dispersal effectiveness is influenced by ambient availability of other seed types when scatterhoarders are making caching decisions regarding a focal species. In our experiments, however, we only modified seed availability by pairing acorns with dense patches of other seed types. Thus, several implications follow. It is likely that comparable variation in masting patterns or forest composition would have a similar effect on dispersal. For example, I would predict that in a forest with American chestnuts, which tends to produce nut crops annually, white oaks would be consistently at a dispersal disadvantage and unlikely to be dispersed and cached by rodents. In a forest dominated by white oaks, red oaks would have a significant dispersal advantage, but red oaks in a forest dominated by the sawtooth oak may be at a significant disadvantage and less likely to achieve successful dispersal. The few studies in other systems that have tested the effects of availability of other seed types (Hoshizaki and Miguchi 2005; Klinger and Rejmánek 2009) or forest composition (Hoshizaki and Hulme 2002) on variation in dispersal success support these claims. However, general studies in this area are few, and in oaks, these questions have not yet been adequately addressed.

I now move to Chapter 8, where I focus on the effects of both competitors and predators, which appear to interact in some unexpected ways to influence the outcome of oak dispersal.

THE OAK DISPERSAL PROCESS: EXTRINSIC FACTORS II

8

INTRODUCTION

Chapter 7 examined how extrinsic factors such as acorn abundance, availability of other seed types, forest structure, and forest composition affect the oak dispersal process. In this chapter, I shift the focus on how the cache site relates to the probability of the seedling establishment and the long-term survival of the oak. This depends on the likelihood of the cache being pilfered by competitors, the suitability of the cache site for germination, and the location of the cache with respect to the oak's likelihood for establishment. I first focus on those of our studies that have quantified rates of pilferage by conspecific competitors of eastern gray squirrels and the ability of this species to reduce pilferage by means of deceptive behavior when caching. I then present a conceptual model of caching for eastern gray squirrels that deviates a bit from that of the optimal spacing model introduced by Stapanian and Smith (1978). Finally, I close with reference to the above and the specific cache sites chosen by eastern gray squirrels as a classic example of directed dispersal, in which cache locations and specific cache sites chosen by these rodents maximize the probability of seedling establishment.

SCATTERHOARDING AND THE THREAT OF SEED PILFERAGE BY COMPETITORS

Spacing of caches: Why does it matter? As discussed in Chapter 9, scatterhoarding behavior is all about the spacing of caches. And, it is, in turn, the spacing of caches that is critical for facilitating seed establishment from the caches of scatterhoarders. The behavior of scatterhoarding differs significantly from larderhoarding in that caches are generally not defended by scatterhoarders because the spacing of caches itself presumably reduces the need for defense against conspecifics and other competitors. Working from this assumption—seeds are scatterhoarded to reduce pilferage by other rodents and birds—early researchers developed some simple, but very compelling models to predict the manner in which seeds should be spaced by scatterhoarders (Stapanian and Smith 1978; Clarkson et al. 1986).

First articulated by Stapanian and Smith (1978) and later Clarkson et al. (1986), these models, sometimes referred to as the optimal density models, were based on the assumption that scatterhoarders are faced with a trade-off between the benefits of spacing caches and the costs of recovery. Greater distance between cache sites translates into reduced pilferage rates, but with this increase in inter-cache distance also comes an increase in recovery costs. Stapanian and Smith (1978) proposed that naïve competitors conduct area-restricted searches, and if successful at recovering one seed, continue searching. However, at lower densities (i.e., greater spacing), a threshold will be reached, and naïve competitors will cease searching without successfully recovering a seed. Stapanian and Smith (1978) argued that for each scatterhoarder there is an optimal density at which a particular seed type should be cached, one that balances both advantages and costs of cache spacing. This optimum density could change when seed quality changes. Higher seed profitability would allow animals to decrease cache densities; lower seed quality would allow higher cache densities. Avian studies on scatterhoarding suggest a slightly modified strategy, one in which seeds are hyper-dispersed to reduce cache pilferage (Male and Smulders 2007, 2008). Although such studies on birds argue against an optimal spacing of caches, they do support the argument that cache spacing reduces pilferage by competitors.

Pilferage of caches by competitors has long been assumed to be a major driver in the evolution of hoarding behavior (Andersson and Krebs 1978; Smith and Reichman 1984; Vander Wall and Jenkins 2003). Recent studies further suggest that high rates of reciprocal pilferage between conspecifics may be the rule rather than the exception (Vander Wall and Jenkins 2003; Hirsch et al. 2012). Drawing on a number of studies that determined pilferage rates from artificial, observed-made caches, Vander Wall and Jenkins (2003) concluded that pilferage is common for many scatter-hoarding rodents and birds and sometimes occurs at rates as high as 2–3% per day of an individual's total food stores. Vander Wall and Jenkins (2003) justifiably reasoned that such rates would be catastrophic for any species that depends on food storage for periods of several months or more, unless each individual in a population has an equal opportunity to pilferage. This led Vander Wall and Jenkins to propose a rather ingenious model of reciprocal pilferage—a potentially stable evolutionary strategy—in which each individual in a population sustained heavy losses to pilferage, but each of these hoarders also was able to pilfer from its neighbors.

Vander Wall's and Jenkin's (2003) synthesis was not only logical, but it also accounted for a wide array of published observations on food-hoarding animals, and thus seemed a reasonable depiction of the competitive environment in which most food-hoarding animals find themselves. Moreover, since its publication, several studies seemed to corroborate the model. Hirsch et al. (2012), for example, showed that territorial agoutis, common scatterhoarders of palm seeds in the tropics, regularly pilfer from neighbors' territories. This occurred so frequently, in fact, that the researchers followed the re-caching of individual fruits as they were moved across numerous agouti territories and were repeatedly cached, recovered, and re-cached.

CACHE PILFERAGE AND DECEPTIVE CACHING IN EASTERN GRAY SQUIRRELS

It was these early models on cache spacing, coupled with growing evidence for high rates of cache pilferage, that led us to investigate whether reciprocal pilferage was common among eastern gray squirrels and whether such interactions could help us to better understand patterns of oak dispersal. Our field notes from studies at our Kirby Park site included incidents in which squirrels would observe a conspecific caching and then move directly to the cache site to excavate and steal the cached acorn.

Early in these studies, we also observed an interesting behavior that appeared to be directed specifically at would-be pilferers. Hereafter referred to as deceptive caching, this behavior suggested a high potential for cache robbery. Through a series of first observations, and later more focused field experiments, we successfully demonstrated that in the presence of neighboring conspecifics, squirrels would often excavate and cover an empty cache site (Steele et al. 2008). Carrying an acorn in its mouth, a gray squirrel would dig a potential cache site, and then thrust its body and head forward several times into the excavated hole, appearing the entire time to bury the acorn (Fig. 8.1). The animal would then follow this "burying" behavior with 3–5 stereotypic swipes of the front paws that appeared to cover the exposed cache site. The result was a display that convincingly suggested to either a researcher or neighboring squirrel that an acorn had just been cached. In fact, on several occasions, we observed the immediate investigation of the empty cache by a neighboring squirrel. The potential pilferer, though, came up empty-handed while the caching animal, with the acorn still in its mouth, moved off to carefully cache the acorn in a more secure location. Through a series of experiments, we were able to demonstrate that while the frequency of this deceptive behavior was relatively rare,

8.1. An eastern gray squirrel (*Sciurus carolinensis*) caching an acorn. Squirrels that engage in deceptive caching (Steele et al. 2008) engage in the same sequence of behavioral acts required for caching, without ever depositing a food item in the cache. *(Michael A. Steele)*

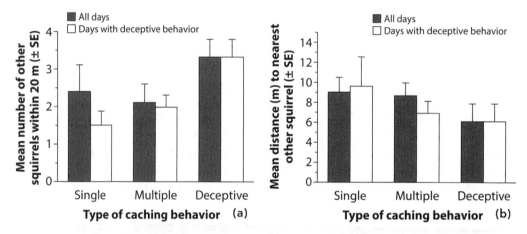

8.2. Frequency of deceptive caching events versus other caching events (single and multiple caching events) in relationship to **(a)** the number of neighboring squirrels and **(b)** the proximity to the nearest squirrel during each caching event. Single and multiple events refer to the number of cache sites excavated when an acorn was actually buried. *(From Steele et al. 2008)*

it was performed significantly more often in the presence of conspecifics (Fig. 8.2). Moreover, we were able to induce this deceptive behavior, along with several other pilferage averting behaviors, when we actually engaged in cache pilferage ourselves (Steele et al. 2008). This study clearly demonstrated a strategy that squirrels employ to reduce pilferage rates.

Whereas deceptive caching suggested a possible strategy by which our squirrels were reducing pilferage, a separate line of investigation at the same study area seemed to further support the concept of reciprocal pilferage posited by Vander Wall and Jenkins (2003). In an attempt to map caches and follow cache fates for individual squirrels, we presented individual squirrels with acorns

(a)

(b)

8.3. Shown are **(a)** Passive Integrated Transponder (PIT)-tag. **(b)** A single PIT-tag was implanted in an acorn by drilling a small hole in the basal end of the acorn, inserting the tag, and sealing the hole with ground acorn cotyledon and an odorless wood glue (Suselbeek et al. 2013). *(Shealyn A. Marino)*

8.4. Student technician using a PIT-tag reader to confirm that a cached acorn was still in the ground. *(Michael A. Steele)*

in which we had carefully inserted PIT (passive integrative transponder) tags (Suselbeek et al. 2013; Fig. 8.3). These tags allowed us to detect and follow the fate of cached acorns without ever disturbing the cache (Fig. 8.4). At a minimum, we could repeatedly revisit each mapped cache site and at least determine when the cached acorn had been recovered. In many cases, we were able to recover the naked PIT tag near the empty cache site and determine that the recovered acorn had been eaten. After following the cache fates of more than 40 animals over two summers and one autumn, a striking pattern emerged. These tagged acorns only remained in the original cache site for a few days; in many cases they were recovered within 24 hours. We assumed these removal rates were due to high rates of cache pilferage, which would likely follow from the artificially high density of squirrels in this seminatural, urban park (Steele et al. 2001a). Perhaps, then, our site was an ideal arena for investigating the phenomenon of reciprocal pilferage?

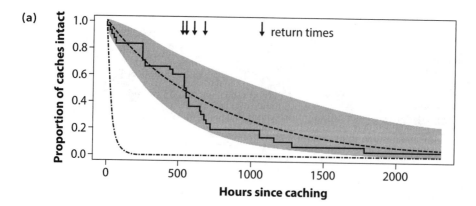

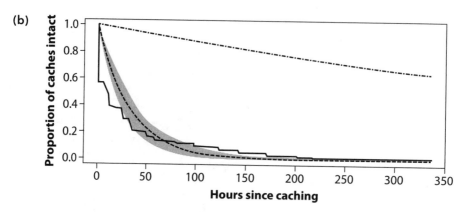

8.5. Survivorship of cached acorns when **(a)** cache owners were experimentally removed from their home range immediately after caching, compared with **(b)** cache survivorship when squirrels remained on their home range during the same time period. Shaded area represents 95% confidence intervals and the line outside confidence intervals shows survivorship for acorns in the opposite treatment for comparison. Arrows indicate when displaced squirrels were returned to the study area. (*From Steele et al. 2011*)

RECIPROCAL PILFERAGE OR HOARDER'S ADVANTAGE?

To determine if these high removal rates were the result of exceptionally intensive rates of conspecific pilferage, we needed to determine which squirrels were responsible for recovering caches: Was it the cache owners or the naïve conspecifics (pilferers)? Our hunch and the literature supported the latter. In an attempt to address this, we continued to map PIT-tagged acorns immediately after they were cached and followed the longevity of caches by revisiting them every twelve hours. However, this time, we selected a subsample of squirrels in which we sought to simulate their mortality immediately following the placement of their caches. We mapped the caches of an individual squirrel and then, when-

164 OAK SEED DISPERSAL CHAPTER 8

ever possible, live-trapped the same animal immediately after it had cached. The animal was then transferred to a large holding facility where it was housed and fed for several weeks before it was returned to the study site. This approach—effectively a simulation of mortality of the cache owners—allowed us to estimate the actual rates of cache pilferage. The results were striking, and not what we expected (Fig. 8.5). When squirrels were removed from their home range, their cached acorns remained in the ground. Rates of cache removal dropped by tenfold. True pilferage by naïve competitors still occurred, but at a much lower rate than expected. And after squirrels were returned to their home ranges, as many as six weeks later, many of the cached acorns again disappeared. These results qualitatively suggested the possibility of long-term memory of cached acorns by the gray squirrel. However, more importantly, and certainly more definitively, we documented clear evidence of a hoarder's advantage (Steele et al. 2011). The cache owner had priority over its own caches compared with neighboring naïve squirrels. This result, at least for our study animals, seemed contradictory to the model proposed by Vander Wall and Jenkins (2003).

AN ALTERNATIVE VIEW OF SCATTERHOARDING

Throughout our studies in the Kirby Park study area, and in several forests across North America, as well as several other studies conducted in oak forests worldwide, another compelling pattern of rodent scatter-hoarding behavior began to emerge. All of these studies showed that whenever controlled experiments were used to determine the hoarding preferences of rodents, preferred nuts were carried significantly greater distances from sources than those that were less preferred. For example, in our early studies at Kirby Park, when we presented acorn pairs to individual squirrels, the animal consistently dispersed preferred acorns farther from the source (i.e., the point of presentation). Regardless of whether it was sound versus infested acorns, acorns of red oak versus those of white oak, or large acorns versus small acorns (see Chapter 5), the former preferred acorn in each of these acorn pairs was dispersed farther (Hadj-Chikh et al. 1996; Steele et al. 1996). In the forested sites where we presented acorns at a central seed source, this was also the case (Steele et al. 2001b; Moore et al. 2007). Perhaps preferred acorns were being moved to preferred caching sites or, alternatively, as predicted by Stapanian and Smith (1978), maybe rodents were spacing preferred seeds farther apart, thereby optimizing the dispersion of their caches? The literature, for example, is now replete with evidence that larger and more profitable seeds (including acorns) are consistently dispersed farther from their sources than smaller seeds (Jansen et al. 2004; Xiao et al. 2004, 2005; Galvez et al. 2009). Many of these authors conclude that this is done to optimize the spacing of caches. It is also assumed that this often in-

8.6. Differential Geographic Positioning System (GPS) used to map caches (± 10 cm) made by individual gray squirrels. *(Dale Bruns)*

creases the probability of seedling establishment. Both assumptions, although reasonable, remain untested.

Throughout our experiments at the Kirby Park study area, we observed the same pattern, regardless of where the animal was encountered within its home range. In other words, it did not seem that animals were necessarily dispersing preferred acorns to a preferred caching area within their home ranges. Why consistently disperse larger seeds farther? With this question in the back of our mind, we sought to determine if acorns were simply spaced as predicted by the optimal density models, or if there was another explanation for this response.

We began by repeating past trials that involved the presentation to individual tagged gray squirrels. Each spring, after live-trapping several dozen squirrels and using nyanzol dye to apply individual marks, visible from a distance of 25 m or more, we systematically presented the animals with PIT-tagged acorns of three types: large northern red oak acorns (*Q. rubra*), smaller acorns of northern red oaks, and pin oak (*Q. palustris*) acorns. Our goal was to vary acorn size but control for other seed characteristics (e.g., lipid and tannins). Pin oaks, although a different species than Northern red oak, belong to the red oak group and possess similar levels of tannin and lipid. The smaller northern red oak acorns were larger than pin oak acorns but substantially smaller than the larger red oak acorns. We were therefore confident that we were presenting acorns that varied in both size and energetic profitability.

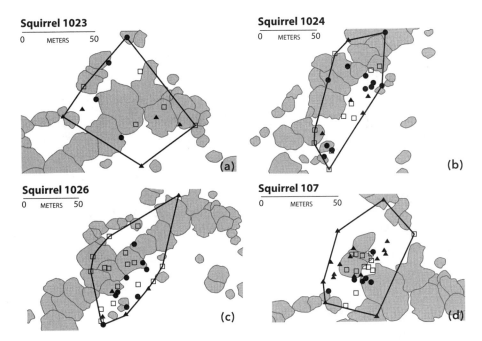

8.7. Schematic of the caches and caching area of four individual gray squirrels in relation to tree canopy. Shaded areas indicate tree crowns drawn from aerial photos. Solid triangles = cached large red oak (*Q. rubra*) acorns, Solid circles = cached small red oak acorns, and Open squares = cached pin oak (*Q. palustris*) acorns. Lines delineate caching area for each squirrel. *(From Steele et al. 2014)*

We randomly and sequentially presented individual squirrels with one acorn of each of the three acorn types and recorded whether the acorn was eaten or cached, and, if cached, the caching time and the distance the seed was dispersed (defined as the distance from the point of presentation to the cache site). We then temporarily marked the cache site with a pin flag and recorded the GPS coordinates with a differential GPS (Fig. 8.6). And, because our study area was only one kilometer from our geo-referenced GIS center, this high-end, differential GPS allowed us to determine the precise GPS location of each cache with considerable accuracy (±30 cm). Our goal was to map all the caches and test the predictions of the optimal density models by comparing the distribution and cache density of each of the three acorn types.

Our hypothesis was simple—larger acorns should be spaced farther apart. Early in our study, however, we observed a potentially interesting behavior and a plausible alternative hypothesis. Student technicians and our postdoctoral associate, Thomas Contreras, suspected that squirrels were relying on an alternative strategy. These researchers noted that they consistently encountered squirrels

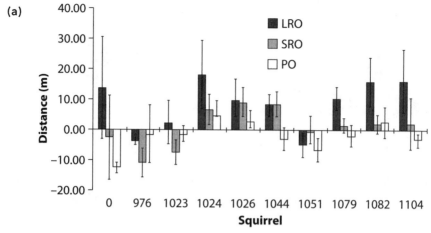

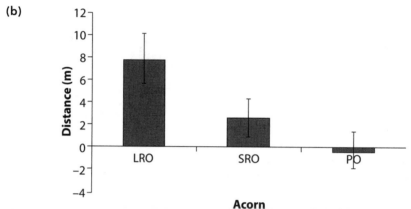

8.8. Mean distance (meters ± SE) of caches from canopy edge for each of the three acorn types presented to gray squirrels (*Sciurus carolinensis*) for **(a)** 10 individual squirrels and **(b)** the same data for all squirrels combined (F = 4.615; DF = 2, 196; P = 0.011). The three acorn types are the same as in Figure 8.8. *(From Steele et al. 2014)*

near the base of trees when attempting to present acorns. The squirrels then dispersed the acorn out in the open, beyond the canopy of the tree. Larger acorns appeared to be carried out in the open farther, whereas the smaller pin oaks were stored under the canopy or near the canopy edge. On occasion, competitors in the tree above would quickly move to the ground and attempt to pilfer recently cached acorns. This introduced the alternative hypothesis that more profitable acorns may be cached farther from cover.

We decided to test both hypotheses, the first by measuring cache densities of each acorn type. The second hypothesis was tested by comparing the position of each acorn in relation to canopy cover. To do this, we plotted acorn caches

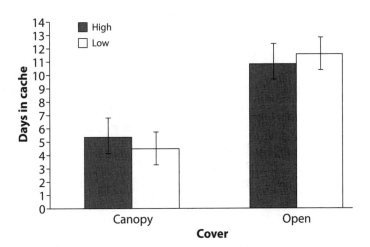

8.9. Cache survival (days in the cache) of red oak acorns artificially cached by researchers under two conditions: varying cache density (high versus low) and under or outside tree canopies. *(Michael A. Steele, unpublished data)*

on a digitized aerial photograph on which we could measure the distance from the canopy that an acorn was cached. We were also able to compare the relative abundance of habitats (canopy cover, canopy edge, or open habitat) with the distribution of different acorn types.

The results were compelling. In our initial analyses, cache densities between different acorn types did not vary significantly. However, larger acorns were consistently dispersed outside of canopy cover, acorns of intermediate size were often cached just beyond the canopy, and the smallest acorns (those of pin oak) were consistently cached under canopy or at the canopy edge (Steele et al. 2014; Figs. 8.7 and 8.8).

To further test if cache spacing and canopy cover influenced cache pilferage, we designed a second experiment in which we systematically made artificial caches of red oak acorns at high and low densities, both under canopy cover and out in the open where squirrels often cached the larger acorns of northern red oak. Results indicated that it was the habitat structure, not the cache density, that influenced the pilferage rate. Pilferage rates were lowest in the open habitat regardless of cache density (Fig. 8.9).

As is often the case in science, at the same time we were conducting the above experiment, a former student (Leila Hadj-Chikh) from our lab, then a doctoral candidate at Princeton University, was closing in on the same question. Also working with gray squirrels, Hadj-Chikh designed a similar experiment to determine how canopy structure influenced cache placement. She presented individual squirrels with acorns of two species, northern red oak and the scarlet oak (*Q. coccinea*), both red oak species, the first (mean = 7.13 g) approximately twice the size of the second (3.52 g). In her experiments, however, Hadj-Chikh sequentially presented the same individual squirrel with one acorn of each type from the base of the same tree. Following the presentation of one acorn, she waited for the same squirrel to return and then presented the second acorn. She randomly varied presentation order between squirrels. As in our study, the larger red oak acorns were dispersed farther into open habitat (about 18 m on average) as compared with those of scarlet oak (10 m). However, in this study, Hadj-Chikh was able to show that squirrels' investment in the dispersal process (the caching distance per unit mass) was nearly identical for the two species (2.9 versus 2.6 in red oak and scarlet oak, respectively). Squirrels were carefully investing more energy in the dispersal of larger acorns but also dispersing them farther into open habitat. This investment in the dispersal process was, however, directly proportional to acorn size.

PREDATION RISKS AND THE HABITAT STRUCTURE HYPOTHESIS

In 2013, we combined our results with those of Hadj-Chikh in one publication and offered what we consider an important alternative model of rodent scatter-hoarding. We call this conceptual model the habitat structure hypothesis. With this hypothesis, we propose that the squirrels, and possibly other rodents, select cache sites in more open vegetation where the likelihood of pilferage is lower. We also propose that in this process, rodents effectively trade-off higher risks of predation for a lower risk of cache pilferage. A growing body of literature, much of which follows from the studies of J. Brown and B. Kotler, demonstrates that rodents are highly sensitive to predation risks and that their perception of such follows most directly from the structure of their habitat. Even the scent of predators is reported to influence patterns of oak dispersal by rodents (Sunyer et al. 2013).

If we are correct that rodents risk some level of predation for more secure cache sites, and that the selection of such cache sites also results in a hoarder's advantage, then two critical implications for seed dispersal likely follow, both of which may be particularly relevant to the dispersal of oaks. First, it is likely that these higher risks of predation will, on many occasions, result in the actual predation of the hoarder. With a hoarder advantage operating at many of these cache sites, seedling establishment is likely to follow when the hoarder is killed. Thus, the dispersal process may depend not only on the scatter-hoarding rodents but also on the meso- and top carnivores that prey on the rodents, especially at the time of caching.

Secondly, because these cache sites are located in open vegetation, seedling establishment is further enhanced. In the shade intolerant oaks, for example, successful seedling establishment is likely to be favored in these more open habitats. Hence, this entire process of caching in more open habitats to reduce cache pilferage may constitute a form of directed dispersal (see below), one in which cache site selection by squirrels and other rodents results in the deposition of seeds at sites that disproportionately favor seedling establishment.

Evidence for this potential mechanism of oak dispersal is even more compelling when one examines the relationship between shade tolerance across numerous tree species and their seed quality from a squirrel's perspective. As discussed in Chapter 9, squirrels evaluate seeds, not just on their energetic profitability, but on their overall utility for storage and long-term use (Sundaram et al. 2015). Fruits of higher utility than even the most preferred acorns are dispersed and cached at greater distances from sources and more often in open vegetation. Several of my colleagues are now investigating how this relates to shade tolerance in a range of hardwood species. Habitat structure may therefore drive the interplay of conspecific competition for cached food but also the spatial arrangement of established oak seedlings. However, it is important to emphasize that this habitat structure hypothesis is only one account by which scatterhoarders may disperse seeds and manage their caches; it is in no way mutually exclusive of other strategies (e.g., the optimal density approach). It does, however, appear to apply well to the oak dispersal system.

DIRECTED DISPERSAL OF THE OAKS

The habitat structure hypothesis strongly suggests that gray squirrels may contribute to the directed dispersal of the oaks. Directed dispersal is the concept that particular seeds or diaspores are disproportionately dispersed to sites best suited for establishment with respect to special habitats, light conditions, edaphic factors, or the density of conspecifics. Suggested first by Howe and Estabrook (1977) for dispersal of fruits by tropical birds and by Vander Wall and Balda (1981) for dispersal of pine nuts by corvids, the concept was first formally defined by Howe and Smallwood (1982). However, the first study to fully document the phenomenon, especially in tropical systems, is credited to Wenny and Levey (1998). They studied the dispersal of the fruits and seeds of a montane, shade-tolerant tree, *Ocotea endresiana*, in Costa Rica by five species of frugivorous birds. Four of the birds dispersed seeds a short distance from the parent tree to sites similar to random locations where survival was limited. In contrast, the three-wattled bellbird (*Procnias tricarunculata*) dispersed seeds farther from their source to song perches in canopy gaps where seedling survival was highest due to lower rates of fungal damage. The bellbirds, by chance, consistently directed seeds to sites where the probability of establishment and recruitment were highest.

In the oaks, directed dispersal has been strongly suggested across a number of studies but has not been rigorously tested in most cases (Table 8.1). A review

Table 8.1. Overview of studies investigating evidence of directed dispersal oaks by rodents and corvids. Shown are the location, oaks studied, the potential dispersal agents, the primary findings, whether data provides some support for the directed dispersal hypothesis (DDH), and the source.

LOCATION	OAK SPECIES	DISPERSAL AGENT(S)
Forests and other landscapes of Europe	Pedunculate oak (*Q. robur*) and sessile oak (*Q. petraea*)	Avian (primarily the European jay [*Garrulus glandarius*])
European forests	Pedunculate oak	European jay
European forests	Pedunculate and sessile oak	European jay and the European wood mouse (*Apodemus sylvaticus*)
Mediterranean forests	Holm oak (*Q. ilex*)	European jay
Eastern deciduous forests of North America	Pin oak (*Q. palustris*)	Blue jay (*Cyanocitta cristata*)
Central Japan	Jolcham oak (*Q. serrata*)	small Japanese field mouse (*Apodemus argenteus*)
Spain	Holm oak (*Q. ilex*)	Algerian mouse (*Mus spretus*)
Santa Cruz Island on west coast of the United States	Island scrub oak (*Q. pacifica*)	Island scrub jay (*Aphelocoma insularis*)
Eastern deciduous forests of North America	Red oak, white oak (*Q. alba*)	Rodent caching behavior
Midwestern US	White oak, black oak (*Q. velutina*) and pin oak (*Q. palustris*)	Avian dispersers, primarily blue jays (*Cyanocitta cristata*) and tufted titmice (*Baeolophus bicolor*)
Eastern deciduous forests (park system) of North America	Red oak	Eastern gray squirrel (*Sciurus carolinensis*)
Eastern deciduous forests (park system) of North America	Red oak and pin oak (*Q. palustris*)	Eastern gray squirrel
Mediterranean coastal forest of Spain	Holm oak and Downy oak (*Q. pubescens*)	Wood mouse (*Apodemus sylvaticus*) and Algerian mouse
Northeastern China	Mongolian oak (*Q. mongolica*)	Siberian chipmunk (*Tamias sibiricus*)

PRIMARY FINDINGS	DDH	REFERENCES
Review of 234 papers show a majority of studies find that oaks are dispersed by birds to open, non-forested sites that include both natural and anthropogenic habitats suitable for establishment	Y	Bobiec et al. 2018
In-depth study arguing for close mutualistic relationship between the jay and oak; strong evidence for long- distance dispersal to open sites and significant survival in cache if not recovered	Y	Bossema 1979
Jays frequently disperse acorns 100–1000 m to open areas, whereas mice tend to disperse acorns in more dense forests	Y	den Ouden et al. 2005
Jays selectively disperse > 95% of acorns long distances to afforested or open pine stands suitable for establishment	Y	Gómez 2003
Acorns cached several kms from source in open sites suitable for germination; acorns cached singly; caches covered with leaves	Y	Darley-Hill and Johnson 1981
Occasional selective transport of acorns to canopy gaps with higher densities of woody debris	?	Iida 2006
Acorns selectively cached in open habitats to reduce pilferage; more acorns establish in the open habitats than under vegetation	Y	Muñoz and Bonal 2011
Acorns selectively dispersed to chaparral and coastal sage brush habitat where seedling densities were highest	?	Pesendorfer et al. 2017
All acorns aided in germination and establishment when cached; however, when probability of caching was also included relationship between red oaks and rodents was mutualistic and that of white oaks and rodents antagonistic	Y & N	Sawaya et al. 2018
Artificially stored acorns near natural avian cache sites showed lower rates of germination and seedling establishment than randomly stored acorns	N	Sipes et al. 2013
Squirrels disperse acorns to open sites and maintain hoarder's advantage even in high density populations	Y	Steele et al. 2011
More profitable acorns dispersed beyond tree canopy to open habitats where pilferage rates are lowest	Y	Steele et al. 2014
Rodents limited recruitment by pilfering cached acorns; rooting boars also increased rates of acorn consumption by rodents	N due to complex interactions	Sunyer et al. 2015
Selectively cached in soil of higher water content, which improves germination and seedling establishment	Y	Yi et al. 2013b

of such studies strongly suggests that directed dispersal of the oaks may occur at three stages in the dispersal process: dispersal of the acorn to sites less likely to be visited by competitors (potentially due to higher predation risks), selection of cache sites in open vegetation where seedling survival is higher, and preparation and use of cache sites most suitable for germination. Nine of the 15 studies reviewed in Table 8.1 suggest dispersal to open sites and to sites where competition with conspecifics is lower. Two of the studies suggest that cache site selection or cache preparation increases the probability of seed germination when not recovered. Only one study carefully compares the likelihood of germination and establishment from avian prepared caches with randomly prepared artificial caches. In this case, avian caches showed lower rates of seed and seedling survival than artificial caches (Sipes et al. 2013). Some studies infer successful dispersal by showing dispersal near sites with high seedling densities.

The goal of the disperser is to store the acorns in sites where the fruit is likely to remain intact until time of recovery. If this fortuitously translates into high rates of establishment, it indeed constitutes directed dispersal. But how often do dispersers fail to recover acorns dispersed to sites suitable for germination and seedling establishment? This question still remains largely uninvestigated.

DO PREDATORS DRIVE THE DISPERSAL PROCESS?

If a scatterhoarder invests considerable energy moving the seed and preparing a cache site unavailable to pilferers, and it is also well adapted for remembering the location of caches (see Chapter 12), why would it fail to recover the seed? I suggest the answer to this follows largely from the fact that acorns are often dispersed to open vegetation by both rodents and corvids (Table 8.1). If this behavior indeed reduces pilferage rates because of higher predation risks, then it follows that predation of cache owners, even if infrequent, would translate into regular patterns of oak establishment.

Future studies should investigate the likelihood that predation of scatterhoarders contributes to the dispersal of oaks. In the closing chapter, I revisit this topic as a promising area of future investigation, one that several of my colleagues and I are now pursuing.

I now move on to a broader examination of the rodents and jays as both acorn predators and dispersers (see Chapters 9 and 10, respectively), the mind of the avian and rodent scatterhoaders in Chapter 11, and the principal predispersal predators (insects) of the oaks in Chapter 12.

THE RODENTS: SEED PREDATORS OR SEED DISPERSERS?

9

INTRODUCTION

The relationship between the rodents and oaks is a delicate one, with rodents frequently acting as both catastrophic agents of seed predation but also, under many circumstances, moving acorns, and often placing them in sites optimal for germination and establishment. Under some circumstances rodents fail to recover these caches, which ultimately results in germination and seedling establishment. In this chapter, I focus on the broader aspects of the rodent-plant relationship and the many factors that can influence it. Herein, my goal is to (1) characterize the nature of these interactions between rodents (in particular, squirrels) and plants with a special emphasis on a few species, (2) review how this relationship likely varies in different communities of rodents in forests worldwide, (3) explore how evolutionary interactions serve to strengthen the potential for this mutualism (even when achieved through diffuse coevolutionary processes), and (4) discuss further how various environmental conditions can shift the balance between mutualism and seed predation in most communities.

The rodents, the largest order of mammals, represent approximately 40% of all mammal species (Martin et al. 2011). Identified by the chisel shaped incisors (both an upper and bottom pair) and the wide space (the diastema) between the incisors and the other cheek teeth where canines and other premolars are absent, the rodents occur nearly worldwide (Fig. 9.1). Rodents are not endemic to most oceanic islands, New Zealand, the Arctic, and the Antarctic; however, many rodents have been introduced throughout the world, especially species of the Muridae. Members of this

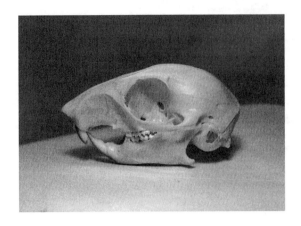

9.1. Skull of the eastern gray squirrel (*Sciurus carolinensis*). Note the sharp, continuously growing upper and lower incisors, the cheek teeth (for grinding), and the gap (diastema) between the two, all characteristics of the Order Rodentia. (*Tyler Brzozowski*)

family of rodents pose significant problems for other species, such as humans (e.g., agricultural crops, stored grains, bark stripping, and excessive burrowing in fields and lawns), ground nesting birds, and other small mammals (Harris 2009). One invasive rodent, a squirrel (Sciuridae), is the eastern gray squirrel (*Sciurus carolinensis*). Introduced to the United Kingdom in the early 1900s, this species is now recognized as one of the top 100 invasive species in the world (Steele and Wauters 2016; Shuttleworth et al. 2016).

Rodents are well noted for opportunistic diets with many species feeding on everything from plant material to seeds to animal tissue; yet, many are considered primarily herbivores, granivorous, or even exclusively carnivorous at some time during the year (Thorington et al. 2012). It is those that are granivorous that are most relevant to my discussion. Granivores feed primarily on fruit, seeds, and nuts, and this assemblage of food items characterizes the diet of many rodents. However, even rodents that feed extensively on seeds during much of the year, such as tree squirrels, often shift to other items, even animal tissue, when little else is available (Koprowski 1994).

The extensive diet of seeds and nuts of some rodents places significant selective pressure on various tree species. The trees, likewise, have evolved characteristics that either defend against seed predation by rodents, or, as I discuss throughout this book, in the case of oaks, both defend against predation or partial predation and promote seed dispersal through the process of scatterhoarding.

THE SQUIRRELS AND CONIFERS

Perhaps our best understanding of rodent-tree interactions comes from a long line of studies on the tree squirrels, especially in North America where numerous studies help us to understand their ecological and evolutionary interactions

(West) Cascade Range:

- Wet maritime climate
- Lightning rare
- No forest fires
- Open = nonserotinous
  Lodgepole pine cones:
  1. Soft cone surface
  2. Weak attachment
  3. Symmetrical shape
  4. More seeds/cone

- *Tamiasciurus douglasii*
  (Douglas squirrel):
  1. Lighter weight
  2. Weaker jaw muscles
  3. Weaker lower jaw

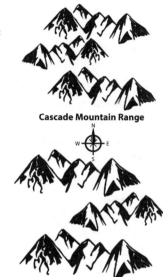

Cascade Mountain Range

(East) Cascade Range:

- Dry continental climate
- Lightning common
- Forest fires common
- Closed = serotinous
  Lodgepole pine cones:
  1. Hard cone surface
  2. Strong attachment
  3. Asymmetrical shape
  4. Fewer seeds/cone

- *Tamiasciurus hudsonicus*
  (red squirrel):
  1. Heavier weight
  2. Stronger jaw muscles
  3. Stronger lower jaw

9.2. Conditions found west and east of the Cascade Mountain Range in southwestern British Columbia. Cause and effect are much more logical going from the top to the bottom of the list of conditions rather than in the opposite direction. In other words, the independent variable is the rain shadow effect of the Cascade Range, not the difference in strength of the squirrels' jaws. (*Redrawn from Smith 1998 with permission from the Virginia Museum of Natural History*)

with trees (Steele and Koprowski 2001; Steele et al. 2005). In many parts of the world, especially in the northern continents, the tree squirrels are critical agents of both seed predation and dispersal, sometimes serving as the only agent of dispersal for some tree genera such as *Juglans* (the walnuts) (Steele et al. 2005). One of the first to explore the close evolutionary relationship between rodents and trees was Dr. Christopher C. Smith in his 1965 dissertation and key papers that soon followed (Smith 1970, 1981). It was Smith's early work that inspired and shaped some of my thinking during my undergraduate and graduate education. Although largely not experimental, this comparative work was exceptionally well-reasoned. By comparing two of the three species of tree squirrels in the genus *Tamiasciurus* (red squirrel [*T. hudsonicus*] and Douglas squirrel [*T. douglasii*]) and their relationship with the cones of the lodgepole pine (*Pinus contorta*) in southwestern British Columbia, Smith (1965, 1968, 1970, 1998) was able to make a compelling case for the effect of climate, tree, and cone characteristics on the biology of two closely related species of squirrel (Fig. 9.2). East of the Cascade Range, where the red squirrel is found, the rain shadow of the mountains produces a dry climate, where lightning frequently results in forest fires. There, lodgepole pines produce tough "serotinous" cones that are characterized by hard, secure bracts, fewer seeds per cone, and an asymmetrical shape compared with cones of lodgepole pine found elsewhere, especially west of the range where the cones are seemingly more easily exploited by *T. douglasii*. The serotinous cones in the east retain their seeds, sometimes for a decade, until triggered by fire, which improves conditions for seed germination and seedling establishment.

Smith showed that the red squirrels in the east had a relatively large body mass, stronger jaw muscles (e.g., masseter), and a more pronounced lower jaw, all characteristics that aided the squirrel in removing cones, stripping cone bracts, and feeding on the protected seeds of these serotinous cones. Smith's first inclination, which he later admits followed from his focus on the squirrel's biology, suggested that, east of the Cascades, the cone characteristics were driven by that of the squirrel. He later realized that it was far more logical to assume that the abiotic conditions likely drove the cone characteristics and, in turn, those of the squirrel (Smith 1998). West of the Cascades, the wet coastal climate meant fewer lightning strikes, limited forest fires, and lodgepole pine cones with very different characteristics (e.g., soft surface, a weak point of attachment to the stem, and symmetrical cones). The squirrels (*T. douglasii*) west of the range have a smaller body mass, weaker jaw muscles, and a smaller lower jaw. Lodgepole pine cones on the western slopes are not serotinous, and they mature and are shed within a few months each year after they first develop.

In studies of individual *Tamiasciurus* behavior, Smith also showed that individual squirrels are highly sensitive to their efficiency when feeding on and storing individual cones and different cone species (1970). In boreal conifer forests, a single squirrel of *Tamiasciurus* can larderhoard several thousand cones in their central midden in a single season (Steele 1998, 1999). The midden is a central larder in the squirrel's territory where, each year, thousands of cones are stored in a massive pile that also contains cone bracts that accumulate as the cones are consumed (Fig. 9.3). The midden keeps cones moist and prevents them from opening and shedding their seeds until eaten by the squirrel that vigorously defends its single midden, usually near the center of the squirrel's territory. As Smith (1970) found, provisioning of cones to the midden is not done haphazardly. Starting in late summer, as the seeds in the cone mature, the individual red or Douglas squirrel first retrieves the cones from the species of tree in their territory with the highest seed density per cone. After the cones of this species are larderhoarded, the squirrel moves on to the next species of highest seed energy per cone (Smith 1965, 1970, 1998).

This intensive predation by the squirrels places strong selective pressure on seeds and cones of conifers found in the range of *Tamiasciurus*, which occurs throughout boreal forests of North America. Smith suggests the squirrel may have a different effect on different tree species (Smith 1970, 1975). Whereas

9.3. A red squirrel (*Tamisciurus hudsonicus*) midden of conifer cones. *(Warren Johnson)*
SEE COLOR PLATE

seed size and seed number relative to overall cone tissue influence feeding efficiency, the seed size is tied closely to germination strategies. Thus, the response of trees to squirrel selection may result in tree species with small-seeded cones producing fewer seeds per cones. In contrast, larger-seeded species are not advantaged by producing fewer seeds but instead by releasing seeds earlier in the season (Smith 1970, 1975).

C. C. Smith's work on the relationship between red squirrels and conifers was only the beginning. Related studies by Elliot (1974, 1988) further demonstrated the efficiency with which these squirrels harvested and stored cones in the midden and the strong selective pressure exerted on the lodgepole pine.

Additional long-term studies on the effect of conifer mast crops demonstrate close evolutionary interactions between tree squirrels and squirrel populations in North America and Europe. Dr. Stan Boutin of the University of Alberta, who oversees the Kluane red squirrel project in the southwest Yukon of Canada, has monitored the close interactions between red squirrels and white spruce (*Picea glauca*) for over 30 years. Likewise, parallel studies by Drs. Luc Wauters and André Dhondt in Italy and Belgium have followed similar interactions between conifers and the Eurasian red squirrel. Among their numerous findings are their joint efforts demonstrating that both squirrel species are able to anticipate a masting event well before it occurs and time an increase in reproduction to coincide with this peak in seed abundance (Boutin et al. 2006).

In conifers south of the boreal forests, warmer and drier climates mean that middens are not as critical for survival. Although the range of *Tamiasciurus* extends below these preferred boreal habitats (Steele et al. 1998, 1999), in these areas, middens are far reduced in size or nonexistent. In fact, red squirrels are known to scatterhoard in many parts of their southern range (Koprowski et al. 2016).

Other species of squirrels (*Sciurus*) found in conifer forests exert different selective pressures on the trees on which they feed. In Ponderosa (*Pinus ponderosa*) pines of the west and southwestern United States and parts of Sierra Madre Occidental of Mexico, the Abert's squirrel (*Sciurus aberti*), for example, is associated exclusively with Ponderosa pine (Keith 1965; States and Wettstein 1998), feeding heavily on cones from summer to autumn during and after seed maturation but before seeds are released from the cones (Keith 1965). However, the Abert's squirrel rarely stores these seeds or cones (Snyder 1998). During the remainder of the year or in between mast cycles (every 3–5 years) (Snyder 1998), the squirrel depends heavily, sometimes exclusively when cones are not available, on the inner bark (cambium and phloem) of the terminal twigs of the Ponderosa pine. Moreover, the squirrel feeds selectively on individual trees based on different chemical attributes of the phloem and xylem, as well as the flow rates of each, all genetically based characteristics. One result to follow is lower reproductive output (of both male strobili and female cones) in these preferred trees. And the squirrels also place strong selective pressure on the trees (Snyder 1992, 1993, 1998; Snyder and Linhart 1994). Interestingly, other species such as the dwarf mistletoe (*Arceuthobium vaginatum*), the North American porcupine (*Erithrizon dorsatum*), and the pine bark beetle (*Dendroctonus ponderosae*) all feed selectively with respect to other characteristics of the phloem and xylem of the Ponderosa pine and, as a result, likely exert diversifying selection on the tree (Snyder 1998, and references therein).

SEED PREDATION IN SOUTHEASTERN FOX SQUIRRELS

Elsewhere in North America, in most conifer forests, other tree squirrels are also well-adapted seed predators that feed heavily on the seeds of the immature cones of these trees. In my doctoral research in the coastal plain of North Carolina, I observed just how adept the southeastern fox squirrel (*Sciurus niger niger*) is at exploiting the cones of the longleaf pine (*Pinus palustris*). Longleaf pine forests, characterized by an open understory of fire-prone wire grass and towering longleaf pines were once extensive throughout the southeast, stretching from southeastern Virginia to eastern Texas. However, a boom in timber harvest for ship building and other resources (turpentine and tar), coupled with more recent fire suppression, have resulted in the replacement of longleaf with loblolly pine (*Pinus taeda*). Today longleaf pine occupies a small fraction of its original range and an extensive biodiversity originally associated with longleaf forests is now threatened or endangered, with southeastern fox squirrels as one such species.

Although now sometimes found in loblolly stands, the threatened southeastern fox squirrel was historically tied closely with longleaf pine. My studies, and those of my graduate advisor, Dr. Peter Weigl, on the interactions between this subspecies of fox squirrel and longleaf pine suggested a very close relationship between the two. Pete has long argued that the large body size of the southeastern fox squirrel (often > 1000 g), which makes it the largest of all the 10 subspecies of fox squirrels found in the eastern half of the United States, is closely tied to its dependence on the large cones of longleaf pine. Western fox squirrels (*S. niger rufiventer*) in the midwestern US and gray squirrels farther east both show a typical Bergmann's cline in which body size increases with latitude (Barnett 1977). Nearly 60% or more of mammals exhibit such a cline. Originally attributed to an increase in energetic savings as a result of lower surface to volume ratios in colder, northern climes, we know that this trend is more likely attributable to an advantage in dealing with periods of food scarcity more common in the north (Lindstedt and Boyce 1985; Geist 1987).

Despite the evidence for a Bergmann's cline in other squirrels, Weigl et al. (1998) reported a significant opposite trend (a reversed Bergmann's cline) for fox squirrels in the eastern US, with the largest fox squirrels occurring in the southeast (e.g., *S. niger niger* in Virginia and the Carolinas, and the largest of the 10 fox squirrel subspecies in Georgia and Florida, *S. niger shermani*). Weigl et al. (1998) attributed this to the subspecies' long association and dependence on longleaf pine and the advantage that larger size provided for moving across open stretches of the pine-savannah habitat and for handling the large cones (up to 400 g) of this pine (Fig. 9.4a).

Feeding on the seeds in conifer cones, rodents follow a somewhat stereotypical behavior common to many species. They start at the base of the cone by removing one bract at a time to access the seeds at the base of these bracts. This requires rotating the cone as each bract is peeled away from the cone with the incisors. The first few spirals of bracts are often sterile with no seeds, but after several spirals of bracts are removed, the animal begins to encounter viable seeds, which it then consumes.

In my studies of fox squirrels feeding on longleaf pine cones, described in far greater detail in Steele and Koprowski (2001; Steele 1988; Steele and Weigl 1992), I uncovered several strategies employed by these squirrels that demonstrate the tremendous efficiency with which cone predation is executed. First, I offer some context. As revealed by extensive radiotelemetric studies, we showed that fox squirrels in southeastern longleaf pine forests enter a midsummer shutdown in activity characterized by a significant drop in home range size and a reduction in fine grain patterns of space use, as well as daily patterns of activity (Weigl et al. 1989; Steele and Koprowski 2001). This striking reduction in activity and movement through the habitat coincided with a period of both high temperatures and food scarcity. Weigl et al. (1989) described this as the disappearance period when animals are at their poorest condition and lowest body mass. We eventually concluded that the animals were limiting time out

9.4. Shown are **(a)** southeastern fox squirrel (*Sciurus niger niger*). (Barbara A. am Ende) **(b)** Longleaf pine cone (with bracts stripped and seeds consumed) and cone core dropped at the base of the tree. *(Mark Witwer)*   SEE COLOR PLATE

of the nest mostly in response to this period of food scarcity. But by late July and early August, when temperatures still exceeded 35°C, we found the animals began to move throughout the longleaf forests. Telemetry studies indicated an extensive fine grain use of the habitat, suggesting that the squirrels were spending short periods of time in numerous trees across their now expanding home range. And, a closer investigation within each home range revealed why.

By late July or early August, fox squirrels were beginning to feed on the longleaf cones. Cones were cut in the treetops and eaten at the base of the trunk. The core of each cone was then dropped to the ground beneath the tree canopy (Fig. 9.4b). This obvious record of food consumption gave me an ideal opportunity to follow patterns of cone use—and follow them I did. By marking > 900 longleaf pine trees at eight spatially independent study sites, I was able to follow cone consumption through two field seasons. I tagged each tree in midsummer at the time I observed the first cones eaten, and then revisited each tree every 10–14 days, recording any additional cone predation, until those cones remaining on trees opened and the seeds were dispersed by the wind. During each visit, cones were collected and counted, and, at the end of the season, I counted the cones remaining in each tree just as they were beginning to open and the season of harvest was concluding.

This detailed record of cone consumption by fox squirrels revealed some very interesting patterns of seed predation. Early in the season, I was able to first document that the increase in space use following the distinct summer lull in activity was initially the result of squirrels consuming a few cones in many trees within their home range. Many of the trees had only a single cone eaten and often these cones were not entirely stripped of bracts. Subsequent caloric analysis of the seeds across the growing season, combined with measures of feeding rates of captive squirrels, showed that by early August, seeds were mature enough for the squirrels to maintain a positive energy balance when feeding on the cones. Thus, early in the season, squirrels were visiting trees and assessing seed maturity, thereby identifying trees in which seeds were ripe enough to harvest (Steele and Koprowski 2001).

Immediately upon seed ripening, the pattern of cone predation began to change. By the middle of August, a few trees in each home range had many to nearly all the cones harvested from the tree. And, although all seeds were fully mature at this time, the behavior I had earlier declared sampling behavior had also continued to take place until cones opened and shed their seeds. Within each established home range, it appeared that sampling behavior continued along with intense seed predation in just a few preferred trees. This pattern was evident at all sites, although more intensive at some. As reported by Steele and Koprowski (2001; Steele 1988) with the exception of one site with a particularly heavy cone crop, sampling accounted for 25% to 48% of the tree visits during any collection period between mid-September and early October. Sampling behavior at the site with the heaviest cone crop varied between 12–30% of the trees visited.

With the aid of some rappelling equipment and a friend who was adept at climbing, we were able to collect a small sample of cones from many of the trees in which the squirrels fed. We then measured cone mass, seed number, and total seed mass, and then estimated profitability (energy/time) of cones for each tree. Combining these estimates with total cone number allowed further estimates of patch (tree) profitability. We then could compare tree and cone profitability in trees preferred by the squirrels with those they sampled and subsequently avoided. Many preferred trees simply had higher cone numbers, but in others, cone profitability was lower, but the higher density of cones meant the overall patch (tree) profitability was higher (Steele and Weigl 1992). Squirrels were selecting the trees, not necessarily based on individual cone quality or profitability, but on the tree that produced the highest feeding rate. And, upon selection of these preferred trees, squirrels repeatedly returned to the tree to remove nearly all of the cones in the tree with the exception of those that were most inaccessible.

Two additional comparisons of the preferred and sampled trees raised yet other questions about the squirrels' ability to assess tree or cone quality. The first was a comparison of volatile chemicals released from the cones (Steele 1988; Steele and Koprowski 2001). Cones from selected trees had significantly higher levels of two terpenes (i.e., pinenes) emitted from the cones. In the second, I conducted a simple experiment in which I presented free-ranging squirrels with patches of cones from both preferred and sampled trees on the ground. These feeding stations were established away from cone sources and thus the context from which the cones were harvested. When I controlled for cone size/profitability, I found that cones from preferred trees were still selected and eaten more often than those from sampled trees (Steele 1988; Steele and Koprowski 2001).

These results illustrate how well equipped the fox squirrel is at exploiting the cones of longleaf pine yet raise numerous other questions about the interaction between the squirrel and this tree species. From the tree's perspective, these collective results mean that in any population of longleaf pine subject to seed predation by fox squirrels, those trees that invest the most in seed and cone production are also those likely to have the fewest cones escape predation (i.e., those with the lowest reproductive success). Indeed, this raises the tantalizing question of how has the species responded to such intensive selection? This remains an open question that begs a serious look.

EXCEPTIONS TO THE RULE: SECONDARY DISPERSAL OF CONIFER SEEDS BY RODENTS

This discussion regarding conifers and squirrels—a brief but necessary diversion from the main focus of this book—serves to illustrate the common role of squirrels and other rodents in general as potentially devastating seed predators. From Smith's and Elliott's work on *Tamisciurus*, to studies of Abert's squirrel, to my own work on fox squirrels and longleaf pine, it is clear that in many conifer forests across North America and Europe, the arboreal squirrels exert significant selective pressure on conifers as intense seed predators. If I stopped here, however, I would seriously err in providing the complete picture.

With seeds set inside the green and relatively accessible female cones of most gymnosperms, they are indeed highly susceptible to both insect and vertebrate predation. And, even if the seed is able to escape seed predators until the end of the growing season when cones dry, the bracts open and the typical winged-seed of most conifers is dispersed by the wind, the single seed, now on the ground, seems even more susceptible to a community of avian and mammalian seed predators—or is it?

Although most rodents that feed on conifer seeds function nearly exclusively as seed predators, studies across the globe have shown that some rodents, in some circumstances, do disperse the seeds of some conifer species (Steele et al. 2005). Species that have been shown to positively influence conifer seed fate include Abert's squirrel of the southwestern US (Hall 1981), the Eurasian red squirrel in both Europe (Wauters and Casale 1996) and Asia (Miyaki 1987), and the Japanese squirrel (*Sciurus lis*) in Japan (Kato 1985). However, it is the research of my colleague, Dr. Stephen Vander Wall, that most clearly demonstrates a modified seed fate for some wind dispersed pines and suggests the need for more serious investigations on the subject (Vander Wall 1992, 1994, 2002, and references therein; Vander Wall and Joyner 1998). Vander Wall (2002) shows that in the Carson Range of western Nevada, seeds of the wind dispersed Jeffrey pine (*Pinus jeffreyi*) are first moved by the wind from their relatively large cones by late September and October. As these seeds (~132 mg) then accumulate on the ground in the vicinity of the tree, they are eaten by rodents. Two of these rodents, the yellow pine chipmunk (*Tamias amoenus*) and the golden-mantled ground squirrel (*Callospermophilus lateralis*), also regularly scatterhoard the seeds. All but one species in the rodent community at Vander Wall's study sites, which include the deer mouse (*Peromyscus maniculatus*), Douglas squirrel, California ground squirrel (*Otospermophilus beecheyi*), and the Golden-mantled ground squirrel all prey heavily on the seeds, larderhoard them, or, in the case of the latter species, scatterhoard them too deep to aid in establishment. The yellow pine chipmunk, however, places seeds at soil depths optimal for germination when not recovered and thus serves as an important agent of secondary dispersal for this pine.

Yellow pine chipmunks initially scatterhoard the seeds to sequester them from competitors and may cache and recover seeds up to six times in a single season. In the process, chipmunks move the seeds up to nearly 100 m from the parent tree. Although the chipmunk's ultimate goal is to move the seeds into its larder, in the process many of these cached seeds escape recovery and germinate. Vander Wall has successfully demonstrated that this process of secondary dispersal is a critical means by which this pine achieves seed dispersal and seedling establishment.

Vander Wall and Longland (2004) argue that this process of secondary dispersal of pine seeds by rodents is not a simple coincidence but likely one stage in the evolution of the seed dispersal process. Although the seeds of Jeffrey pine are winged and first dispersed by the wind, the relatively large seed-to-wing ratio means a high wing loading and thus a relatively short dispersal distance by the wind. The trade-off to follow, however, is that larger seeds are consistently removed and scatterhoarded by rodents and, to a lesser extent, by jays (Vander Wall 1994). Small-seeded pines, which are dispersed more efficiently by the wind, are generally less sought after by rodents and jays. One exception to this rule is the small-seeded white pine (*Pinus strobus*) in the eastern US, which appears to achieve some level of secondary dispersal by white-footed mice (*Peromyscus leucpous*) (Abbott and Quink 1970). Vander Wall and Longland (2004) further discuss secondary dispersal of some conifers by rodents in the context of other secondary dispersal processes, such as those involving endozoochory and dung beetles (Andersen and Feer 2005), endozoochory and rodents, endozoochory and myrmecochory, and ballistic dispersal and myrmecochory (see Chapter 1).

DOES FRUIT MORPHOLOGY DRIVE FORAGING, FOOD STORING STRATEGIES, AND THE SOCIAL STRUCTURE OF SQUIRRELS IN NORTH AMERICA?

C. C. Smith (1998) suggests that differing patterns of wind pollination in boreal forests versus temperate forests drove "optimum flowering patterns," which, in turn, influenced fruit morphology and hence patterns of food storage in arboreal squirrels. He argues further that patterns of food storage—larderhoarding versus scatterhoarding—directly led to different types of social structure in *Tamiasciurus* and *Sciurus*, respectively. For example, the ability, as well as the need of *Tamiasciurus* spp. to harvest and larder large numbers of cones in a single midden requires that tremendous energy be invested in both provisioning and subsequently defending these central stores. Because cones typically house total seed masses of usually less than 30% of the woody cone, seeds are less accessible to many competitors as long as cones remain closed (Smith 1998). By storing cones in these moist middens, often under heavy snow, cones remain closed and unavailable to many competitors that would otherwise seek these seeds. However, many competitors, especially conspecifics, frequently raid the middens (Gerhardt 2005; Donald and Boutin 2011). Red squirrels, for example, appear to regularly raid their neighbors' middens, although some individuals are more prone to this behavior than are others (Gerhardt 2005).

This intense competition for the contents of middens has led to a strict system of territorial behavior, in which each midden, located in the approximate center of each squirrel's territory, is vigorously defended with frequent vocalizations and chases during daylight hours. In more northern and boreal forests, the result is a social system based on this intense territorial behavior (Steele 1998, 1999) in which individual mothers bequeath a territory to one of their young at the time of dispersal of the young (Price and Boutin 1993; Larsen and Boutin 1994; Boutin et al. 2000, 2006). Reproduction in both *Tamiasciurus* and Eurasian squirrels (*Sciurus vulgaris*) is clearly tied to cone availability but often in unpredictable ways, such as the ability of these species to switch on reproduction in anticipation of conifer mast crops and, in the case of *Tamiasciurus*, to even provision the cache in anticipation of future offspring well before conception (Boutin et al. 2000, 2006).

In contrast, in temperate forests, larderhoarding is less likely even for *Tamiasciurus hudsonicus*, whose range extends well out of boreal forests to mixed deciduous/coniferous forests and even deciduous forests (Steele 1998). In these forests, the territorial behavior is often relaxed and the provisioning and maintenance of middens is not as pronounced. Moreover, as C. C. Smith (1998) argues, *Sciurus* species, such as the gray squirrel and fox squirrel, especially in North America, are associated closely with temperate forests. Here, squirrels feed on the fruits of beech, oak, and chestnut (Fagaceae), and walnuts and hickories (Juglandaceae), all of which are characterized by single-seeded fruits with a higher relative proportion of nut kernel/shell ratio, thus rendering them of considerable value to larger competitors (Smith 1998). As a result, a central larderhoard is not effectively defended by an individual squirrel. Consequently, *Sciurus* species typically store food by means of scatterhoarding in which individual nuts or food items are stored in widely dispersed cache sites. Although these scattered caches allow the animal to distribute food supplies in a way that reduces competition, it also means that it is not possible for the tree squirrels to defend these widely spaced caches.

Smith further argues that whereas the fruit types have necessitated scatterhoarding behavior by the tree squirrels, this food storing technique, in turn, translates into a different social system for *Sciurus*. The inability of *Sciurus* spp. to defend stored food means there is no need for territoriality. The tree squirrels instead exhibit overlapping home ranges and their social system takes the form of a dominance hierarchy, most often expressed when they converge on a central food source, nest cavities, or in competition for mates (Steele and Koprowski 2001).

RODENT SCATTERHOARDING AND OAK DISPERSAL

Scatter-hoarding behavior in mammals (mostly rodents) is an involved process by which these animals store food during critical periods of food shortage and, in the process, disperse seeds (Vander Wall 1990). Here, I briefly highlight some of the major players—both the rodents and the investigators who have uncovered these interactions (Table 9.1). Many, but not all, of these studies, have focused on scatterhoarding of acorns and are thus featured throughout this book. Significant research on the behavior of scatterhoarding include the pioneering studies of W. Hallwachs in Costa Rica on the Central American agouti (*Dasyprocta punctata*) and the more recent studies by P. Jansen and collaborators on the agouti in Panama as well as the red acouchy (*Myoprocta acouchy*) in South America. In North America, much of the study of scatterhoarding has been defined by the extensive, groundbreaking research by S. Vander Wall and collaborators, which now extends over three decades in various ecosystems of Nevada; the early work of M. Stapanian and C. C. Smith on fox squirrels in the midwestern forests; the recent studies by R. Swihart, myself, and colleagues in the midwestern forests and eastern deciduous forests; and limited research in central Mexico, some of it my own. The past two decades have also seen an explosion of studies in the Mediterranean dry forests by A. Muñoz, R. Bonal, J. M. Espelta, and their collaborators as well as A. J. Gómez, R. Pérea, and others. The majority of this research is also focused on oak dispersal. In Asia, there has been extensive research on various aspects of scatterhoarding and seed dispersal by mice (*Apodemus* spp.) and squirrels of several genera by a major cohort of scientists who studied under the direction of Zhibin Zhang at the Chinese Academy of Sciences and a disproportionate effort on the islands of Japan, much of it focused on seed dispersal of several woody plant species by scatterhoarding rodents (Table 9.1). Finally, any discussion on scatter-hoarding behavior would not be complete without highlighting the seminal contributions of my colleague Tad Theimer whose research in both Australia and the southeastern US helped to highlight how context dependent the scatter-hoarding process can be in its contributions to the dispersal of seeds.

Table 9.1. Sample of granivorous, scatter-hoarding rodents that potentially contribute to seed predation and dispersal of oaks and associated tree species. This summary is not comprehensive but intended only to highlight the diversity of rodent taxa involved in dispersal and predation of the fruits of oaks and other tree species.

LOCATION	GRANIVOROUS RODENTS *That Potentially Scatterhoard and Disperse Seeds and Nuts*	TREE GENERA OR SPECIES	REFERENCES
Dry forests of NW Costa Rica	Central American Agouti (*Dasyprocta punctata*)	*Quercus oleoides*	Hallwachs 1994
Central Hardwoods Forests and eastern deciduous forests of North America	Eastern gray squirrel (*Sciurus carolinensis*), fox squirrel (*S. niger*), Eastern chipmunk (*Tamias striatus*), and white-footed mouse (*Peromyscus leucopus*)	Black oak (*Q. velutina*), chestnut oak (*Q. montana*), red oak (*Q. rubra*), and white oak (*Q. alba*)	Hadj-Chikh et al. 1996; Steele et al. 2001b; Moore et al. 2007; Lichti et al. 2014
Central Hardwoods Forests of North America	Fox squirrel and eastern gray squirrel	Seeds and nuts of numerous hardwood species (e.g., *Quercus*, *Carya*, *Castanea*, *Fagus*, and *Juglans*)	Stapanian and Smith 1978, 1984, 1986; Sundaram et al. 2017
Mediterranean dry forests of Spain	Algerian mouse (*Mus spretus*) and wood mouse (*Apodemus sylvaticus*)	Holm oak (*Q. ilex*), pedunculate oak (*Q. robur*), Pyrenean oak (*Q. pyrenaica*), and sessile oak (*Q. petraea*)	Perea et al.2011a,2011b, 2014; Muñoz and Bonal 2007, 2008ab; Bonal et al. 2012b
Tropical forests of Panama	Central American agouti	Palm (*Astrocaryum standleyanum*)	Hirsch et al. 2012; Jansen et al. 2012
Tropical forests of French Guiana, SA	Red acouchy (*Myoprocta acouchy*)	Carapa procera (*Meliaceae*)	Jansen et al. 2004
Harwood forests of Eastern Europe	Bank vole (Myodes glareolus) and yellow necked-mouse (*Apodemus flavicollis*)	European beech (*Fagus sylvatica*)	Wróbel and Zwolak 2013

LOCATION	GRANIVOROUS RODENTS *That Potentially Scatterhoard and Disperse Seeds and Nuts*	TREE GENERA OR SPECIES	REFERENCES
Harwood forests of central Europe	Yellow-necked mouse and wood mouse (*Apodemus sylvaticus*)	Invasive northern red oak (*Q. rubra*)	Bieberich et al. 2016
Temperate hardwood forests of Central Mexico	Mexican gray squirrel (*Sciurus aureogaster*)	Numerous oak species	Steele et al. 2001b
Tropical forests of southwest China	Chinese white-bellied rat (*Niviventer confucianus*), red spiny rat (*Maxomys surifer*), and yellow-bellied rat (*Rattus flavipectus*)	Evergreen broadleaf tree (*Castanopsis hystrix*)	Cao et al. 2018
Temperate broadleaf forests of north and northeast China, subtropical evergreen and broadleaf forests of south China, and subtropical rainforest of southwest China	16 rodent species of 10 genera: *Apodemus* (5 spp.), *Leopoldamys* (1 spp.), *Maxomys* (1 spp.), *Niviventer* (2 spp.), *Rattus* (2 spp.), *Tscherskia* (1 spp.), *Myodes* (1 spp.), *Tamias* (1 spp.), *Sciurotamias* (1 spp.), and *Tamiops* (1 spp.)	Different rodents or combination of rodents disperse the following genera and species of woody plants: *Amygdalus* (1 spp.), *Armeniaca* (1 spp.), *Camellia* (1 spp.), *Castanopsis* (5 spp.), *Corylus* (2 spp.), *Juglans* (2 spp.), *Lithocarpus* (4 spp.), *Pinus* (1 spp.), *Pittosporopsis* (1 spp.), and *Quercus* (4 spp.)	Zhang et al. 2016
Temperate deciduous forests of northern China	Korean field mouse (*Apodemus peninsulae*), striped field mouse (*A. agrarius*), Pére David's rock squirrel (*Sciurotamias davidianus*), and Siberian chipmunk (*Tamias sibiricus*)	Wild apricot (*Armeniaca sibirica*) and wild peach (*Amygdalus davidiana*)	Zhang et al. 2016
Subtropical broadleaf evergreen forests of southwest China	Predominantly Bower's rats (*Berylmys bowersi*), chestnut-bellied rats (*Niviventer fulvescens*), Chinese white-bellied rats (*N. confucianus*), and Edwards's long-tailed rat (*Leopoldamys edwardsi*)	Jolcham oak (*Q. serrata*)	Xiao et al. 2004
Subtropical broadleaf evergreen forests of southwest China	Asian red-cheeked squirrels (*Dremomys rufigenis*)	Chinese cork oak (*Q. variabilis*) and qinggang oaks (*Cyclobalanopsis glaucoides* and *C. stewardiana*)	Xiao and Zhang 2012
Deciduous broadleaf forests of central China	Pére David's rock squirrel	Qinggang oaks (*C. multinervis* and *C. breviradiata*) and White oaks (*Q. aliena* and *Q. serrata*)	Xiao et al. 2013a

LOCATION	GRANIVOROUS RODENTS That Potentially Scatterhoard and Disperse Seeds and Nuts	TREE GENERA OR SPECIES	REFERENCES
Secondary broadleaf and mixed conifer forests of northeast China	Eurasian red squirrel (*Sciurus vulgaris*) and Korean field mouse	Manchurian walnut (*J. mandshurica*)	Yi and Wang 2015
Forests of Japan	Eurasian red squirrel, Japanese squirrel (*S. lis*), Siberian chipmunk, large Japanese field mouse (*A. speciosus*), and small Japanese field mouse (*A. argenteus*)	Different rodents or combination of rodents disperse the following genera and species of woody plants: *Abies* (3 spp.), *Aesculus* (1 spp.), *Castanea* (1 spp.), *Castanopsis* (1 spp.), *Corylus* (1 spp.), *Fagus* (1 spp.), *Juglans* (1 spp.), *Kalopanax* (1 spp.), *Larix* (1 spp.), *Lithocarpus* (1 spp.), *Pinus* (4 spp.), *Prunus* (1 spp.), and *Quercus* (5 spp.)	Review by Tamura et al. 2005

Table 9.1 is a snapshot of investigations and scatter-hoarding rodents across the globe. It is not intended to be a comprehensive list. Nevertheless, there are some important patterns to follow from this overview. First, it is clear that scatter-hoarding rodents occur widely across the globe in temperate, subtropical, and even tropical forests. Oak dispersal by scatter-hoarding rodents, although more common in temperate forests of the Holarctic region, occurs in subtropical and tropical regions as well. Where there are oaks, there are scatter-hoarding rodents. Second, the distribution of none of these rodents across the globe, however, is tied exclusively to the oaks as most species typically depend on a diversity of other fruit and seed sources and therefore often occur outside the range of oaks.

Finally, rodent-mediated seed dispersal of the oaks is accomplished by relatively few taxa of rodents, but the impact of some species on seed and nut dispersal and forest regeneration appears disproportionately significant (Lichti et al. 2017). Among the scatter-hoarding rodents responsible for oak dispersal, it is several genera of the squirrels (Sciuridae) that are especially important. This includes the tree squirrels (*Sciurus*), especially in North America, isolated areas of Central America and Eurasia, and other squirrel genera (*Callosciurus*, *Sciurotamias*, and *Dremomys*) of southeast Asia. As discussed in Chapter 5, several species of these four genera of squirrels appear to have evolved behavioral responses to early germinating oaks. Also, in North America and Eurasia, *Tamias*

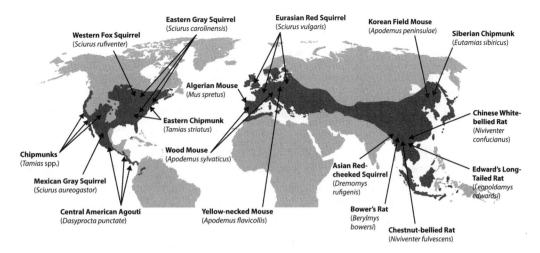

9.5. Global distribution of acorn scatter-hoarding and dispersing rodents. The species shown is not an exhaustive list but includes those most highlighted in the literature. Distribution of oak (shown in blue) is from Chapter 2. *(From Nixon 2006; Shealyn A. Marino)*

species (the chipmunks) likely contribute to oak dispersal, especially *Tamias sibiricus* in China and Japan. Other species of *Tamias*, may on occasion, scatterhoard acorns that eventually establish, but many species of this genus are likely to move initial scatterhoards into larders where acorns will perish (Fig. 9.5).

Other contributors to oak establishment include both the New World mice, primarily *Peromyscus* spp. in North America belonging to the family Cricetidae and a few genera of Old World mice and rats belonging to the family Muridae. Among the murids, oak dispersal is frequently attributed to several species of Eurasian field mice (*Apodemus* spp.), other field mice (*Mus* spp.), and some rat species distributed across Eurasia and/or North Africa. In some systems, mice can contribute significantly to oak dispersal over short distances, but under many conditions, mice are also devastating seed predators. Finally, I mention briefly the agouti (*Dasyprocta* spp.) that regularly feeds on and scatterhoards seeds and nuts. Although the distribution of this genus extends well outside that of the oaks, where their range overlaps that of *Quercus*, at least one species (*D. punctata*) appears to be well equipped to scatterhoard acorns (Hallwachs 1994). My limited observations in this system suggest that this rodent is able to thwart early germination of *Q. oleoides* acorns suggesting it likely evolved closely with this oak (see Chapter 6). More studies are needed to understand the extent to which this species disperses the oaks. Throughout the range of oaks worldwide, the rodents rely on acorns as a major part of their diet. Yet, the species of rodents that seem to have entered into a close evolutionary relationship with oaks seems relatively limited.

MECHANISMS OF RODENT
SCATTER-HOARDING BEHAVIOR

Only recently have we accumulated enough information and insight into rodent scatter-hoarding behavior to begin to synthesize these results in a way that best informs how scatterhoarding drives the seed dispersal process. Recently two of my colleagues and I attempted a synthesis of these many studies (Lichti et al. 2017). Spearheaded by Dr. Nate Lichti, this review sought to integrate current findings in a way that provides foundation for a mechanistic model of seed fate that logically follows from this synthesis of rodent scatter-hoarding behavior. Although a comprehensive model is still under development, the original review identified six rodent behaviors involved in linking the seed to seed dispersal, and ultimately, to seedling establishment. These six processes, or rodent behaviors, in the order in which they would naturally occur are exposure to seed crops, harvest of seeds, allocation to cache, preparation for caching, placement in the cache, and cache recovery (or failure of recovery). In addition, the review outlined nine key variables that drive decision-making by rodents at different stages in the scatter-hoarding process. These include motivation, food value, costs of secondary metabolites in seeds, handling time, missed opportunity costs, metabolic costs, perishability of the seed, pilferage risk, and predation risk.

Many factors can further influence these six stages of rodent scatter-hoarding behavior and the nine key variables. I point you to the online supplements attached to Lichti et al. (2017), which provide an exhaustive overview of the hundreds of studies that demonstrate how four key seed traits (seed size, energy and nutrient content, secondary metabolites, and seed perishability) influence rodent behavioral decisions when handling seeds. As an example, the simple question of how intraspecific and interspecific variation in seed size influences seed removal behavior, food selection, caching, dispersal (distance seeds are moved), or cache recovery has been covered by well over a hundred papers, the vast majority of which were published in the past 15 years.

Table 1 in Lichti et al. (2017) provides an overview and synthesis of results of 151 papers reviewed in these four supplements. From this overview, several conclusions can be drawn regarding how these four seed traits generally influence rodent scatter-hoarding behavior: *use* or *removal* of seeds are favorably influenced by larger seed size (when size varies within species but not across species), total energy content, and carbohydrate levels, but negatively influenced by secondary metabolites and seed perishability; both lipid and protein levels do not appear to increase patterns of *use*, *removal*, or *caching* of seeds; and a high majority of studies show that *consumption* of seeds decreases significantly in response to seed size and secondary metabolites. In contrast, nearly all studies investigating the effects of perishability on seed consumption (19 of 20 studies) report that rodents consume seeds in response to perishability. *Caching*, in contrast to *use* or *consumption* of seeds, significantly increases in response to seed size across studies, both when seed size varies within a species and across species. In many studies, *caching* often increases in response to total energy in the seed as well as the presence of secondary metabolites. I refer you to Lichti et al. (2017) for a more thorough analysis of these patterns, but emphasize that we are just now beginning to capture a comprehensive perspective on how seed characteristics influence patterns of seed use by rodents.

THE VARIABLE NATURE OF THE SCATTERHOARDER-PLANT RELATIONSHIP

Although the discussion above clearly delineates the patterns of behavior exhibited by scatter-hoarding rodents when handling seeds, these responses are never universal. For every seed characteristic examined in the studies reviewed by Lichti et al. (2017), there was always at least one, and sometimes several, that showed a significant departure from that observed in the majority of investigations. In other words, depending on the many specific conditions that can vary in a study, rodents may show very different behavior than expected. Yet, another way to put this is that the response of rodents toward seeds, either as mutualists (seed dispersers) or antagonists (seed predators), will always depend on the context in which each foraging decision is being made.

The scatterhoarder-plant mutualism is highly dependent on a number of factors that can cause this relationship to shift from one of mutualism to antagonism. The potential instability of mutualistic interactions in general was first reviewed by Bronstein (1994), but it was Theimer (2005) who so clearly articulated the conditionality of the mutualism between scatter-hoarding rodents and seed plants. Theimer (2005) identified two major factors that spatially or temporally define this mutualism. The first factor is the relative abundance of seeds to scatterhoarders. Put simply, as the relative abundance of seeds of a particular species available to each individual scatterhoarder increases, the likelihood of the seed surviving and germinating in the cache also increases. When this ratio is low, the probability of all seeds being eaten increases.

The second factor is the relative importance of the particular scatter-hoarding species at contributing to dispersal and establishment to the plant species. If, for example, the seed plant depends exclusively on the scatterhoarder for establishment, then mutualism is not likely to vary conditionally (Theimer 2005). However, for most species, this is not the case and several scatter-hoarding species may contribute to plant recruitment. This certainly appears to be the case for the oaks.

The first factor varies in response to both seed abundance and rodent abundance. Above all else, masting drives this relationship and influences the seed/rodent abundance in two ways: masting cycles cause rodent populations to fluctuate; they crash following mast failures and increase significantly soon after mast crops (see Chapter 4). As a result, rodent populations tend to be lower at the time of a heavy mast crop, which elevates the seed-to-rodent ratio thereby providing conditions that favor dispersal, scatterhoarding, and the failure of cache recovery. So important is masting to this process that it is considered by Vander Wall (2010) to be one of the four critical adaptations of tree species to drive scatter-hoarding behavior (see below).

UTILITY MODELS FOR UNDERSTANDING SEED USE BY RODENTS AND OTHER SEED CONSUMERS: A NOVEL APPROACH FOR BEHAVIORAL ECOLOGISTS

As discussed previously, numerous seed characteristics contribute to foraging decisions in rodents, and many of these seed traits appear not to act independently of one another. This led my colleague, Dr. Mekala Sundaram, to adopt utility models (i.e., economic choice models) to evaluate the perception of squirrels to these collective seed traits when deciding to eat or cache a seed or nut (Sundaram 2016). Borrowed from economics and psychology, utility models allow one to evaluate a consumer's perspective with respect to a product's (seed's) many traits and the user's collective preference for each product. Utility is a way of evaluating one's comprehensive perspective of a particular choice over another—in this case, the different seeds that are to be eaten, discarded, or cached.

The use of economic models to address questions in evolutionary ecology and animal behavior is not new. From the use of game theory that plays a critical role in our understanding of the evolution of selfish versus cooperative behavior to the marginal value theorem (MVT) that allows us to predict how an animal should likely forage in a patchy environment, these economic models have had a dramatic impact on our predictive power in behavioral and evolutionary ecology (Futuyma and Kirkpatrick 2017). Game theory, now employed across several fields (economics, psychology, and computer science), was introduced to biology by Axelrod and Hamilton (1981) to tackle the question of how cooperative behavior likely evolved.

The MVT, first articulated for behavioral ecology by Eric Charnov (1976; see also MacArthur and Pianka 1966) became a central model in optimal foraging theory, which allowed one to predict the time a forager should remain in a food patch based on the cumulative rate of energy return in the patch and the travel time between patches. As the animal forages in the patch, the feeding rate reaches a point of diminishing returns, and the animal must decide when to leave the patch and move to another to forage. This powerful model allowed us to predict the optimal solution to a wide range of patch use problems, from the foraging decisions of horned lizards (Iguanidae; *Phrynosoma*) when feeding on ant hills (Munger 1984) to the patch choices and residence times of hummingbirds when feeding on flower nectar (Pyke 1978). As I discuss in Chapter 6, even the partial consumption of an acorn by a squirrel can be assessed with the MVT to predict when it is best for the animal to drop the acorn and move on to another. And the tannin and lipid gradients in the acorn change this patch dynamic by reducing the optimal time in the patch (in this case, the acorn). This, in turn, increases the possibility that the partially eaten acorn can germinate.

As in the MVT, other foraging models, such as those that predict optimal diet breadths (Emlen 1966; Stephens and Krebs 1986), are based on the costs and benefits of foraging. Their derivation from economic models follows from little more than changing the currency from dollars to energy. In the diet models, for example, the value of each food item (e.g., nut species) can be assessed based on the energy extracted from the nut minus the energy to open the nut and consume it, and the relative abundance of the nut compared to that of other nuts. These simple measures could then be used to predict the optimal diet breadth for a foraging species.

Although the concept of optimal foraging has received considerable criticism for advancing the misconception that animals evolve to perform optimally, behavioral ecologists often argue that the best or only way to evaluate foraging behavior is to first predict the optimal solution to a foraging problem. This notion misses the important fact that, while the rate at which energy can be acquired from a food item is a key factor influencing animal foraging decisions, there are numerous other factors that may vary along with energy content and feeding time. When one considers that foraging decisions may include the decision of what to cache, as well as what to eat, these other factors increase even more. This is where the concept of utility increases the researchers' predictive power. And, this is why the use of utility is so clearly novel (Sundaram 2016; Lichti et al. 2017; Sundaram et al. 2018).

Considering the many characteristics that are known to influence scatter-hoarding and seed-handling decisions by rodents (Lichti et al. 2017), the calculation of utility allows integration of these diverse variables into a single measure of utility—a multivariate, unit-less predictive measure. A model that maximizes utility, rather than optimality, will not just account for the seed's energetic value but also for the many other factors that influence eating and caching decisions (Lichti et al. 2017). Sundaram et al. (2018) measured seed utility from the perspective of eastern gray squirrels by conducting a series of 426 discrete choice experiments with free-ranging squirrels. They presented pairs of nuts of several seed types (e.g., oak, hickory [*Carya*], walnut [*Juglans*], and chestnut [*Castanea*]) that varied with respect to energetic value, tannin levels, shell thickness, protein level, kernel mass, and dormancy period. They found that among these different nuts, squirrels selectively cached those of larger kernel size, heavier shells, and higher tannin levels. And, squirrels preferentially consumed nuts of shorter dormancy and thicker shells (Sundaram et al. 2018). This study demonstrated how squirrels balance numerous seed traits when making decisions about whether to cache or to eat a seed. They found that squirrels often trade-off multiple seed traits when caching seeds but are less flexible when consuming seeds.

Studies by Sundaram et al. (2018) offer a novel approach to understanding the complex decision process that rodents engage in when eating or dispersing seeds. And, as the authors emphasize, these models and procedures from the field of economics also have important applications for other areas of research in ecology and evolution. Unfortunately, many behavioral ecologists today are still fully entrenched in the traditional optimal-foraging way of thinking and are slow to embrace utility models to evaluate complex behavioral decisions. I thus urge students in the field of behavioral ecology to explore this powerful new approach for evaluating animal decision-making.

RELATIVE FREQUENCY OF SEEDS AND CACHING DECISIONS OF RODENTS

Previous discussions on the conditional nature of scatter-hoarding decisions shows how various seed characteristics may influence rodent behavior and ultimately seedling dispersal and establishment. One of the most significant factors that can shift the scatterhoarder-plant interaction to one of mutualism is the relative abundance of various seed types. In Chapter 4, I discuss in detail how masting alone can drive oak dispersal and establishment by manipulating the behavior of scatterhoarders. In addition to heavy seed crops (Hoshizaki and Hulme 2002; Moore et al. 2007; Xiao et al. 2013b), which cause temporal variation in individual seed types, several other factors can contribute to differences in the relative frequency of seed types. For example, forest composition can contribute to spatial variation in the frequency of seed types, as two or more species of trees may produce variable, unsynchronized crops (Lichti et al. 2014). And, some forest stands may include multiple tree species and others more of a monoculture (Hoshizaki and Miguchi 2005).

In addition, we are now beginning to understand that even seed traits within a single species may vary considerably within a single seed crop. I have observed acorns of red oak (*Q. rubra*) and white oak (*Q. alba*), for example, to vary in mass as much as 2 g to 6 g and 1 g to 4 g, respectively, between individual trees within a season. Likewise, investigators in other systems have observed similar degrees of variation in seed size within and between individual oak trees. Other characteristics of acorns that drive dispersal may also vary within and between trees during an annual seed crop. My lab, for example, is in the process of trying to uncover how critical measures of seed chemistry may vary between trees and within trees during the narrow period of seed fall. Although variation in many of these seed traits have not yet been quantified, they seem likely to interact in subtle but significant ways to influence rodent scatter-hoarding behavior.

The first studies to investigate how the frequency of subtle differences in seed traits of the same species influence dispersal decisions in a rodent was conducted by my colleagues Zhishu Xiao, Zhibin Zhang, and I (Xiao et al. 2010). We examined the response of Pére David's rock squirrel (*Sciurotamias davidianus*) to variation in germination morphology of *Q. aliena* and *Q. serrata* acorns in central China. Although both of these oak species are white oak species and thus germinate in the autumn during seed fall, germination phenotypes vary considerably between individual trees, with varying proportions of acorns per tree showing either immediate germination or delayed germination over a five to six week period (Fig. 9.6). As discussed in detail in Chapter 5, because of early germination in white oak acorns, several tree squirrels (including *S. davidianus*) selectively eat these acorns over those of red oak or excise the embryo, permanently arresting germination prior to caching. In Xiao et al. (2010), however, we found that acorns of these two white oaks often escape this fate because of variation in germination morphology. By presenting free-ranging squirrels with tagged acorns of varying frequencies of these two germination phenotypes, we were able to demonstrate that dispersal success of both phenotypes increased when the relative frequencies of either phenotype increased relative to the other. We found that 21% and 41% of the dispersal success of *Q. serrata* and *Q. aliena*, respectively, followed from frequency of the germination phenotype presented to squirrels. Such frequency-dependent selection by the squirrels could selectively maintain these two germination phenotypes and further account for dispersal success in white oak species (Xiao et al. 2010) that is otherwise considered unlikely due to early germination in these species (Fox 1982; Steele et al. 2001b; Steele et al. 2007).

In another study to investigate frequency-dependent hoarding in rodents, Sundaram et al. (2016) explored how the frequency of different seed types influenced caching decisions of gray squirrels. We predicted that, when presented with a seed type of high caching value (utility) and another seed type of markedly lower value, individual squirrels would consistently cache the high-value seed regardless of the frequency of the two seed types. And, indeed they did. For example, when presented with different frequencies (2:8, 4:6, 6:4, 8:2) of high-valued black walnuts and lower valued American hazelnut (*Corylus americana*), captive squirrels cached black walnuts in 25 of 27 caching events. Only two ha-

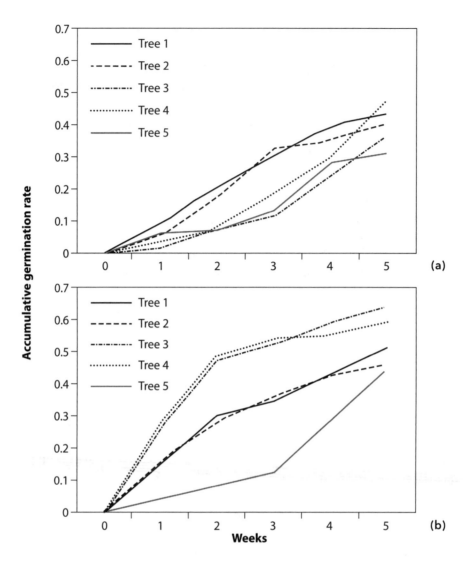

9.6. Immediate versus delayed germination in acorns of two white oak species/ variants from southeast Asia, **(a)** *Quercus aliena* var. *acutesevata* and **(b)** *Quercus serrata* var. *breviptiolata* over a 5–6-week period. *(From Xiao et al. 2010)*

zelnuts were cached by one squirrel. The strong caching preference for black walnuts meant that there was no frequency dependence in these caching decisions. In contrast, when squirrels were presented with English walnut (*J. regia*) and Chinese chestnut (*Castanea mollissima*), two nuts of nearly equal value from the squirrel's perspective, the frequency of the two nut types significantly influenced the caching decisions. Captive squirrels showed no preference for either nut and consistently cached the nut of lowest frequency, thus demonstrating a negative frequency-dependent selection. In identical trials with free-ranging squirrels, Sundaram et al. (2016) found similar results for both experiments: no evidence of frequency-dependent hoarding when nut value was different, but negative frequency-dependent hoarding when nut value was similar. Sundaram et al. (2016) also tested if the squirrels' caching decisions were influenced by the number of existing caches. Interestingly it did when squirrels were presented with seeds of similar value, but in the black walnut-hazelnut experiment, the statistical model predicting caching events was weakened when we accounted for cached items.

Studies on the effects of relative frequency of seed types on dispersal and caching decisions clearly illustrate just how complex the interactions can be between seed traits and the relative frequency of seed types. That food-hoarding rodents can account for such variation speaks to their ability to process complicated information when making caching decisions. Moreover, the sensitivity of squirrels to the relative frequency of similar nut types has important implications for understanding how such caching decisions may influence seed establishment and seedling survival of rarer seed types and potentially the promotion of tree diversity (Sundaram et al. 2016).

HAVE SEED-CONSUMING RODENTS COEVOLVED WITH TREES?

It has long been assumed that many of the characteristics of seeds have evolved in response to rodent and bird seed predators. While the simple fact of this is inescapable, the strength of this evolutionary relationship between seed tree and the seed consumer is not well understood nor is the precise manner in which such evolutionary trajectories have taken place. In other words, little is known about the degree to which these relationships are driven by the plant, the animal, or both as in coevolutionary interactions. Nevertheless, some clear examples, beyond the early studies on *Tamiasciurus* by Smith, document the selective pressures between seed consumers and seed trees. Steele et al. (2005) review the strong negative selective pressures that squirrels (both *Sciurus* and *Tamiasciurus*) exert on conifer reproduction, emphasizing the potential for these species to have a negative keystone effect on conifer forests (Fig. 9.7). And, whereas these authors recognize the limited circumstances in which squirrels may disperse conifer seeds (see earlier discussion in this chapter), they identify three major selective pressures that significantly constrain conifer seed development, maturation, and dispersal. They include bark and twig clipping, which significantly reduces seed set; pre-dispersal seed predation when squir-

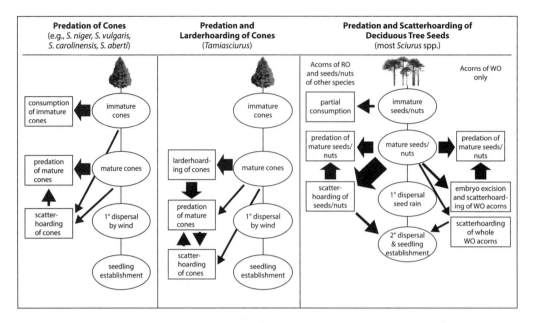

9.7. Overview of the three most significant ways in which tree squirrels influence seed fates in temperate and boreal forests. Ellipses indicate primary stages in the development and establishment of seeds or cones; the path from seed development to seedling establishment progresses from top to bottom. Rectangles and arrows indicate the influence of tree squirrels on seed fates. Arrow width is proportional to effect. RO = red oak species (section *Lobatae*) and WO = white oak species (section *Quercus*). *(From Steele et al. 2007)*

rels feed on immature cones; and the extensive larderhoarding of cones performed by pine squirrels (Smith and Balda 1979; Steele et al. 2005).

Perhaps some of the most persuasive evidence of the evolutionary impact of squirrels on conifers follows from the research of Craig W. Benkman and members of his laboratory at the University of Wyoming. Benkman and his collaborators have documented a series of strong cascading ecological and evolutionary effects to follow from the intense seed predation inflicted by pine squirrels on conifer cones in the western US. By comparing sites where pine squirrels are present with those where they have not occurred for more than 10,000 years, Benkman has shown, through these careful comparative studies, how conifer seed predation by pine squirrels has led to the evolution of seed defenses in six species of western conifers (Siepielski and Benkman 2008, and references therein). In limber pine (*Pinus flexilis*), for example, the presence of pine squirrels that readily feed on the immature cones has led to the evolution of cones that invest in fewer seeds per cone, smaller seeds, larger cones, more resin per cones, and greater seed-coat thickness (Benkman 1995a). Trees where pine squirrels are present (Rocky Mountains) invest half as much energy per cone in reproductive (seed) tissue than those where squirrels are absent (Great Basin).

Benkman (1995b) demonstrates further that selection on pines by pine squirrels can interfere significantly with avian dispersal of pine seeds. Seed dispersal in pines depends largely on seed size. Pine species that produce seeds <90 mg are typically wind dispersed, whereas those that produce larger seeds (>90 mg) are dispersed by jays and nutcrackers (family Corvidae). Many pine species, but not all, that produce large seeds exhibit traits such as a reduction in seed-wing size, longer periods of seed retention, and cone orientation that allows access by birds—all characteristics that favor dispersal by corvids (Lanner 1996; Siepielski and Benkman 2007). Hence, cone and seed characteristics that result from pre-dispersal predation by pine squirrels results in cone and seed traits not suitable for corvid dispersal (Siepielski and Benkman 2007).

This apparent conflict between the squirrel seed predators and the avian dispersers appears to have far reaching effects on ecosystem processes. Siepielski and Benkman (2008) provide strong evidence that this selection of pine squirrels on both limber and whitebark pine (*P. albicaulis*) results in a significant reduction in seed availability to these trees' primary dispersal agent, the Clark's nutcracker (*Nucifraga columbiana*). As a result, forests where pine squirrels were present exhibited 50% lower stand densities, suggesting that tree recruitment was significantly depressed in these forests (Siepielski and Benkman 2008).

Another outcome of the differential selective pressure to follow from the presence/absence of pine squirrels, coupled with the complementary presence/absence of another avian agent of conifer dispersal, the common crossbill (*Loxia curvirostra*), are definitive biogeographic mosaics of selection. The crossbill species complex is well-documented to be coevolved with lodgepole pine (Benkman 1999, 2003; Benkman et al. 2001) as well as black spruce (*Picea mariana*) in Newfoundland (Parchman and Benkman 2002) but only where *Tamiasciurus* spp. are absent. Similar mosaics of selection involving crossbills and tree squirrels (*Sciurus* spp.) have been reported in ponderosa pines of the western US (Benkman et al. 2001) and Allepo pine (*P. halepensis*) in, and adjacent to, the Iberian Peninsula (Mezquida and Benkman 2005).

WHAT ABOUT THE EVOLUTION OF TREES AND RODENT DISPERSAL AGENTS?

Just as pre-dispersal seed predation of conifers by pine and some tree squirrels drive the evolution of cone and seed characteristics, so too has the evolution of the nuts and trees of hardwood species been shaped by scatter-hoarding rodents. It would seem obvious that many nut characteristics—from the hard shells of hickory and black walnut to the chemical defenses such as tannins in acorns—are likely evolutionary responses to defend against seed predators, such as rodents. Such a conclusion, however, would greatly underestimate the complexity of these relationships. Take the tannins, for example. The presence of tannins alone is likely an adaptation that predates the rodents and instead evolved to protect acorns from pathogens, such as fungi and microbes that plague acorn survival (Steele et al. 1993). Thus, the tannin gradient that correlates with partial acorn consumption by rodents may have evolved long before the rodents. However, when you consider the multiple gradients (lipid, tannin, and Na) in *Q. rubra* acorns (see Chapter 6), for example, it is far more plausible to argue that these interacting gradients represent an adaptation to promote partial consumption of the basal portion of the seed. And, because these gradients direct partial predation by weevils, birds, and rodents, they are likely the result of diffuse evolution exerted by this entire suite of seed predators, rather than selection by just the rodents.

Perhaps one of the most definitive examples of evolutionary exchange between rodents and the nuts that they consume and scatterhoard is the relationship between early germinating acorns of white oak and the behavior of embryo excision by the squirrels (see Chapter 5). This behavior—performed by at least four genera of sciurids—allows squirrels to arrest germination when caching acorns that would otherwise germinate immediately and escape predation in the cache (Fox 1982; Steele et al. 2001b; Steele et al. 2006; Xiao and Zhang 2012). Evidence that this is in fact an evolutionary response of the squirrels follows from documentation that at least two of these squirrel species show an innate tendency for this behavior (Steele et al. 2006; Xiao and Zhang 2012). Yet, despite this likely evolutionary response of the sciurids to early germination, there does not seem to exist a strong evolutionary counterpunch by the tree or acorn to embryo excision.

Several rodent species in North America and Asia, many of which do not perform embryo excision, harvest acorns from early germinating seedlings (see Chapter 6). In response to this selective pressure, it appears that some oak species, especially in the white oak group, exhibit a pronounced cotyledonary petiole, which separates the acorn from both the epicotyl and the radicle. This anatomical arrangement allows the acorn to be pruned without disturbing the developing seedling, which is then able to survive acorn removal. Thus, the acorn too exhibits traits that appear to have evolved in response to rodent behavior. But even these interactions do not appear to involve the tight coevolutionary interactions observed in other systems (Ehrlich and Raven 1964; Janzen 1980).

As I describe in detail in Chapters 5, 7, and 8, scatter-hoarding gray squirrels selectively disperse and cache larger, more energetically profitable acorns (of higher utility), farther into more open habitat (Steele et al. 2011) where pilfer-age rates are lower (Steele et al. 2015) and cache longevity is significantly higher (Steele et al. 2014). To the extent that cache owners then fail to recover these acorns, it seems likely that this behavior would select larger acorns. Because larger seed size also enables acorns of some species to escape predation by in-sect and vertebrate seed predators (Bartlow et al. 2018a; Bogdziewicz et al. 2019, and references therein), it would seem that there is a strong selection on acorn size. However, for some oak species (*Q. palustris*), smaller acorn size is favored for dispersal by corvids (Bartlow et al. 2011). This illustrates how the diversity of acorn predators and scatterhoarders can exert conflicting selective pressures on the oaks.

One tree-scatterhoarder interaction that most likely involves coevolution is that of the fox squirrel and black walnut. The squirrels are the only vertebrate seed predators of black walnuts. Thus, the walnut's heavy, impenetrable shell most likely evolved in response to predation by these rodents. Stapanian and Smith (1978) argued that the fox squirrel, in particular, is engaged in a coevolu-tionary relationship with black walnut. By producing large, highly valuable, but storable nuts, the walnut tree offers a significant reward that ensures the nuts will be dispersed and widely spaced in scatterhoards, reducing the probability of cache pilferage by other seed predators. This behavior ensures seedling survival when these nuts are not recovered by cache owners. Couple this with masting patterns in walnuts and the result is evolutionary manipulation by the tree to ensure optimal spacing of individual nuts in a way that maximizes seedling es-tablishment and tree regeneration (Stapanian and Smith 1978).

Vander Wall (2010) has extended the argument of Stapanian and Smith (1978) by reviewing similar relationships for a broader range of plant species that rely on scatter-hoarding vertebrates for dispersal and seedling establish-ment. Based on this synthesis, he concludes that plants rely on four character-istics that collectively manipulate the behavior of scatterhoarders in a way that promotes seed dispersal. These include the production of (1) large, high quality propagules that encourage scatterhoarding, (2) seeds or nuts that have a high handling time due to either physical defenses (e.g., size and shell thickness) or physiological defenses (e.g., chemical traits), (3) masting behavior which further encourages scatter-hoarding behavior, and (4) seeds that produce weaker olfac-tory cues that would otherwise increase cache pilferage.

Although some of these studies of nut-producing trees and the rodents that scatterhoard their propagules suggest strong evolutionary relationships be-tween the two (e.g., Stapanian and Smith 1978; Steele et al. 2005; Vander Wall 2010), many treatments fall well short of a clear argument for coevolution (Janzen 1980). I suggest there are likely several reasons for this. Successful seed dispersal of hardwood trees by scatter-hoarding animals is a highly conditional mutualism. Even the most optimal dispersal agent may switch to strict seed predation under many conditions. In addition, several hardwood tree species, especially the oaks, achieve successful dispersal and establishment by multiple mechanisms and several dispersal agents (i.e., several species of both rodents and birds) under a range of different situations. The result is often multiple avenues of favorable selection. The combination of both insect and vertebrate seed predators and the vertebrates that may contribute to dispersal, establishment, and recruitment results in a suite of selective pressures that reduces the likelihood of coevolution between one tree species and one or a few dispersal agents.

Despite the lack of evidence for a tight coevolutionary relationship between scatterhoarders and nut producing trees, there has been little effort to rigorously test for one. Sundaram et al. (2015), however, conducted one of the first such studies by examining the evolutionary relationship between a rodent dispersal agent and the multiple tree species with which it interacts. In this study, we sought to determine how a set of seed traits of several hardwood species influenced feeding and caching decisions of eastern gray squirrels and how these two sets of variables, in turn, correlated with the phylogenetic relationship among these tree species. The first step was to measure 11 physical and chemical characteristics of the seeds and nuts of 23 hardwood species (e.g., tannic acid, shell strength, seed size, utilizable levels of protein, carbohydrate and fat, and number of days of cold stratification required for germination [i.e., perishability]) with which gray squirrels are typically associated. This included 10 oak species, two species of chestnut and their hybrid (*Castanea* spp.), four hickories, three walnuts, the tanoak (*Notholithocarpus densiflorus*), the American hazelnut, and the American beech (*Fagus gardifolia*). This included primarily species found within the Central Hardwoods Region of the eastern US, which includes much of the range of the gray squirrel. Sundaram et al. (2015) then randomly presented free-ranging squirrels with these seeds to determine their feeding and caching decisions and the energy invested in processing each seed type.

With these data in hand, correlations were run between seed traits and the squirrels' responses to each seed type, while controlling for phylogenetic relationships between trees. In this analysis, seed traits alone contributed to 27% to 73% of the squirrels' responses. This was then followed with a multivariate analysis (a phylogenetic, principal components analysis [pPCA]) that identified three key axes that explained 30–70% of the squirrels' behavioral responses (i.e., energy investment). Three vectors of this multivariate analysis explained this variation. The first was defined by a negative relationship between the time invested in caching seeds and seeds with thin shells, low lipid levels, and high carbohydrate levels. The second identified a relationship between consumption times of squirrels and seeds with higher protein levels, low tannin, and shorter periods of dormancy. The third axis was defined by the kernel mass, which was correlated with all measures of squirrel foraging but was not significantly associated with phylogeny.

The significance of Sundaram et al. (2015) follows from both the range of traits investigated and the added phylogenetic analysis which demonstrates that while some of the traits influencing squirrel behavior were defined by phylogeny, others were not. This led the authors to conclude that the relationship of the gray squirrel's behavior (nut predation and scatterhoarding) and the hardwood trees is unlikely explained by a strong coevolutionary interaction between the two. Instead, the squirrel's responses are best explained by a diffuse, or weaker, collective evolutionary relationship between hardwood trees and the squirrels.

In closing, rodents are universally distributed wherever there are large seeded plants, and in many of these plant communities are key agents of seed and nut dispersal. In some forest ecosystems, an entire guild of rodents may contribute to the process, whereas in others, often a single keystone rodent species may function as the primary dispersal agent of one or a few tree species. And, while rodents are often keystone dispersal agents in many oak forests, and have evolved closely with oaks, evidence for strong coevolutionary relationships between the rodents and oaks are limited. Chapter 10 now turns to the birds, primarily members of the Corvidae, which are also important scatterhoarders, and, in some forests, contribute significantly to oak dispersal.

THE JAYS: 10
A DIFFERENT KIND
OF OAK DISPERSAL
AGENT

INTRODUCTION

I now shift the discussion to the various ways in which birds—primarily a few members of the family Corvidae—facilitate oak dispersal. I review how their contribution to oak dispersal differs considerably from that of rodents and how these contributions may have profound consequences for some oak species and far less for others. Although the corvids include a relatively diverse group of birds (crows, ravens, rooks, jackdaws, jays and magpies, nutcrackers, and a few other specialized groups), several of which feed on and cache acorns, dispersal of the oaks is largely attributed to the activity of jays. A few other non-corvid bird species may also aid in dispersal in some systems. Here I discuss findings from my own research and that of others showing (1) how jays selectively disperse smaller-seeded oaks, (2) the susceptibility of jays to rodent pilfering, (3) strong habitat preferences by jays when scatterhoarding, (4) the potential for long-distance dispersal of some oaks by jays, and (5) dispersal limitations imposed by jays on many oak species. This review will demonstrate how gape limitations and central-place foraging in jays interact to influence seed size selection and dispersal of several oak species. I also suggest how susceptibility of jays to rodent pilferage influences patterns of directed dispersal by jays—an outcome that may result in a disproportionate rate of oak establishment.

As reviewed in Chapter 1, the birds contribute significantly to the dispersal of woody plants with avian-mediated dispersal, explaining as much as 40% and 90% of the dispersal of woody plant species in temperate and tropical forests, respectively (Herrera 2002). In Steele et al. (2010), my coauthors (Nate Lichti and Rob Swihart) and I reviewed species accounts for all breeding birds in Pennsylvania (~190 species) and estimated that 83 (43.7%) of these species dispersed seeds in some capacity. Thirty-nine of these 83 species, representing 15 families of birds, were ranked as significant agents of seed dispersal (Steele et al. 2010).

Birds have played a major role in our understanding of animal-mediated seed dispersal, shedding considerable light on a number of central factors related to the dispersal process. Some of the most important of these include the effects of avian-mediated dispersal on plant recruitment and demography (Herrera et al. 1994; Wenny and Levey 1998), recruitment limitations (Clark et al. 1998a), long-distance dispersal (Clark 1998; Clark et al. 1998b; Powell and Zimmermann 2004; Levey et al. 2008), directed dispersal (Wenny and Levey 1998), dispersal of invasive species (White and Stiles 1992), context dependent dispersal (Sargent 1990; Carlo 2005; Schupp 2007), the effects of landscape structure and forest fragmentation on dispersal (Tewksbury et al. 2002; Gómez 2003; Levey et al. 2005), and the effects of dispersal processes on the evolution of tree species diversity (Terborgh et al. 2002).

Whereas avian-mediated dispersal is widely studied (but see Clark et al. 1998b and Gómez 2003), much of it focuses on frugivory rather than scatterhoarder-mediated dispersal. One reason for this is that the number of avian species that scatterhoard seeds, fruits, or nuts is considerably limited, as are the woody plant species these scatterhoarders disperse (Steele et al. 2010). Yet their influence, albeit restricted to a few systems, appears immeasurably significant where it occurs.

Among this limited pool of scatterhoarders, it is the few species of the family Corvidae (the crow family) that are responsible for oak dispersal, primarily several species of jays. And with a significant worldwide distribution of oaks in temperate and subtropical forests, many, but not all, of the more than 500 species of oaks are regularly dispersed by one or a few species of corvid in each region. However, before focusing on the work of the key scatter-hoarding corvids, I briefly mention a few other avian species reported to feed on and move the acorns—species that generally can be ruled out as important dispersers of the oaks.

AN OVERVIEW OF OTHER AVIAN CONSUMERS OF OAK

Acorns are often dispersed and eaten by a number of bird species, although the activity of only a few contribute to oak establishment. For example, in deciduous forests of midwestern and eastern North America, acorns of several oak species are frequently consumed by the tufted titmouse (*Baeolophus bicolor*), white-breasted nuthatch (*Sitta carolinensis*), and red-bellied woodpecker (*Mel-*

anerpes carolinus), along with the blue jay (*Cyanoccita cristata*). Although all regularly disperse and scatterhoard acorns, only the tufted titmouse, nuthatch, and jay cache acorns on the ground where establishment is even possible (Richardson et al. 2013). The tufted titmouse will cache acorns and select sites suitable for germination (Lichti 2012), but, like other parids, will often cache above ground level. Poor cache site selection and the short duration of caching cycles for the nuthatch further eliminate this species as a viable dispersal agent of oak (Vander Wall 1990). This leaves primarily the blue jay, which is a significant contributor to oak dispersal throughout North America.

Pérea et al. (2011b) examined acorn movement by guilds of avian and rodent nut consumers in oak-beech forests of central Spain. Although guild diversity was generally low overall, it was higher in oak stands and mixed oak-beech stands compared with pure beech stands. As shown in many other studies in this region, Eurasian jays (*Garrulus glandarius*) and the wood mouse (*Apodemus sylvaticus*) were abundant in oak forests and were the primary dispersal agents of oak. Although the Eurasian nuthatch (*Sitta europaea*) was also reported to eat oak, this species showed a distinct preference for beech seeds and was significantly more abundant in beech forests than mixed forests or those dominated by oak. Great tits (*Parus major*) occurred more often in oak-beech and oak forests where they regularly fed on acorns. However, this species does not store acorns and is thus unlikely to contribute to oak dispersal.

The acorn woodpecker (*Melanerpes formicivorus*) is another species in the western hemisphere that relies extensively on acorns but is likely to have a limited effect on oak dispersal (see below) because it stores acorns above ground. Across much of its range in California, the southwestern US, Mexico, Central America, and northwestern South America (Colombia), the species' distribution is closely associated with the oaks. As communal food hoarders, acorn woodpeckers live in tight social groups that larderhoard acorns in trees and other wooden structures where they place each acorn in an individually excavated hole (Koenig and Mumme 1987) where these stored acorns can be viewed by all members of the social group (Vander Wall 1990). This highly visible food store, or granary, serves as the focal center of the social group, which is defended against other groups of conspecifics (Fig. 10.1). Although acorns in many ways represent a marginal food source for these woodpeckers due to their high tannins and low protein levels, they are critical for survival and breeding success (e.g., clutch size, and hatchling and fledgling success; Koenig and Mumme 1987). As these authors note, the acorn woodpecker's dependence on these oaks is especially puzzling given that the acorns are stored six months before the breeding season, and the acorns constitute only a small percentage of the overall diet of the woodpeckers, particularly the nestlings.

Each social group of acorn woodpeckers can move and store tens of thou-

10.1. An acorn woodpecker
(*Melanerpes formicivorus*) at a
typical granary. *(Ingrid Taylar)*
SEE COLOR PLATE

sands of acorns (upward of 50,000 or more in a single year). Thus, it is generally assumed that this specialized behavior renders this species an intense acorn predator. However, it is not known how often accidental dispersal may occur as a result of woodpeckers moving and attempting to store acorns. This led Scofield et al. (2010) to determine the genetic relationship between source trees and acorns of *Q. agrifolia* stored by these woodpeckers. They found, with few exceptions, source trees were located within 150 m of granaries usually within a social group's territory. Maternal tree sources were also located within 90 m of each other, suggesting that, even if accidental dispersal occurs, the effects are likely to have a localized effect on the genetic structure of these oak stands (Scofield 2010). Nevertheless, it would be interesting to know the frequency with which such accidental dispersal occurs.

Finally, a few species of pigeons have been implicated in dispersing oaks in both North America and Europe, though limited direct evidence is available to support these claims. The most salient argument has been advanced by Webb (1986) who maintained that rapid range extension of several nut species of beech, oak, and hickory, following the last glaciation event in North America, was the likely work of the now extinct passenger pigeon (*Ectopistes migratorius*) (see below). Webb (1986) also cites early natural history observations that implicate the band-tailed pigeon (*Patagioenas fasciata*) in the dispersal of Garry oak (*Q. garryana*) in British Columbia and the wood pigeon (*C. palumbus*) in dispersal of several oaks in Europe. It should be noted that pigeons are not reported to cache acorns, so, if they contribute to oak dispersal and establishment, it is likely as a result of incidental loss and/or partial consumption of acorns (see Chapter 6).

10.2. The Rook
(*Corvus frugilegus*).
(*Michael A. Steele*)
SEE COLOR PLATE

THE CORVIDS AND OAK DISPERSAL

The key avian agents of oak dispersal include primarily the jays, and, under limited circumstances, some of the other members of the Corvidae, such as the rook (*Corvus frugilegus*), Eurasian magpie (*Pica pica*), jackdaw (*Coloeus monedula*) (Lockie 1956; Waite 1985; Källander 2007), and possibly the nutcrackers (*Nucifraga* spp.). The nutcrackers include only three species worldwide, one in North America and two in Eurasia. The nutcrackers are conifer specialists, dispersing large quantities of seeds of numerous species record distances (>20 km) and regularly contributing to patterns of dispersal and establishment (Vander Wall 1990). I mention the nutcrackers because of their adaptations for moving seeds so closely parallel those of the jays (Pesendorfer et al. 2016a). Although no formal reports of acorn use are reported for this species, the nutcrackers are known to shift their diet to small nuts when conifer seeds are less available. Thus, it is possible that oak dispersal occurs under some circumstances.

Among the rooks, magpies, and jackdaws, all are known to feed on acorns, but it is only the rook (Fig. 10.2) that is reported to disperse and cache acorns on a regular basis. Found in the British Isles and other parts of northern Eurasia, this species has a broad diet of both animal material and fruits, including acorns. Based on observations made in southern Sweden, Källander (2007) reported that rooks dispersed acorns of *Q. robur* from 50 m to 4 km and then cached acorns in open, grassy areas. Rooks often dispersed multiple acorns (2–7), but, after arriving at a dispersal site, cached acorns individually. During this process, scatter-hoarding rooks were often susceptible to pilferage by other rooks. Few other studies have examined oak dispersal by this species.

10.3. Avian dispersers of acorns. Shown are **(a)** blue jay (*Cyanocitta crystata*). *(Jeff Flinn)* **(b)** Steller's jay (*C. stelleri*). *(Scott Vail)* **(c)** California scrub jay (*Aphelocoma californica*). *(Eric Lu)* **(d)** Florida scrub jay (*A. coerulescens*). *(John Macnamara)* **(e)** Island scrub jay (*A. insularis*). *(Shijo Joy)* **(f)** Mexican jay (*A. wollweberi*). *(Roger Dietrich)* **(g)** Eurasian jay (*Garrulus glandarius*). *(Jaroslaw Stefanoff)* SEE COLOR PLATE

The jays are the primary avian agents of oak dispersal. A polyphyletic group of corvids found nearly worldwide, the jays include three groups: Old World jays, New World jays, and the grey jays found primarily in northern North America and northern Eurasia. Although these three groups include more than four dozen species, it is a relatively small number of jays that are responsible for oak dispersal. As reviewed by Pesendorfer et al. (2016a), the jays that contribute to the dispersal of oaks include the blue jay (*Cyanocitta crystata*), Steller's jay (*C. stelleri*), western scrub jay (*Aphelocoma californica*), Florida scrub jay (*A. coerulescens*), island scrub jay (*A. insularis*), Mexican jay (*A. wollweberi*) in North America, and the Eurasian jay (*Garrulus glandarius*) in Europe and Asia (Fig. 10.3). I also mention the white-throated magpie jay (*Calocitta formosa*), which I have observed handling acorns of *Q. oleoides* in northwest Costa Rica, although studies of avian dispersal of oaks are rare in Mexico and Central America. Despite this limited diversity, the jays contribute significantly to the dispersal of several oak species across the globe (Fig. 10.4). Here I review the key

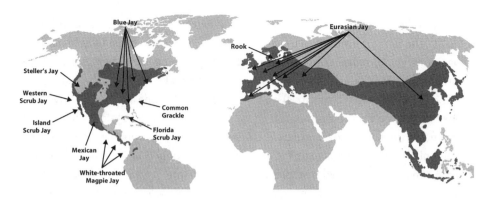

Avian Species	Oak Species Dispersed	Geographic Region of Research
Western Scrub Jay (*Aphelocoma californica*)	*Quercus douglasii* *Quercus agrifolia* *Quercus lobata* *Quercus gambelii*	Southwestern United States
Island Scrub Jay (*Aphelocoma insularis*)	*Quercus pacifica* *Quercus agrifolia*	Northwestern Mexico
Steller's Jay (*Cyanocitta stelleri*)	*Quercus garryana* *Quercus lobata*	Northwestern United States
White-throated Magpie Jay (*Calocitta formosa*)	*Quercus oleoides*	Central America
Common Grackle (*Quiscalus quiscula*)	*Quercus phellos*	Eastern United States
Mexican Jay (*Aphelocoma wollweberi*)	*Quercus emoryi*	Central Mexico
Florida Scrub Jay (*Aphelocoma coerulescens*)	*Quercus geminate* *Quercus inopina* *Quercus myrtifolia* *Quercus chapmanii*	Southeastern United States
Blue Jay (*Cyanocitta cristata*)	*Quercus palustris* *Quercus phellos* *Quercus velutina* *Quercus alba* *Quercus muehlenbergii*	Central and Eastern United States
Eurasian Jay (*Garrulus glandarius*)	*Quercus robur* *Quercus petrea* *Quercus ilex* *Quercus coccifera* *Quercus faginea* *Quercus suber* *Quercus rubra* (limited dispersal of this invasive species)	Northeastern Europe and Western Russia, the Iberian Peninsula, and Dongling Mountains of China
Rook (*Corvus frugilegus*)	*Quercus robur*	Southern Sweden

10.4. Global distribution of avian dispersers of oak. Shown are the oak species they likely disperse and the general location of primary studies. *(Map modified from Nixon 2006; Shealyn A. Marino)*

studies on oak dispersal by these corvids and highlight several aspects of avian-mediated dispersal of the oaks that differ markedly from dispersal by rodents.

The ability to fly is one of the major differences between rodent and avian agents of dispersal. That adaptation alone translates into longer dispersal events, the ability to move quickly through different habitats and across landscapes, and the ability to harvest and disperse localized seed crops faster than many rodent competitors. From the early work of Darley-Hill and Johnson (1981), the blue jay's ability to move acorns of some species considerable distance was quite evident. They observed a small group of blue jays—with some of the birds tagged for visual identification—feeding on and dispersing the acorns in a stand of 11 pin oaks (*Q. palustris*) on the campus of Virginia Technical University. They estimated that in one season these few jays dispersed 133,000 acorns and consumed another 49,000 acorns of the estimated acorn crop of 246,000 acorns. Most of the remaining acorns (26% of the crop) dropped to the ground by jays, were considered rejected because of insect infestation and fungal damage. At this site, jays were observed carrying from one to five acorns per dispersal event, which they then moved approximately 1 km before caching.

Soon after joining the faculty at Wilkes University, I was reminded of Darley-Hill's and Johnson's paper when I noticed a group of about seven blue jays descending on two pin oaks on campus. The two mature trees were loaded with acorns as were most of the other pin oaks in the area. I immediately recruited some eager undergraduates and put them to work. I assigned each student to one or more two-hour shifts to observe and record jay visits at the two neighboring trees. The observation shifts extended from sunrise to sunset on three consecutive days. Students were equipped with a lawn chair, binoculars, notebook, and a water bottle and asked to record each jay-visit and the approximate number of acorns that each jay harvested and placed in the esophagus or bill before departing.

Although I knew from earlier work that this single observational exercise lacked some originality, I was driven by my simple curiosity to estimate just how fast these jays could move the acorns. It was staggering. In the three days, jays made a total of 3229 jay-visits to the two trees and, based on observations of the average number of acorns dispersed by each jay (~2.5), we estimated that in excess of 8000 acorns were moved to cache sites during that three-day period (Fig. 10.5).

Estimates of acorns moved by other jays, summarized by Vander Wall (1990) and Pesendorfer et al. (2016a), are comparable to those we observed. For example, individual Eurasian jays are estimated to cache as many as 8000 acorns per season (Bossema 1968, 1979), Florida scrub jays about half that many (DeGange et al. 1989), and western scrub jays about 5000 acorns annually (Pesendorfer et al. 2016a, and references therein). Although I have witnessed 8000 acorns removed in a 48-hour period by rodents during one low mast year, I still find this work of the jays far more impressive, especially given that they were moving these acorns approximately 1 km (Steele et al. 2010). Elsewhere, maximum observed estimates for jay dispersal of oaks is approximately 1 km for Steller's jays and Eurasian jays, 1.9 km for blue jays, and 400 m for island scrub jays (Pesendorfer et al. 2016a, and references therein).

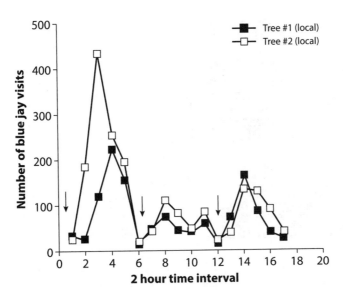

10.5. Number of visits by blue jays observed at two adjacent pin oak trees (*Quercus palustris*) over a three-day period in Northeastern Pennsylvania (October 1993). Arrows indicate the beginning of each day of observations. Although they occasionally consumed acorns, these birds (approximately 7–12 individuals) were harvesting and dispersing the acorns > 0.5 km from the site. Based on these and other observations, we estimate the birds were dispersing 2–3 pin oak acorns per visit. *(Results from unpublished data; Steele et al. 2010)*

MULTIPLE-PREY LOADING BY AVIAN DISPERSERS OF OAK

The rapid dispersal of acorns by jays is accomplished in part by the ability of these birds to move more than one acorn at a time. Vander Wall (1990) describes in detail the unique adaptations that corvids exhibit for carrying multiple seeds or nuts. The jays, for example, possess a flexible esophagus (a crop) that allows them to swallow and carry multiple seeds or nuts. Other corvids, such as the nutcracker, have a pouch either under the tongue (sublingual pouch) or in front of it (antelingual pouch) that accomplishes the same (Vander Wall 1990, and references therein). This translates into the ability of corvids to carry heavy seed loads—as much as a 28.0 g load in the case of Clark's nutcracker (Vander Wall 1990; Fig. 10.6). For dispersal of conifer seeds, this means multiple seeds can be moved considerable distances.

From the jays' perspective, this also means multiple acorns can be carried, at least when acorn size is relatively small (<2 g). The realization of this led some of my students and colleagues to consider further how acorn size, multiple prey loading, and investment in dispersal events may influence scatterhoarding of

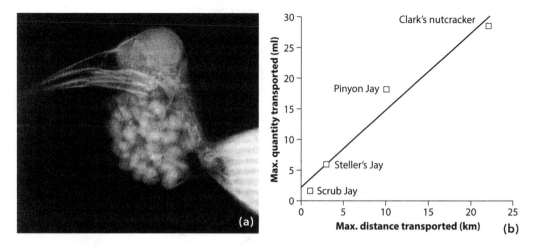

10.6. Shown are **(a)** radiograph of sublingual pouch of a Clark's nutcracker (*Nucifraga columbiana*) showing the position and load of 28 seeds (30.6 g) of single leaf piñon pine (*Pinus monophylla*). *(From Vander Wall 1990)* **(b)** Relationship between maximum load (quantity of conifer seeds) and maximum dispersal distance of four corvids. *(From Vander Wall 1990; Vander Wall and Balda 1981)*

acorns by blue jays. Bossema (1979) suggested that single acorns were dispersed shorter distances by Eurasian jays than loads of two or more acorns. We therefore hypothesized that, if larger loads with multiple acorns would be dispersed farther then smaller acorns may encourage even larger loads. To test this, we presented jays with acorns of different sizes and recorded the estimated prey load when dispersing the acorns and the distances they carried these acorns when caching them (Bartlow et al. 2011).

During each year in which we conducted the experiment, we collected pin oak acorns from several individual tree sources and from these sources created composite samples of both small pin oak acorns (<1.5 g per acorn) and large pin oak acorns (>2.0 g per acorn). We then conducted individual trials in which we presented a small group of jays (1–5) during eight trials in 2008 and another eight trials in 2009 with either small or large acorns in each trial. Acorns were presented to jays on a 1 m² feeding platform positioned a few meters above the ground with two Reconyx camera traps positioned directly over the feeder (see Fig. 7.7). The feeding platform, designed by my colleague, Nate Lichti, consisted of a heavy piece of plywood with a grid of 13 cm × 13 cm and small depressions approximately every centimeter. Each depression could hold a single acorn and was uniquely numbered so in some experimental trials we could follow the fate of individual acorns. In each trial, we randomly pulled 100 acorns from one of the composite samples and placed the acorns on the tray. Observers then waited

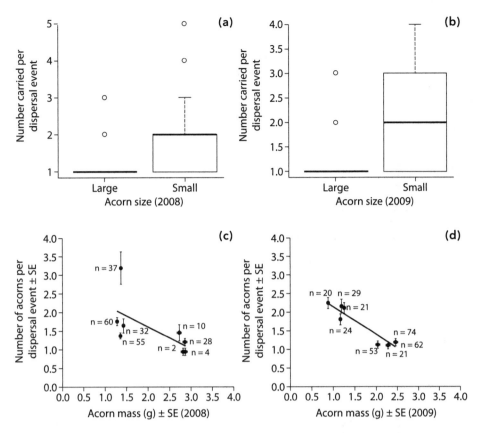

10.7. Box plots showing the distribution of the seed loads (number of pin oak [*Q. palustris*] acorns carried per dispersal event) carried by blue jays (*Cyanocitta cristata*) for large and small acorns in **(a)** 2008 and **(b)** 2009. Seed loads were significantly larger for smaller acorns in both years of the study (see text). The relationship between seed mass (g) and seed load (see text) across all trials in **(c)** 2008 and **(d)** 2009. Shown are the means per trial (± SE); number of dispersal events is shown for each experimental trial. *(From Bartlow et al. 2011)*

until jays arrived and began either feeding on or dispersing the acorns to sites where they then cached them. Researchers were positioned about 70 m away with a spotting scope where they could observe each jay and record the number of acorns retrieved from the tray.

When feeding on the acorns, our blue jays usually carried single acorns to the top of nearby trees and immediately ate the acorns. However, in many of trials, jays swallowed two or more acorns and then travelled up to 130 m and then regurgitated the acorns to cache them. Across the two-year experiment, jays cached at seven primary sites and occasionally at more than a dozen secondary sites. Although much of the surrounding forest was dominated by hard-

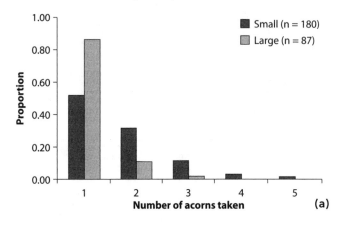

Prey Load per Visit in 2008

Small (n = 180)
Large (n = 87)

(a)

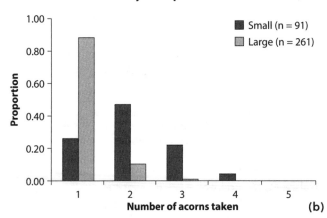

Prey Load per Visit in 2009

Small (n = 91)
Large (n = 261)

(b)

10.8. Proportion of small and large acorns dispersed in single or multiple loads by individual blue jays (*Cyanocitta cristata*) in **(a)** 2008 and **(b)** 2009. *(From Bartlow et al. 2011)*

woods, primarily oak, maple (*Acer* spp.), and some hickory (*Carya* spp.), jays often cached in a conifer stand approximately 120 m from the feeder, along the edge of the hardwood forest or on the edge of a nearby pond.

Overall, when jays dispersed smaller acorns, they carried multiple acorns (1–5 acorns, mean = ~2 acorns). Mean load size (acorn number) was consistently higher for smaller acorns than larger acorns and mean mass of individual acorns was negatively correlated with the number of acorns dispersed (Bartlow et al. 2011; Fig. 10.7). The proportion of small and large acorns dispersed in single and multiple loads is shown in Figure 10.8. In the second year of the study, but not the first, smaller acorns were dispersed farther than larger acorns (Fig. 10.9), suggesting that, in some circumstances, multiple seed-loading of smaller acorns by blue jays may favor dispersal. This assumes that greater dispersal distance favors establishment, which is not always the case. However, in this system,

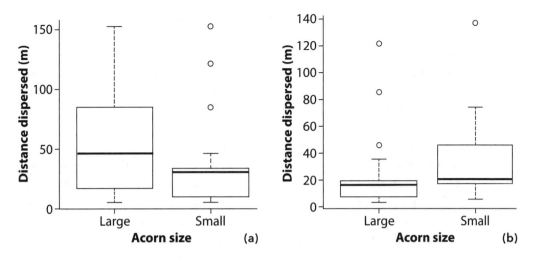

10.9. Dispersal distances of small and large acorns of pin oak (*Q. palustris*) by blue jays (*Cyanocitta cristata*) in **(a)** 2008 and **(b)** 2009. *(From Bartlow et al. 2011)*

greater dispersal distances were usually associated with forest edges and open conifer stands where establishment is likely higher (see above).

The study by Bartlow et al. (2011; see Gómez 2004 for European jays) demonstrates that jays may, in some cases, exert directional selection toward smaller intraspecific acorn size, or perhaps a disruptive selection in some circumstances. This is, nevertheless, a markedly different pattern of selection than that of most rodents (but see Muñoz and Bonal 2008a), which usually selectively disperse large acorns of the same species (see Chapter 4). Such acorn size constraints also influence patterns of acorn selection by jays between oak species in several ways that appear not to apply to acorn selection in many rodent species.

OTHER ACORN PREFERENCES OF JAYS

Scarlett and Smith (1991), working in the vicinity of Fayetteville, Arkansas, during autumn seed fall sought to determine acorn preferences by observing free-ranging blue jays visiting trees of six oak species. This included two species of early germinating white oak (subgenus *Quercus*), the white oak (*Q. alba*), and post oak (*Q. stellata*), and four species of red oak (subgenus *Lobatae*), the northern red oak (*Q. rubra*), pin oak, willow oak (*Q. phellos*), and black oak (*Q. velutina*). In the spring, these researchers also presented individual piles of all six of these acorn species to jays on the ground, with the exception of black oaks, which they replaced with larger-seeded acorns of the English oak (*Q. robur*), a non-native species that is regularly dispersed by European jays (Bossema 1979; Gómez 2004).

In both sets of experiments, Scarlett and Smith (1991) found that jays preferred the smaller-seeded oaks (especially pin oak but also post and willow oak) rather than the larger acorns of white oak, northern red oak, and the English oak (Fig. 10. 10). Smaller seed size took precedence over other key acorn characteristics, such as tannin levels, lipid levels, and germination schedules (corresponding with oak subgenus). Darley-Hill and Johnson (1981) reported that jays selectively ate and dispersed smaller-seeded oaks (pin oak, willow oak, and black oak, and American beech [*Fagus grandifolia*]) and avoided larger-seeded acorns of northern red oak and white oak. Richardson et al. (2013) found that blue jays consistently selected pin oak acorns over those of black oak, and black oak over white oak, but also found that the assemblage of seed predators and the relative abundance of oaks influenced acorn preferences.

In more controlled experiments with captive blue jays, Moore and Swihart (2006) presented individual jays with acorns and other seed types that varied in size, shell thickness, germination schedule, and seed size as well as background food availability. When alternative foods were available, jays showed a distinct preference for acorns of pin oaks first and to a lesser extent, black and white oak acorns. Jays avoided red oak acorns except when alternative foods were scarce but, even then, only selected small-seeded red oak acorns. Again, seed size appeared to be a key factor driving seed selection by jays, but this study also demonstrated how context (e.g., availability of alternative foods) could modify patterns of acorn selection.

Finally, a more recent study on acorn selection by Pons and Pausas (2007b) involved controlled experiments in the field with European jays and four small-seeded species of Mediterranean oaks (*Q. ilex*, *Q. faginea*, *Q. suber*, and *Q. coccifera*). In this study, the researchers presented jays with different combinations of acorns on elevated feeders much the same way Bartlow et al. (2011) had done. Jays consistently preferred *Q. ilex* acorns, the smallest of the four species, which they attributed in part to this species' higher fat and lower tannin content. The species with the highest tannin levels, *Q. coccifera*, was consistently avoided when other acorns were available. Individual variation in size of acorns of each species also influenced dispersal patterns, but in this case larger individual acorns were removed first, subsequently followed by smaller acorns (Pons and Pausas 2007b). The authors maintain that visual cues allowed the jays to assess acorn size.

These investigations of how acorn characteristics influence acorn selection in jays illustrate several key factors that likely drive avian-mediated dispersal. Most importantly, the preference of smaller-seeded oaks likely follows from the jay's ability to swallow and hold multiple acorns in its modified esophagus. When acorn size prevents swallowing, jays can still carry larger individual acorns in the bill. Pons and Pausas (2007b), for example, found that European jays were limited to swallowing acorns of an estimated area less than 3.6 cm^2.

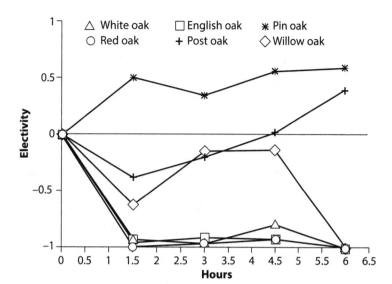

10.10. Mean electivity of acorns of six species of oak. Note the high preference for acorns of pin oak (*Q. palustris*). (*From Scarlet and Smith 1991*) SEE COLOR PLATE

Typically, they could not swallow acorns that were 17–19 mm in width or 31–32 mm in length and instead carried them in the bill (Fig. 10.10). However, there is also a gape limit of the bill for the size of an acorn that can be carried. This is most certainly why red oaks and white oaks of North America are less likely to be dispersed by blue jays (but see Steele et al. 2007).

THE EFFECTS OF TANNIN ON ACORN DISPERSAL BY JAYS

The effects of tannin on acorn selection by jays appears as complicated as it is for rodents (see Chapter 9), although fewer studies have examined this question for jays. Just as laboratory studies demonstrated that rodents could not subsist indefinitely on a diet of only acorns (due to loss of body mass), the same has been reported for both scrub jays (Koenig and Heck 1988) and blue jays (Johnson et al. 1993). However, in the latter case, it was shown that this weight loss was countered when blue jay diets were supplemented with weevil (*Curculio*) larvae, leading Johnson et al. (1993) to suggest a tri-trophic mutualism that resulted from the jays dispersing both acorns and weevils and the jays relying on both for a balanced nutrient source. Fleck and Woolfenden (1997) reported that Florida scrub jays selectively dispersed acorns from individual trees with lower rates of weevil infestation but also selected acorns with intermediate levels of tannin. The acorns of sand live oak (*Q. geminata*) that produced high levels of tannin had the lowest rates of weevil infestation but were also avoided by the jays presumably due to these high tannin levels. A test of such a three-way mutualism, however, was rejected in studies with captive Mexican jays when Hubbard and McPherson (1997) found that these jays selected the densest acorns, free of weevil infestation. Similar evidence for selective dispersal of sound acorns by European jays was reported by Bossema (1979). Dixon et al. (1997b) tested these questions further with blue jays. They presented captive jays with infested and noninfested pin oak acorns in an aviary and found a clear preference for noninfested acorns, and, although jays were observed eating infested acorns, they did not consume weevil larvae. Moreover, by sampling stomach contents of jays that were feeding in the canopy of pin oaks, Dixon et al. (1997b) found that the birds were consuming a variety of insects while harvesting acorns but did not consume weevils. Thus, while jays, compared to rodents, appear to lack the physiological ability to partially deal with tannins, they do not depend on weevils to supplement high tannin diets. The possibility that jays modify tannin levels by first caching them and allowing tannins to leach into the surrounding soil is also highly unlikely given the number of studies that have tested this and found no support for this hypothesis (Dixon et al. 1997a; Smallwood et al. 2001). One study has even found evidence of an increase in tannin activity (protein-binding capacity) during acorn storage (see Chapter 7).

Perhaps one of the most effective ways for jays to thwart the effects of tannin is to regularly consume only the basal half of the individual acorn, where tannin levels are lowest. As detailed in Chapter 6, Steele et al. (1993) reported that when acorns are abundant in the autumn, blue jays, along with other vertebrates (e.g., common grackles [*Quiscalus quiscula*] and gray squirrels [*Sciurus carolinensis*]) and insects (weevil larvae), consistently consume less than 60% of the acorn cotyledon and as little as 30% from the basal (cap end) of each fruit. Whereas lower tannin levels certainly play an important role in this response, higher levels of lipid and nutrients (e.g., Na) in the basal half of acorns are likely to synergistically drive this behavior of partial seed consumption. And the ability of these partially eaten acorns to still germinate and establish suggests that these chemical gradients represent a critical adaptation in the oaks for escaping seed predation by an entire community of potential seed predators, including jays and grackles.

The effects of chemical gradients on partial seed consumption by jays aside, the overall take-home message from these studies on seed selection by jays is not readily forthcoming and certainly not as evident as that for rodent-mediated dispersal. For one, the combination of aviary and field studies leaves many questions open on how tannin and weevil infestation interact to influence dispersal decisions in jays. Whereas captive studies allow controlled experiments, they also remove the option for long-distance dispersal and caching of acorns. And, while jays may frequently select sound, dense acorns for dispersal, much the way rodents do, this does not discount selective consumption of infested acorns along with their weevils when consuming acorns (Steele et al. 1996). Thus, much remains to be done, but as some of my students and colleagues learned, research on free-ranging jays is not, by any means, a simple undertaking.

ACORN ABUNDANCE AND DISPERSAL BY JAYS

Only recently have studies focused on the effects of mast crop sizes on patterns of acorn dispersal by corvids. Pesendorfer et al. (2016b) determined patterns of acorn (*Q. pacifica* and *Q. agrifolia*) dispersal of island scrub jays (*Aphelocoma insularis*) on Santa Cruz Island, part of the Channel Islands of California. These authors followed 26 individually marked birds over three years of varying crop sizes. One of the most important outcomes of this study was evidence for a positive correlation between annual crop size and dispersal distances. This contrasts significantly with an opposite pattern observed in several studies on rodents (Moore et al. 2007; see Chapter 7). Pesendorfer et al. (2016b) also reported that caching rates increased with acorn abundance and decreased with aggressive interactions associated with territoriality. They also concluded that acorn dispersal was highly context dependent.

Just as interannual variation appears to influence acorn dispersal by corvids, so, too, may geographic variation in annual crop size. In one of the first studies to address this question, Pesendorfer and Koenig (2016) determined how within-year variation in acorn crops in the valley oak (*Q. lobata*) influenced the behavior of a high quality disperser (the western scrub jay) as well as the behavior of the poor disperser (the acorn woodpecker). The authors found that western scrub jays concentrated on individual trees with higher acorn crops, thereby increasing quality dispersal rates compared with trees with less abundant acorn crops.

In Norway, long-term patterns of acorn crop size was negatively associated with two measures of Eurasian jay abundance: irruption patterns (erratic large-scale movements of jays) followed over 40 years, and a hunting index followed for 20 years (Selås 2017). These irruptions appeared to be associated with mast failures although heavy snowfall, which interferes with acorn removal, was a likely confounding factor in some years. My colleague, Dr. Jeffrey Stratford, and I are now exploring the potential for blue jays to show a similar pattern of extensive migration in years of acorn mast failure. More research is needed to understand how acorn crops relate to the movement of jays and dispersal patterns.

DISPERSAL KERNELS AND CACHING PATTERNS OF JAYS VS. RODENTS

We know that dispersal kernels created by jays extend far greater than any attributed to rodents, yet detailed comparisons between the two are generally limited. However, a comparative review of dispersal patterns of the wood mouse and Eurasian jay (den Ouden et al. 2005, and references therein), illustrates some of the key differences between these two dispersal agents. Jays prefer forest edges for harvesting acorns, whereas wood mice occurred throughout the forest, especially where vegetation was dense. Jays also maintained larger territories (10–100 ha vs. <1 ha in mice) and consistently dispersed acorns farther than mice (up to several km for jays and <50 m for mice). Compared with wood mice, which will often cache 2–6 acorns in some of their caches (~20% of caches) (den Ouden et al. 2005), Eurasian jays consistently place one acorn per cache that are consistently spaced 0.5–1.0 m apart (Bossema 1979). These jays also bury the acorn in about 1.5 cm of soil (Gómez 2003) sometimes covering the surface with debris—conditions ideal for germination when the acorn is not recovered. Many of these differences are similar to those observed for rodents and blue jays in eastern deciduous forests of North America, with the exceptions that tree squirrels often disperse acorns to open areas in sites suitable for germination and bury only one acorn per cache (Steele and Smallwood 2002).

HABITAT USE AND ACORN DISPERSAL BY JAYS

Among the most distinguishing characteristics of jay scatterhoarding is their consistent preference for successional, disturbed habitats or conifer forests for caching (Steele and Smallwood 2002; Steele et al. 2011). The majority of studies suggest that closed forests are generally avoided for caching. Instead, jays frequently cache acorns in abandoned fields (Deen and Hodges 1991; Pons and Pausas 2007c) or other open habitats (den Ouden et al. 2005, and references therein) such as conifer stands, which are the most common habitat selected by Eurasian jays. Kurek and Dobrowolska (2016), for example, found that in Poland, jays cached 76% of acorns in Scots pine plantations and only 8% in clear cuts and other open habitats. Similarly, Leverkus et al. (2016) reported that in southern Spain, 63% of acorns cached by jays were placed in conifer forests compared with 37% that were cached in more "open areas." On Santa Cruz Island, California, Pesendorfer et al. (2017) found that the island scrub jay showed a marked preference for chaparral and coastal sage-scrub habitat for scatterhoarding over cleared areas and grassland habitat. However, the widely advanced assumption that jays simply prefer open habitat for caching underestimates specific selection patterns. Gómez (2003) found that the sites Eurasian jays selected for caching holm oak acorns, as well as the sites where oak seedlings established, were especially homogenous with respect to light conditions compared with the more heterogeneous conditions characteristic across the oak's distribution. Jay selection for caching thus favored the shade tolerant holm oak, suggesting possible directed dispersal.

Johnson et al. (1997) assessed patterns of habitat selection based on the available habitat types found within the range of blue jays in Iowa and reported that they cached in a diversity of habitats, with disproportionately more caches made in dense regenerating woodlands, forest edges, and canopy gaps. Habitats avoided by jays included woodlands and grasslands, except after a recent burn. In contrast, Lichti (2012) found that blue jays showed no statistical preference for a specific habitat when caching acorns in Indiana and two sites in Pennsylvania. These studies collectively suggest that jays frequently, but not always, cache in open habitats, but even when it occurs, it is not the only distinguishing characteristic defining cache site selection. The literature thus suggests that cache site selection by jays may happen at the level of the landscape, habitat, or microhabitat, or a combination of all three.

I identify four features of habitat selection by jays when dispersing acorns that have not been well investigated and perhaps deserve additional attention. First, open vegetation may reduce predation risks when jays are caching by increasing their ability to remain vigilant. Second, soil type and, in particular, soil moisture may differ significantly between preferred and less preferred sites. Optimal soil conditions may influence cache longevity: too much moisture and the acorn may germinate; not enough moisture and the seed may dry out and deteriorate. Too much moisture in the soil may also enhance acorn odors and increase pilferage rates by rodents (Vander Wall 1995a). In sites that are too open, stored acorns may not survive in the cache. A third feature that may be critical is that sufficient ground cover may be necessary for covering the cache. Jays frequently insert acorns in the soil and then cover caches with leaves and other plant material and sometimes specific items (e.g., twigs) that may serve for remembering cache locations (Bossema 1979). In addition, structural features in the habitat, such as trees, stems, or shrubs, may be important for relocating caches via triangulation on caches much the way fox squirrels (*S. niger*) are reported to recover caches (Lavenex et al. 1998).

Finally, I suggest that jays may select specific habitats where rodent densities are lowest. This may, in part, explain why jays consistently avoid more mature hardwood forests when caching, but prefer successional sites with varying amounts of ground cover. Certainly, when conifer stands are surrounded by hardwood forests as they were in Bartlow et al. (2011), the relative abundance of rodents is likely to be lower in conifers where jays regularly cached. I am reminded of a simple observation I made when observing one of our radio-tagged jays caching acorns. It selected a single pin oak acorn that dropped to the ground and cached it in a thicket of common blackberry (*Rubus allegheniensis*). Out of curiosity, I marked the site and placed a camera trap focused on the cache. A week later, I retrieved the memory card and learned that within 24 hours of caching the acorn, the jay's cache was pilfered by an eastern gray squirrel. This simple field observation has long reminded me of a looming question that still requires further research. How much are these two species regularly pilfering from one another's scatterhoards?

Although we know little of the competition between the jays and rodents, it is no doubt a significant factor that deserves further investigation. While preparing this book, I watched both blue jays and gray squirrels harvest pin oak (*Q. palustris*) acorns from a single tree. Whereas squirrels will often harvest acorns from the ground, moving them one at a time to cache sites, on this particular day about six squirrels were moving up and down the tree quickly cutting terminal twigs that each contained a number of acorns. A few of the squirrels were then removing the acorns and moving them to nearby cache sites. Although Jeffrey Stratford, an established avian ecologist, was quick to doubt my explanation that the squirrels were racing the jays at acorn removal, I look forward to the day that I can present to him the published study that says otherwise.

Schmidt and Ostfeld (2008a) demonstrated that jays and squirrels interact at food sources when they found that foraging squirrels eavesdrop on foraging jays. These authors experimentally found that squirrels perceived a lower value of cached food when researchers played recordings of jay vocalizations near the cache site. Squirrels distinguished sites where playbacks were delivered with other sites where recordings were not played, indicating that squirrels were expecting higher pilferage rates in the presence of jays. To date, the effects of these interspecific interactions on the outcome of dispersal events has not been further explored.

JAYS AND OAK DEMOGRAPHY: SURVIVAL, GERMINATION, AND OAK ESTABLISHMENT

Studies on the fate of acorns dispersed by jays are surprisingly limited. Yet, despite sparse data on the subject, many authors, including myself, have suggested that because of the tendency to select less dense vegetation for caching, conditions for establishment and growth are generally higher in many of the habitats selected by jays (Johnson et al. 1997; den Ouden 2005; Steele et al. 2010). When artificial caches are used to simulate the fate of cached acorns, some authors have found significantly higher rates of germination and establishment in open habitats compared with woodland sites (Crawley and Long 1995). However, others suggest that acorns dispersed to more open habitats by jays may suffer high rates of seed predation by wild boar (Gómez 2003; Pérea et al. 2014) or rodents (Johnson et al. 1997). Additionally, when dispersed to sites where rodent predation is lower, acorns are more susceptible to desiccation (Kollmann and Schill 1996). Lichti (2012) followed radio-tagged acorns of red, white, and pin oak acorns, the majority of which were dispersed by blue jays, as well as several other birds. For each seed type and dispersal agent, Lichti followed the time to recovery of the acorn (disturbance rate), survival time, and its movement to a new location. Although he found no habitat preference for jays when caching acorns, Lichti did find that acorns cached in old fields were disturbed less often and survived longer than those stored in open, plowed fields or forested sites. However, in related studies in which artificial caches were created near jay caches and then compared with random caches, Lichti and colleagues (Sipes et al. 2013) found lower survival and establishment at the former. This is clearly the opposite of what is predicted if jays are, as frequently suggested, engaging in directed dispersal of acorns and exhibiting a relatively strong mutualism with the oaks (Bossema 1979; Johnson et al. 1997; den Ouden et al. 2005; Pesendorfer et al. 2016a). Although jays consistently cache acorns individually (Gómez 2003) in a few centimeters of soil often below the leaf litter, many studies have assumed this behavior results in a greater establishment rate. More data on seedling establishment and survival of acorns cached by jays are sorely needed to better understand the nature of this mutualism and the many factors that can strengthen or weaken this relationship. Despite the arguments that jays regularly engage in directed dispersal of the oaks (Pesendorfer et al. 2016a), more careful, quantitative analyses are needed.

Another aspect of oak establishment potentially involving jays is the exceptional ability of early germinating white oaks to sustain acorn removal following germination and early establishment. First reported by Bossema (1979), Eurasian jays frequently remove the acorns of germinating seedlings, potentially leaving the seedling intact (Korstian 1927). As reported in detail in previous chapters, my lab has reported strong experimental evidence that many of these oaks can sustain acorn removal, even exhibiting a specific trait (cotyledonary petiole) that appears to facilitate acorn removal (Yi et al. 2012b, 2013a, 2019). It may very well be that many white oak species are first dispersed and cached, the acorn later removed, and the seedling able to establish and grow. Thus, the ability of some oak species to sustain acorn removal speaks to an even stronger mutualism between oaks, jays, and rodents than previously considered. More field studies that explore this relationship are necessary.

LONG-DISTANCE DISPERSAL OF OAKS BY JAYS

One of the few studies to map individual home ranges of Eurasian jays was conducted by Rolando (1998) in which he radio-tracked jays and conducted other observational surveys of banded jays in Tuscany, Italy. Rolando reported that home-range sizes were smallest in spring and winter and largest from summer through autumn. In spring and winter, however, the largest ranges were associated with greater habitat heterogeneity. Patterns of space use within individual home ranges varied considerably between jays, but no signs of territoriality were observed throughout the study.

Rolando's (1998) fundamental study on home-range sizes and long-term movement patterns of jays serves as an important baseline for understanding the potential for long-distance dispersal events. At a minimum, Rolando's (1998) research suggests long-distance dispersal events are relatively rare. Documented dispersal events of 1–5 km reported by some authors (Johnson and Adkisson 1985) are certainly not inconsistent with these estimates of home-range size. However, evidence of significant long-distance dispersal events at the hand (bill) of jays comes not from watching jays disperse acorns but from paleo-botanical records indicating rapid movement of large-seeded trees, especially the oaks, following the retreat of the glaciers during the end of the Pleistocene, approximately 11,000 years ago.

As reviewed by Clark et al. (1998a), the early work of the British botanist, Reid (1899), presented an apparent disconnect between the rate at which the oaks and other trees, including beech and hickory, were dispersed and any plausible explanation of how this could be achieved—an enigma that Clark et al. (1998a) coined Reid's Paradox. For Reid, this disconnect followed from his effort to track the spread of early oak populations by dating fossil remains and leaf and acorn deposits in bog sentiments, a process today that is accomplished by dating pollen deposits (Delcourt and Delcourt 1987ab). Based on contemporary fossil pollen records, average range extensions for several tree taxa are estimated to vary from 100 m per year for American chestnut (*Castanea dentata*), and 350 m and 400 m per year for oaks and some of the wind-dispersed pine (*Pinus* spp.), respectively (see Webb 1986 for an overview, and references therein). Although such migration distances seem well within the normal capability of a jay, it is important to realize that these average range extensions are based on the pollen produced by adult trees. When an acorn is dispersed, it must germinate, establish, and grow to reproductive age before it can be detected in the fossil record and before it can produce acorns that can again be dispersed. This means that range expansions of these various tree taxa must involve significant long-distance dispersal events. The actual dispersal distances required to achieve the observed rates of range extension can be calculated by multiplying estimated dispersal distance by the generation time (time-to-reproduction) for each tree taxon. For the oaks, with an approximate generation time of 20 years, this means it would take an average dispersal distance of 7 km per year to achieve the estimated migration rate of 350 m per year (Webb 1986; but see Johnson and Webb 1989). As Webb (1986) points out, these dispersal distances are considerably greater for tree taxa (*Quercus*, *Carya* [hickories] and *Fagus grandifolia* [American beech]) with the largest seed sizes. Hence, this is the essence of Reid's Paradox.

Clark et al. (1998a) explained how Reid's Paradox was largely resolved by theoretical modeling of seed dispersal and additional contributions of paleo-ecologists. Above all else, the theoreticians' recognition that a dispersal kernel (the distribution of seeds around a parent source following dispersal) is best described as leptokurtic in which seed densities are highest at the source and then extend out with a "fat tail" to the right (Clark et al. 1998a; see Chapter 1). This fat tail in the dispersal kernel (distribution) in effect accounts for rare, long-distance dispersal events and further increases the rate at which a plant population could move (Clark et al. 1998a). Further contributions from paleo-ecology followed from the realization that more rapid tree generation times and greater reproductive output increase the rate of population spread. The obvious question to follow is what dispersal agent(s) were responsible for the rapid and long-distance movement of oaks following the retreat of the ice masses at the end of the Pleistocene. Several mammalian and avian candidates have been pro-posed—from black bears (*Ursus americanus*) (Clark et al. 1998a, and references therein) to passenger pigeons (*Ectopistes migratorius*) (Webb 1986).

Several years ago, my colleague, Peter Smallwood, and I shared a lengthy dis-cussion about the possibility that black bears, which consume large quantities of acorns when they are abundant, may disperse acorns as frugivores. We hy-pothesized that bears were likely to pass whole acorns in the feces, intact and able to germinate. We decided to try to test this with some captive bears at the Philadelphia Zoo. We planned to lace their meal with some acorns and then ex-amine their feces for evidence of whole intact acorns that we could then collect for germination trials—simple enough we reasoned. However, before we even sought permission to do this, I thought it might be wise to call my former post-doctoral advisor, Dr. Roger Powell, one of the world's leading carnivore experts. Roger had spent years studying black bear behavior so it seemed likely he might have examined just a few piles of bear feces in his career and would have an opinion on our plan. Indeed, he did. I asked him simply what is the probability that a whole acorn could be passed through the digestive system of a bear. His response was simple: "zero percent; acorns would come out looking like corn-meal." And that was the end of our experiment with bears.

Among the birds, two groups in particular are the most likely candidates for the rapid northward migration of the oaks: the jays and the passenger pigeon. Webb (1986) advanced a compelling argument that the movement of the oaks, and several other species, may have been the work of the now extinct passenger pigeon. In her review, Webb (1986) notes that passenger pigeons, once the most common bird species in North America, were estimated to number as many as five billion individuals. Given their extensive flock sizes, constant rapid, long-distance movement patterns, cosmopolitan diet of fruits of numerous plant species, including the oaks, and their ability to carry multiple fruits in the crop for long periods, it seems quite likely that this species may have contributed to the rapid advance of oaks (Webb 1986). Although this species never cached acorns and was considered exclusively a seed predator, it is entirely plausible that acorns were moved and dropped or partially consumed as we have shown for numerous other extant seed predators (see Chapter 6). Despite evidence that passenger pigeons would have only accidentally contributed to oak dispersal, the sheer size of their flocks and the number of seeds consumed makes their contribution to the dispersal process quite possible (Webb 1986).

The second most likely contributors to the northward movement of oaks are the jays. Johnson and Webb (1989) argue that the close relationship between the fagaceous tree species and the blue jay may likely explain the northward movement of these trees in North America during the Holocene. From the specific characteristics of the jays for consuming nuts, such as their flexible feet for holding them and a reinforced skull and bill for opening nuts of at least 14 species of four genera of the Fagacaea (see Table 1 in Johnson and Webb 1989), to their behavior of caching nuts in sites often suitable for germination and establishment, they certainly seem a likely candidate. In careful measurement of dispersal kernels created by European jays when dispersing and caching holm oak acorns in a heterogeneous landscape, Gómez (2003) identified two distinct patterns of dispersal committed by the same birds: short dispersal within oak stands and long-distance dispersal to other forest patches. Jays almost always cached under conifer trees and avoided open grassland, but because of the heterogeneous nature of their habitat, long-distance dispersal events occurred frequently. However, the average distance for these longer dispersal kernels averaged 100 m to 400 m with 1 km approximating the longest dispersal event (Gómez 2003).

Concrete evidence for directed dispersal by jays (i.e., higher oak establishment rates at jay cache sites) is still lacking (but see theoretical study by Purves et al. 2007). Evidence is also lacking for the long-distance dispersal events that could possibly account for the average 7 km per year (Webb 1986) that is estimated to be the dispersal rate necessary for the observed rate of the Holocene oak migration. Although more research on both of these important issues is still needed, other studies on the movement of extant oak forests seem to favor the jays. In Scotland, a recent study by Worrell et al. (2014) suggests that oak seedlings in birch (*Betula* spp.) stands and conifer forests are often abundant (400 per ha) with some patches reaching 1500 per ha. DNA analyses comparing potential parent trees and seedlings suggest dispersal distances of at least 1.5 km, still quite short of what is needed to account for historical movement of oak populations. However, far more compelling is the recent genetic work of Hampe et al. (2013) led by molecular geneticist, Dr. Remy Petit, of France. Hampe et al. (2013) examined a recently established population of holm oak, assumed to be the result of jay dispersal, located 30 km from the leading edge of the source stand. Based on these analyses, the authors found that the new population consisted of two nonoverlapping generations that originated from two source trees. These analyses also demonstrated that the first generation of oaks included four mature trees, all full siblings, thus originating from the same source tree. Yet, despite low levels of genetic variation in this first generation, the second generation showed increased variation, thus demonstrating distinct evidence of long-distance pollen dispersal. Hampe et al. (2013) concluded that long-distance establishment of oaks on the leading edge of an oak forest is limited by dispersal of the acorns, presumably by jays, rather than any constraint imposed by pollen dispersal.

The question of what species was responsible for the rapid northward migration of the oaks following the Pleistocene is particularly relevant today because of the strong evidence of the northward migration of many species that is driven by climate change. In most of the studies of the Holocene movement of the oaks, even those as early as Webb (1986) and Johnson and Webb (1989), authors point to the impending concerns of how the oaks will respond to current warming trends and whether jays are up for the task (see Chapter 14).

DISPERSAL OF INVASIVE SPECIES BY JAYS

Because of the distances that jays disperse seeds, they are likely to contribute to colonization and establishment of trees in new areas. Whereas this will certainly benefit native species, it can potentially contribute to the rapid spread of invasive species. Yet, while it is often assumed that jays may be responsible for dispersal of invasive nut species, this conjecture should always be carefully tested. Rodents are known to disperse both invasive sawtooth oak (*Q. acutissima*) in eastern deciduous forests of the United States (Steele, personal observation) and the northern red oak in Europe (Bogdziewicz et al. 2018c), but it appears that, in their invasive ranges, both oak species are sometimes avoided by jays. The size of sawtooth oak acorns appears to limit dispersal by blue jays in eastern North America. Similarly, Bieberich et al. (2016) found that in Germany, Eurasian jays avoided northern red oak acorns when presented in choice experiments with the native pedunculate oak and when presented alone. These authors concluded acorns of the invasive oak were rejected by jays because of their size, mass, and pericarp thickness. In contrast, in western Poland, Myczko et al. (2014) found that, although the probability of dispersal by Eurasian jay was twice as high for the native pedunculate oak than for the alien northern red oak, the latter regularly colonized in new areas, presumably as a result of jay dispersal. It seems, while dispersal of the alien oak occurs less often, the probability of establishment once dispersed may far exceed that of the native species.

Finally, I mention the definitive evidence that Eurasian jays, along with another corvid, the rook, contribute significantly to the spread of the invasive walnut (*Juglans regia*) in parts of Europe (Lenda et al. 2018). Several studies by Lenda and associates detail how the dispersal and establishment of this alien nut species is facilitated by the corvids. Their approach should be considered when exploring the effect of the corvids on the spread of invasive oaks.

THE CHALLENGES OF STUDYING ACORN DISPERSAL BY CORVIDS

To close this chapter, I shift the discussion to some of the challenges involved in following nut dispersal by jays. Just as jays respond differently than rodents to many seed characteristics, they also show markedly different patterns of habitat use when caching. Much of this is certainly due to their greater vagility and the scale at which they perceive their potential habitat compared to most rodents. It also appears that avoidance of rodent competition may also contribute to some of the specific habitat types selected by jays for caching. However, documenting patterns of habitat use by jays and, in particular, specific patterns of acorn dispersal in the corvids in many ways is far more challenging than similar research on rodents.

My colleagues, Nate Lichti and Robert Swihart and I, and the many students that worked with us from 2010 to 2014, learned this firsthand when we launched a study on the effects of forest fragmentation on acorn dispersal by blue jays. In this well-planned study supported by the US National Science Foundation, our goal was to compare patterns of acorn dispersal by blue jays in fragmented forests of the Midwest (central Indiana) with patterns in northeastern Pennsylvania where forests are far less fragmented and forest cover approaches 80% of the landscape compared to approximately 30% of cover in the Midwestern US. Our plan was to first radio-tag blue jays and then present them with native acorns of several species of oak (e.g., pin oak, red oak, and white oak) and then follow patterns of acorn dispersal, survival, and establishment in these contrasting forest landscapes. This was a carefully well-planned study, but as often is the case in science, the surprises, challenges, and procedural failures that usually do not make it into the peer-reviewed literature, kept Nate and the rest of us on our toes.

The first step was to mist-net jays and fit them with a radio-tag. Nate outlined the detailed procedures and carefully selected radio transmitters that were light enough to be carried on the back of a jay. Although the technology for radiotelemetry is rapidly changing, a typical challenge in most radiotelemetry studies is identifying the transmitter that weighs less than 5% of the animal's biomass, but also carries a battery large enough to cover the animal's

10.11. Shown are **(a)** research associate, Shealyn Marino, attaching a transmitter to a blue jay (*Cyanoccita cristata*). *(Michael A. Steele)* **(b)** The antenna extends beyond the bird's tail feathers. Straps on the backpack were designed to drop the transmitter in a few months just before expiration of the batteries. *(Shealyn A. Marino)*   SEE COLOR PLATE

home range and provide an adequate charge for the desired period of tracking. Nate selected a small transmitter that could be carefully strapped to the back of a blue jay with a long whip antenna that ran down and off of the jay's back (Fig. 10.11). The whip antenna increased the range at which the transmitter's signal could be detected. The battery life was approximately three months. The transmitter was first glued to the back of the jay and then carefully strapped around the jay's body. Nate developed a simple method of strapping the transmitter to the back of the jay (a backpack, as he referred to it) that secured it beyond the life of the glue and kept the transmitter snug until which time the straps rotted and the transmitter was shed from the jay. This allowed us to then recover the transmitter at the end of the tracking period.

With the radio-tagging procedures established, my research associate, Shea Marino, several students, and I eagerly set out to mist-net some jays and begin the tagging procedure early in the first year of the study. We started with two major field sites in northeastern Pennsylvania: one was my own property in Luzerne County, Pennsylvania, consisting of a hectare of open grass and meadow surrounded by nearly continuous forests, and a second similar site at Hawk Mountain Sanctuary in Berk's County, Pennsylvania. Study sites in Indiana were very similar in habitat but were surrounded by a markedly different level of forest fragmentation. In Indiana, the goal was also to find continuous forest stands as well as heavily fragmented stands in this larger heavily fragmented landscape.

The mist-netting procedure went flawlessly. Using playback calls of jays on one side of the mist-net, we were able to readily draw adult jays into the net.

They were then quickly removed from the net, fitted with transmitters, and released. We then allowed the typical period of about 10 days for acclimation to the transmitter before we began the formal tracking process. And this is where the trouble began. Shea returned from the field baffled and confused. After two days of searching, first where the jays were caught, and then in a continuously greater circumference of the tagging site, she was not able to detect a single signal of the five jays we had radio-tagged. After lengthy discussions with Nate and Rob and another day of unsuccessful searching, we concluded that our tagged jays had decided to pack up and migrate. Although jays are known to migrate even in large numbers, especially from the northern limit of their range in southeastern Canada, patterns of migration are highly irregular and poorly understood. The most plausible explanation, although not well-documented for blue jays, may be that in years of poor seed and nut crops they migrate in search of better food sources.

When blue jays do migrate, it appears that it occurs in the early part of the fall, and even then, not all individuals engage in the activity—some jays stay behind. So, we reasoned that perhaps we should wait to tag jays until later in the season after the migrants had departed. Perhaps our tagged jays were all destined to migrate, or the tagging process encouraged them to move on. Waiting a few weeks may mean we would catch resident jays that are less likely to migrate. Regardless, weeks after our tagged jays had migrated, we were still observing jays at our study sites, so, we repeated the process, captured a new group of jays, and fit them with radio transmitters. Although an expensive lesson (approximately $200 US per transmitter), it seemed to work. The tagged jays stayed on site and after the requisite acclimation period, Shea and a small gaggle of student technicians set out to begin the tracking process.

The precise methods used for radiotelemetry depend on the specific research question posed. If one is simply attempting to locate an animal in its nest, for example, all that is necessary is a single receiver hooked to an antenna and a pair of earphones (see Chapter 5). The researcher holds the antenna up in the anticipated direction of the tagged animal and moves it in a circular manner parallel to the ground. The transmitter, permanently tuned to a unique frequency (that is associated with that individual animal), releases a pulsed signal (every 1–2 s) that is then amplified by the antenna and transduced into an audio pulse that is detected by the receiver, which can be tuned to the same frequency or many others as needed. The researcher then listens for the strongest signal as she/he is moving the antenna. To identify a specific location where the tagged animal may reside or nest, one only needs to follow the strength of the signal.

However, telemetry is most often used to ask more general questions, such as, what is the individual animal's home range size, and how might that vary in different seasons or varying habitats? When asking these kinds of questions, it is necessary to locate the animal from a distance without disturbing it, and then continuously repeating the procedure to track patterns of movement. These locations can then be superimposed on to aerial photos or digital images of habitat to quantify patterns of habitat use. I know all too well the labor involved in conducting such studies because I spent several years tracking gray squirrels and fox squirrels (*S. niger*) in the southeastern coastal plain of North Carolina (see detailed descriptions in Steele and Koprowski 2001). To identify the location of a radio-tagged animal requires that one work from fixed locations, locate the strongest signal for the tagged animal, and thus the compass bearing from which the signal originates. The researcher must then quickly move to at least one or several more locations and do the same. These fixed locations, identifiable on aerials, must be distributed in relation to where the animal is located so that two of the compass bearings intersect close to a 90° angle to reduce error. Additional bearings would then allow closer estimation of the animal's precise location. After determining these two or more bearings, the researcher would then need to repeat the process a few hours later to determine if and to where the animal may have moved. When tracking my squirrels, I conducted three-day tracking periods in which one or more animals were tracked and located every 2.5 hours from sunrise to sunset for three consecutive days. The approximate 25 to 30 locations during one tracking period for each animal could then be used to calculate the home-range size and patterns of habitat use during that tracking period. For determining home-range sizes, the tracking interval was based on studies (Swihart and Slade 1985) that showed that this interval should be the approximate time the particular species is required to move from one end of its home range to another. This ensures that each observation is statistically independent of any other location during the tracking period.

Although we were not particularly interested in jay home-range sizes, we did want to monitor movement between habitats. This meant we still needed to use the standardized triangulation procedures so we could determine specific locations of jays. However, additional modifications to the procedures I had used for tracking squirrels were also necessary. When radio tracking squirrels, I was able to conduct the entire tracking procedure myself. Because squirrels spent most of their time foraging in one location, often in a particular tree, I was able to move between several tracking stations to take the several bearings necessary to calculate the squirrels' location. Tracking birds, however, is an entirely different endeavor. The birds can move across much of their home range quite rapidly, which makes it nearly impossible for a single researcher to accurately track a bird.

Shea and our team of students knew this and were prepared to overcome this issue—at least they thought they were. On the day we began tracking, we equipped three teams each with two observers, the necessary notebooks, a radio receiver, an antenna, earphones, and a small walkie-talkie so each team could communicate with the others when tracking. Each team was stationed at a location that could easily be identified from digital photographs (e.g., a road intersection) and all three teams were strategically positioned about 2–4 km from where the jay was first located in a way that encircled a much larger potential area of occupancy. We reasoned that each team could simultaneously take a bearing and we could then triangulate on the jay's location—or so we thought. In less than a day of tracking, my team encountered our next challenge in working with jays. We decided we would take simultaneous bearings about every 15 minutes so we could then map jay movements across the landscape. Shea would radio to each of the other two teams and at a specific synchronized time each team would take a bearing on the jay. It seemed to work sometimes, but more often than not, we encountered a common problem. Two of the bearings seemed to point to a common location and then the third bearing was off in a very different direction, sometimes in the opposite direction. We soon learned that within the few minutes of each team attempting to take a bearing, the jay could move, and within a few minutes, it could easily be found at the other end of its range. Handheld telemetry was simply not suitable for mapping patterns of habitat use by jays and most likely other corvids as well.

However, my thoughtful colleagues, Nate and Rob, were not about to let a bird outsmart us—a trio of mammologists. Almost immediately, Rob and Nate developed an alternative approach that had recently been employed at only one other location at the time: the Smithsonian Research Station in Barro Colorado Island (BCI) in Panama, a well-known ecological research station along the Panama Canal. The need for numerous investigators there to radio track a variety of vertebrates led to the development of an automated telemetry system that could regularly track radio-tagged animals instantaneously, while the researchers sat at a computer and followed the animals' movement. This was accomplished by connecting an automated receiver to each of several towers that extended above the canopy of the tropical forest. Each receiver was connected to an array of several antennae that could each provide a signal of varying strength when a particular transmitter was detected. The signal from all antennae on each tower would then be computationally averaged to provide a single direction to the tagged animal, and the signals from all towers were triangulated to provide the precise location of the tagged animal each time the receiver took a reading.

This was great, but certainly well outside our budget and that of most investigators. Moreover, our jays were often likely to move outside of a tracking area, so stationary towers would not likely work. Instead, my colleagues at Purdue came up with a brilliant alternative, which we soon thereafter adopted in Pennsylvania. We switched from handheld receivers to an automated system

10.12. Mobile automated telemetry station designed by colleagues at Purdue University and further adapted by staff at Wilkes University. Automated radiotelemetry unit, housed in sealed container on the trailer, is connected to six antennae on a collapsible tower. *(Shealyn A. Marino)*

much like that employed in Panama, which we housed in a toolbox attached to a flatbed trailer. Also attached to the trailer was a collapsible steel-framed tower to which we could attach a ring of six antennae that were then wired to the receiver (Fig. 10.12). The tower could be raised to 10 m but the whole system was mobile—it was a mobile version of the BCI system. Three such trailers could be positioned around the anticipated range of a jay and moved within minutes, if necessary.

With the receivers tuned to one or more animals' radio frequencies and set to record a single compass bearing for each jay, every minute, 24 hours a day, I felt like my colleagues and I had conquered the world or, at least, solved the latest challenge in our blue jay saga. Perhaps the most gratifying feeling was knowing that I was collecting data while I was in the classroom—or even better—while I slept. All we needed to do was replace memory cards in the receiving unit once a week and send them to Nate.

This tracking system, however, came with its own challenges—from the huge data sets to wade through, to the occasional receiver failure due to ant invasions or high humidity levels. However, we soon learned from researchers at BCI that the technical problems we encountered paled in comparison to those that frequently presented themselves in a tropical forest.

With data in hand, Nate was able to begin to explore movement patterns and, more importantly, acorn dispersal patterns. However, to follow acorn dispersal by tagged jays, we also had planned to place radio-tags in the acorns so we could determine precisely where acorns were cached and then follow recovery patterns by jays and competitors (e.g., gray squirrels). Radio-tagging acorns presented yet another set of challenges. First, we had to develop a method for placing a radio transmitter on or in an acorn. These radio-tags were substantially smaller than the jay tags, but how does one attach a radio-tag to an acorn, especially one the size of a large pin oak acorn? Two challenges in particular needed to be addressed: how to attach the radio so it remains on the acorn until the acorn is eaten, and how to attach the radio in a way that does not discourage the jay from even selecting the acorn. We knew from early attempts in my lab that attaching something to the outside of an acorn would not be effective. In past experiments, we tried gluing small magnets to the outside of acorns that we then presented to free-ranging squirrels. We reasoned that a magnetic locator could then be used to relocate tagged acorns. However, whenever squirrels encountered a magnet, they first peeled it off and then went about their business with the acorn. We did, however, eventually find other uses for the magnetic locator, such as placing magnets underneath artificially cached acorns to determined rates of pilferage.

Again, Nate offered a solution to how to radio-tag an acorn and my crew quickly adopted the procedure. The solution was to drill through the basal end of the acorn and remove enough cotyledon to be able to insert the transmitter. To accomplish this, it was first necessary to carefully wind the whip antenna around the body of the transmitter. Although this significantly reduced the range at which the transmitter could be detected, we accepted this constraint on the assumption that we would learn the approximate location of the cached acorns by following the tagged jays. Once the transmitter was nested in the remaining cotyledon, it was then possible to fill the gaps with the remaining debris mixed with an odorless glue and recreate a cap at the basal end of the acorn. The final product appeared as a sound acorn both to us and to jays, and in preliminary trials, jays readily accepted these acorns (see Chapter 8).

We initiated dispersal trials and within days discovered yet another challenge. When the majority of acorns presented to jays on our feeding platforms were radio-tagged, it took very little time for jays to learn that these acorns contained radios—not cotyledon. And they soon stopped recovering our carefully tagged acorns. After coaxing the jays back to the feeding platforms and conducting additional trials, we eventually learned that fewer than about 25% of the tagged acorns could be tagged without turning the jays away. The smart little buggers made me consider on more than one occasion that their intelligence far exceeded that of the rodents, especially when my wife noticed that the jays that

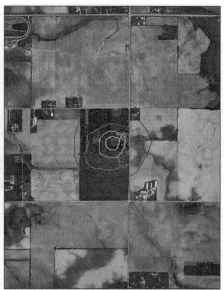

0 125 250   500 Meters

10.13. Home ranges of two blue jays (*Cyanocitta cristata*) in Tippecanoe County, Indiana, based on data collected by automated radiotelemetry: one where forest fragments occurred in relatively close proximity and one in which the occupied forest was isolated from other forests. Concentric circles represent utilization distributions with the highest probability of occurrence in the center of the distribution. *(Nathaneal Lichti)* SEE COLOR PLATE

we were working with on our property would regularly come to the tree outside our back door and vocalize when all the acorns on the feeding tray had been depleted. Although amused, she said it was also a bit unsettling.

Following one challenge after another, we were able to begin to regularly track jays and follow their dispersal of acorns. Over the course of three years in the field, we learned some important patterns of oak dispersal by jays, much of what we discussed previously (Bartlow et al. 2011; Lichti 2012; Richardson et al. 2013; Sipes et al. 2013). Although not yet published, we also found that the use of jay habitat, while feeding on and dispersing acorns, was highly dependent on forest landscapes, ranging from daily range sizes of less than 5 km^2 to greater than 20 km^2 (Fig. 10.13), depending on each jay's ability to move between forest fragments.

Our refinement of our radio-tracking procedures was still instructive in a number of ways, but this experience was an excellent reminder that the investment in time and money for perfecting procedures should always be weighed against the benefit of what is potentially learned. However, some of the answers to that question are still waiting for additional analyses.

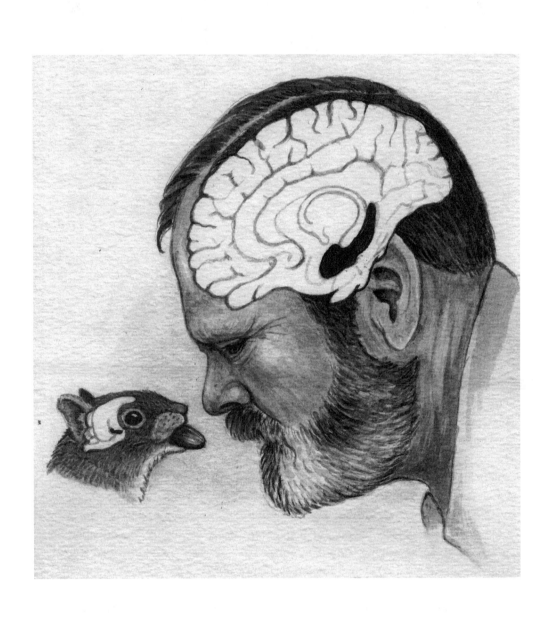

THE MIND OF THE SCATTERHOARDER 11

INTRODUCTION

My goal in this chapter is to provide an overview and synthesis of the growing body of knowledge regarding the cognitive ability of scatterhoarding animals to relocate and manage caches and their ability to thwart pilferage. Much of this research focuses specifically on the neuroscience of scatterhoarding, with far less attention devoted to understanding its relevance to seed dispersal. However, I will attempt a brief synthesis of such studies from which I posit direct implications for understanding the seed dispersal process. Because many of these studies have focused on the cognitive ability of rodents and jays, this research is especially relevant to oak dispersal. I will first dispel the long held notion that scatter-hoarding jays and rodents simply store seeds, forget where many are buried, and, in the process, contribute to plant establishment. Then, through a systematic account of studies from around the world, I will discuss how our current understanding of the cognitive ability of scatterhoarders suggests a different trajectory by which seedling establishment is achieved—one in which some scatter-hoarding birds and mammals consistently maintain priority over their own caches until preyed upon by a predator. In this argument, I will also introduce a novel twist on the predator satiation hypothesis, which suggests that not only will masting promote high rates of dispersal and caching during mast years, it also will increase the number of predators, and the predation of scatterhoarders, thereby promoting seed establishment. I also qualify this claim with the reminder that under certain habitat and environmental conditions, reciprocal pilfering between scatterhoarders may instead prevail (Vander Wall and Jenkins 2003).

A BRIEF HISTORICAL PERSPECTIVE

By this point in my treatment of oak dispersal, it should be abundantly clear that the rodents and corvids, the primary dispersers of acorns, are well adapted for recovering their scatterhoards. It should also be clear that the seed dispersal process in oaks depends on failure of cache recovery via one circumstance or another. Understanding these two conflicting demands—one favoring the scatterhoarder and the other favoring the oak—therefore requires an understanding of how well scatterhoarders are at remembering caches and the specific circumstances in which they fail at recovery.

The traditional view of cache recovery many decades ago was that scatterhoarders cache large quantities of nuts when available and simply forget where many are buried. For many years, the popular view was that squirrels and other rodents simply store more than they need and then they successfully recover only a portion of these stores. One of the first studies to investigate recovery of stored nuts by mammals (i.e., the western fox squirrel [*Sciurus rufiventer*]) was conducted by Cahalane (1942) in 1928–1929. He demonstrated that these squirrels were capable of recovering nearly 100% of man-made caches, which he concluded was accomplished by the detection of olfactory cues. Lewis came to a similar conclusion in 1980. However, it was not until McQuade et al. (1986) explored the relative importance of several cues (visual, spatial, and olfactory) that gray squirrels (*S. carolinensis*) rely on for recovering artificial caches and concluded that visual cues were most important, followed by spatial information, and then olfactory cues. Although the early models of scatterhoarding proposed by Stapanian and Smith (1978, 1984) implied that fox squirrels may have the ability to remember cache locations, their research did not address any specific sensory modality by which this was accomplished.

In his comprehensive review of food hoarding, Steve Vander Wall (1990) provides a detailed account of the various strategies used by the hymenopterans (wasps, bees, and ants), birds, and mammals for relocating storage sites. Any student of animal behavior is familiar with the classic studies by Nikolass Tinbergen (see review and original reference in Tinbergen 1972) on the European bee wolf (*Philanthus triangulum*), a parasitoid wasp that buries its prey (a bee) along with its egg in an underground chamber. The egg then hatches and feeds on this paralyzed prey. As demonstrated by a simple but classic experiment in behavior—one of the many that eventually earned Tinbergen a shared Nobel Prize in medicine—the female first excavates the chamber and then leaves the site to hunt for its prey. By simply manipulating three-dimensional objects around the nest, Tinbergen was able to demonstrate that even this short-lived wasp relies on the spatial distribution of objects around the nest for relocating it when the wasp returns to the nest to deposit its prey (the bee) and egg in the underground chamber.

Certainly, if a short-lived insect is able to do this, then it is likely that mammals and birds are capable as well. However, as Vander Wall (1990) notes, it is one thing to rely on landmarks or spatial information to return to one location (e.g., a nest), but a completely different task to remember dozens or even hundreds of cache sites. By the time of Vander Wall's review, however, it was already evident that at least several species of food-caching birds rely on spatial information for accomplishing this task. By 1990, there was either indirect or direct experimental evidence that the following avian species relied on spatial memory to relocate food caches: the Eurasian jay (*Garrulus glandarius*) (Bossema 1979; Bossema and Pot 1974), Clark's nutcracker (*Nucifraga Columbiana*) (Tomback 1980; Vander Wall 1982; Kamil and Balda 1985), marsh tit (*Poecile palustris*) (Cowie et al. 1981; Sherry et al. 1981; Shettleworth and Krebs 1982), and black-capped chickadee (*P. atricapillus*) (Sherry 1984). Although it was assumed at the time of these studies that birds were unlikely to rely on olfactory cues, by the time of Vander Wall (1990), it had been shown that olfactory cues were used for navigation by some birds. Regardless, many of these early studies on cache recovery by birds provided clear evidence that the four species above were using spatial memory to relocate caches.

In contrast to these early, rather definitive, studies on scatter-hoarding birds, by the time of Vander Wall's review (1990), there was little evidence for the use of spatial memory by mammals for relocating stored food. Exceptions included the study by McQuade et al. (1986) on gray squirrels, and another on red foxes by MacDonald (1976; see detailed review by Vander Wall 1990). In contrast, there was plenty of evidence that mammals relied on olfactory cues for recovering buried seeds and that soil moisture or damp seeds enhanced seed recovery by accentuating olfactory cues (Vander Wall 1998; Vander Wall 2000; Vander Wall and Jenkins 2003; Dittel et al. 2017). As reviewed by Vander Wall (1990), there was even speculation by a number of authors that several mammals, including two species of rodents, may rely on scent marking caches to increase the chances of recovery.

Then, in 1991, my colleague, Dr. Lucia Jacobs, and her coauthor, E. Liman, published seminal findings from Jacobs's doctoral work (1987), showing that when captive gray squirrels were given an opportunity to retrieve nuts they previously cached, as well as nuts cached in the same vicinity by other squirrels, they recovered significantly more of their own caches. This indicated the specific use of spatial information for relocating caches (Fig. 11.1). This was the first study in mammals to demonstrate the use of memory to relocate caches. A year later, Jacobs (1992) showed that Merriam's kangaroo rat (*Dipodomys merriami*)

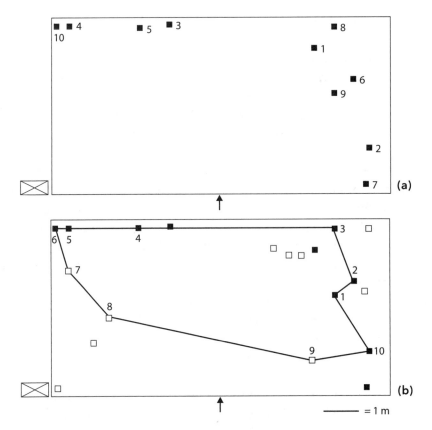

11.1. Schematic representation of **(a)** caching and **(b)** retrieval by the squirrel, Alvin (two-day delay). Solid black squares indicate Alvin's caches. White squares = caches of other squirrels. Numbers refer to the sequence in which nuts were cached or retrieved, the arrow indicates the location of the observer and the source of hazelnuts, and the rectangle in the lower left indicates the squirrels' place of entry into the arena. The polygon in **(b)** defines those caches considered, for the analysis, to be available during retrieval. *(From Jacobs and Liman 1991)*

revisited cache sites made 24 hours earlier even though all odors were removed, and only half of the cache sites still contained seeds. Jacobs's work not only validated the assumptions underlying the scatter-hoarding strategies previously modeled by Stapanian and Smith (1978, 1984), but also, in conjunction with some of the early work on corvids and other birds (Vander Wall 1990), laid the critical ground work for several decades of research from which an excitingly new picture of scatterhoarding emerged.

Today, the image of a scatter-hoarding bird or mammal now can be best generalized as an animal that stores food in individual locations during periods of food scarcity, remembers the location of many of these caches, and regularly engages in a range of manipulative, defensive, and investigative behaviors to manage seed caches in a way that maximizes long-term food availability. This behavior clearly involves a diversity of behavioral strategies that requires the ability to process complex spatial information. Throughout the remainder of this chapter, I discuss some of these strategies and the brain function that enables scatterhoarders to accomplish these tasks. Keep in mind that whereas many vertebrates and even invertebrates have evolved a diversity of sensory modalities for remembering sites and navigational paths, it is the scatter-hoarding birds and mammals that must manage hundreds of locations during a caching season. This capability may therefore reflect superior cognitive ability in these scatter-hoarding species.

COGNITIVE STRATEGIES OF SCATTER-HOARDING BIRDS AND MAMMALS

Among the many strategies employed by both birds and mammals for remembering cache locations, stationary (permanent) landmarks (e.g., logs, saplings, trees, rocks, and so on) seem to offer the most abundant and accurate information. Evidence of the use of landmarks, however, has been observed in captive studies in which temporary landmarks are experimentally manipulated by researchers. Such was the case in the early studies by Jacobs and colleagues.

The use of landmarks can be as simple as one object near the cache site or a number of objects that the animal may use to orient itself to remember the specific location of a cache. Figure 11.2 is a photograph of a young eastern black walnut tree (*Juglans nigra*) of about 15 years of age that established from the cache of an eastern gray squirrel. The animal cached the walnut at the base of a metal pin of an abandoned quoit (horseshoe) field in the author's backyard. The squirrel had likely dispersed the walnut from the nearest mother tree (approximately 30 m from the stake) and cached it at the base of the stake approximately 25 m from the nearest forest edge. The squirrel was most certainly using the metal pin as an obvious landmark at the time of caching.

I learned early in my research that while the occasional use of a specific landmark like the quoit stake maybe helpful, repeated use of identical landmarks would result in significant cache losses for a scatterhoarder. In our oak dispersal experiments, in which we followed the fate of metal-tagged acorns, we found that the placement of pin flags directly next to the cache site of whole intact acorns resulted in rapid removal of these acorns. Although we could not determine to what extend these caches were removed by the cache owner or other pilferers, it seemed almost certain that pilferers could easily associate pin flags with acorn caches. A more natural example in which seed consumers use individual landmarks to locate buried seeds occurs when rodents and jays

11.2. Found on the author's property, this Black Walnut tree sprouted from an unrecovered cache that had been buried next to a horseshoe pin. *(Michael A. Steele)*

learn to locate cached seeds by the emerging plumule of the seedling. (Bossema 1979; Vander Wall 1990, and references therein; Yi et al. 2013a). As my lab has shown, this appears to be relatively common in the early germinating white oak species, which appear to be well adapted for dealing with acorn pruning after germination and seedling emergence (Yi et al. 2013a, 2015, 2019; Zhang et al. 2014b). Seed predators that engage in such acorn pruning are most likely using the emerging seedling to locate seeds buried either by themselves or another scatterhoarder. One study to contradict the idea that naïve foragers would use seedlings or saplings as a visual landmark to pilfer cached seeds was conducted by Ribeiro and Vieira (2016) on the Azara's agouti, (*Dasyprocta azarae*), in the Brazilian Araucaria Forest. These authors reported that this agouti frequently cached individual seeds of *Araucaria angustifolia* (Araucariaceae) directly adjacent to saplings, which they then used as a landmark for recovery. Interestingly, the agouti regularly recovered caches made near these young trees, but when artificial caches were placed near saplings, naïve agoutis were not likely to pilfer these caches. Similarly, Zhang et al. (2016) found that captive Siberian chipmunks (*Tamias sibiricus*) preferred to prepare caches associated with a single landmark and were able to recover caches more quickly from such caches.

The use of landmarks to remember one's own cache locations, however, is often considerably more complex than the examples above. From Jacobs's studies and some of the earlier avian studies, it is evident that animals use several stationary objects to relocate cache sites. This suggests that scatterhoarders may rely on complex spatial information or, in effect, triangulate from three or more stationary objects to locate the cache. They appear to know the cache location in relation to several stationary objects in the environment. In the language of those that study cognitive abilities of food-hoarding animals, a single landmark, such as that described above, is considered a beacon (Gould et al. 2010). It is generally assumed that the use of this kind of landmark does not require spatial information. As reviewed by Gould et al. (2010), most scatter-hoarding birds and mammals, however, appear to depend on the spatial relationship between several landmarks, those that are relatively close to the cache (within a meter; proximal landmarks), or stationary objects that are a considerable distance from the cache (several meters or more; distal landmarks).

Perhaps one of the more impressive studies to demonstrate the importance of spatial information derived from stationary landmarks for relocating cache sites by a rodent was conducted by my friend and colleague, Dr. Pierre Lavenex (Lavenex et al. 1998), now a professor of neuroscience at Lausanne University, Switzerland, and a leading expert on the hippocampus. At the time of this study, Pierre was a postdoctoral fellow in the laboratory of Lucy Jacobs at UC, Berkeley. Working with the habituated western fox squirrels on the campus of the University of Berkeley, he and his coauthors trained individually marked squirrels to retrieve peanuts from concealed dishes on a 1 m × 1 m acrylic learning board. The board contained 25 small dishes distributed evenly across the board (Fig. 11.3). Each of the dishes was covered with a small aluminum cup-shaped cover to conceal the peanut or empty dish. Squirrels were first trained to come to the board and flip the covers to retrieve shelled peanuts from some of the dishes. After adequate training, the researchers conducted experiments to first present the squirrel with four hidden peanuts and allow the squirrels to search, locate, and eat them. The squirrel was then lured away from the board with a larger cacheable nut that was then carried off and either eaten or cached. When the squirrel returned, it was allowed to again search for peanuts that were placed in the same dishes as before. The goal was to determine if the squirrel could remember the location of the four dishes from which it previously recovered peanuts. Although based on just short-term memory, the study was designed to determine the precise type of spatial information used by the squirrels to remember the cache locations.

In one experiment, the squirrel was presented with a peanut in each of the four corner dishes and, at the same time, a black three-dimensional pipe was placed in the center of the board. The idea was to train the squirrel to associate the location of the peanuts by their spatial relationship with the black pipe. In another experiment, the squirrels were presented with peanuts placed in the inner four dishes on the board (Fig. 11.3) and these positions were associated

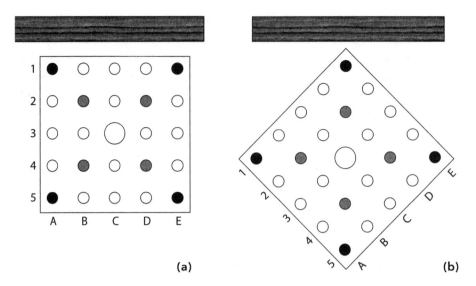

11.3. Shown are **(a)** presentation tray oriented parallel to nearby log and **(b)** presentation tray orientated 90 degrees to nearby log. *(From Lavenex et al. 1998)*

with a red cup in the center of the board. These two proximal (close range) objects in the center of the board were each associated with a different array of peanuts. In the field, squirrels were first tested to determine that they could distinguish between these two-colored objects. Because squirrels are dichromatic, they readily accomplished this task.

Squirrels quickly learned to associate the peanut arrays with the correct object in the center of the board. However, when the board was rotated 45 degrees on its central access in relation to its position when the squirrels were trained, squirrels were first unable to find the location of the peanut. The only explanation was that the squirrels recalled the peanuts' locations based on distant, stationary visual cues such as trees, buildings, and a log that was positioned a few meters from the test board during the experiments (Fig. 11.3). As Lavenex et al. (1998, p. 133) concluded, squirrels "relied on directional spatial information derived from distal environmental cues." The squirrels in effect remembered the spatial arrangement of objects around the food in the previous trial based on stationary objects in the environment, rather than on the obvious proximal cues presented on the array of dishes. This was an important breakthrough in our perspective on how scatter-hoarding squirrels may perceive their caches. And, as is the case in some of the best science, the relatively simple design in this experiment allowed the authors to uncover this impressive capability in squirrels. Lavenex et al. (1998) point out that while the use of distal spatial information was certainly how the squirrels located these artificial caches, they could have easily relied on olfactory cues, which were not masked in this experiment. This raised the possibility that rodents may shift strategies from spatial to olfactory modalities rather than using both simultaneously.

Stephen Vander Wall (2000) demonstrated quite clearly that rodents may shift from the use of spatial information to olfactory cues when locating scatterhoards. He followed patterns of cache recovery by both yellow pine chipmunks (*Tamias amoenus*) and deer mice (*Peromyscus maniculatus*) after caching seeds of Jeffrey pine (*Pinus jeffreyi*), and compared patterns of cache recovery for individuals relocating their own caches, or the caches of other individuals in an outside enclosure (10 m × 10 m), in the Carson Range of western Nevada. He found that both chipmunks and mice relied on spatial information to relocate their own caches, but then switched to olfactory cues to relocate caches of other individuals of both species. However, under dry conditions, rodents of both species could recover their own caches but not those of other individuals of either species. However, when seeds or soil were moistened, which enhances the odor of the cache, naïve rodents successfully pilfered caches of the other two species. Experienced rodents, tasked with recovering their own caches, were capable of doing so successfully under both dry and moist conditions. Vander Wall (2000) hypothesized that during frequent periods of dry conditions, individuals of these two species, which exhibit overlapping home ranges, likely recover and manage their own caches independently. But when conditions are moist, these rodents engage in a rather random pattern of reciprocal pilferage. Vander Wall and Jenkins (2003) argued that reciprocal pilferage, in which competitors cache their own food but also pilfer caches of their neighbors, is common and likely to evolve in many systems. However, when we explored the potential for eastern gray squirrels to engage in reciprocal pilferage behavior in an unnaturally high-density population, we found that individual squirrels had priority over their own caches and that pilferage was unlikely once the cache was established (Steele et al. 2011).

The use of multiple landmarks by mammals for remembering cache sites was first established by Barkley and Jacobs (1998). They tested the ability of Merriam's kangaroo rat (*Dipodomys merriami*) to recover cached food in two different environments: one associated with one landmark or another associated with 16 landmarks. Kangaroo rats showed similar recall ability after one day between caching and recovery in the two environments. But, after 10 days, rats recovered significantly more seeds from the more complex environment (i.e., 16 landmarks), indicating that the additional landmarks extended memory time.

One of the few studies on mammals to explore the potential use of spatial information for the overall long-term maintenance of scatterhoards was conducted by Hirsch et al. (2013) on the Central American agouti (*Dasyprocta punctata*). With the use of camera traps (placed to view individual cache sites) and radio transmitters to monitor free-ranging individuals of this large rodent, the authors were able to determine that cache owners were significantly more likely to repeatedly visit and independently approach their own caches than naïve competitors. The authors suggested by doing so the agoutis were monitoring cache status and at the same time they were recharging their memory regarding cache locations. Although the authors did not know the precise mechanisms by which spatial information was used by this rodent, this field study was a first step toward understanding how a free-ranging, scatter-hoarding animal could possibly manage caches over an extended period of time. In Bartlow et al. (2018b), we reported evidence that cached acorns were repeatedly recovered and re-cached by individual eastern gray squirrels over a five-month period (see Chapter 7). Based on a number of observations, we assumed that most of these cache recoveries were made by cache owners and that by revisiting caches, squirrels could determine the status of caches, consume partially damaged acorns, re-cache sound acorns, and, through the entire process, possibly recharge spatial memory. More studies are certainly needed to investigate the long-term maintenance of caches and the possibility that scatterhoarders refresh memory through this process.

Several of my colleagues and I recently explored the ability of a rodent to use spatial information to distinguish between their own caches and those of a competitor (Yi et al. 2016). We allowed individual Siberian chipmunks (*Tamias sibiricus*) to scatterhoard in 10 m × 10 m seminatural enclosures and then observed individual responses of chipmunks to their own caches and those of naïve conspecifics. Chipmunks recalled the location of their own caches for at least three days but consistently avoided them and instead visited and pilfered caches of competitors. Although this study was only conducted over a short period, it suggests that this scatterhoarder could easily remember its own caches in the context of those of others.

Scatterhoarders may also organize caches in a manner that improves their ability to remember caches and to recover stored items. Delgado and Jacobs (2017) recently demonstrated that western fox squirrels can organize scatterhoards spatially by nut type. These authors presented four nut types either randomly or by one nut type at a time, and from one central location, or from multiple locations. When retrieving nuts from a single location, the squirrels scatterhoarded nuts in separate locations so that each array of one nut species was relatively distinct from that of the others. The authors suggested that this behavior was consistent with chunking behavior—a behavioral strategy cognitive scientists suggest allows subjects to organize behavioral responses to improve memory. Although this ability for chunking seemed to decline when nuts were presented from multiple sources, it certainly suggests a cognitive strategy tree squirrels may use to recall cache types under natural conditions. It is interesting to consider how this behavior may contribute to the spatial arrangement of trees under different situations.

Research on the use of landmarks and spatial information by scatter-hoarding mammals (reviewed above), is not nearly as extensive as comparable inquiries on birds. Table 11.1 summarizes some of the studies on the use of spatial information by food-storing birds, especially the Clark's nutcracker, that was completed since the early breakthrough studies discussed earlier. Much of this research was conducted in the laboratory of Alan C. Kamil in the Department of Psychology at the University of Nebraska. Kamil and his collaborators have shown, for example, the ability to recall the location of hidden food for long periods (183 days), and the ability to rely on geometric shapes derived from the relative positions of landmarks to locate cache sites. And, whereas scatter-hoarding birds have the ability to use both proximal and distal landmarks when remembering caches, the closer cues appear to be more important (Goodyear and Kamil 2004). Moreover, there is evidence that Clark's nutcrackers may form a cognitive map of landmarks in relation to cache sites, or, alternatively, integrate vectors (triangulate) from distal stationary objects depending on the situation in which the birds are challenged. Clark's nutcrackers are also more efficient at relocating hidden locations when these cache sites are located within rather than outside landmark arrays. Finally, there is evidence that Clark's nutcrackers are often highly resistant to stimuli that may interfere with their ability to spatially relocate caches (but see Lewis and Kamil 2006). The collective results from this truly remarkable array of experimental studies demonstrates that Clark's nutcrackers are capable of processing and recalling precise locations of numerous individual caches.

Although many of these studies were conducted in captivity—an essential approach for deciphering the precise mechanisms used in processing spatial information—there is also evidence from the field that the locations of numerous cache sites are regularly revisited by these birds (Tomback 1977; Vander Wall 1990). Thus, the combination of both laboratory and field studies suggest a significant capability of relocating caches by birds, especially the food-storing corvids. And, whereas not all corvids regularly disperse acorns, some do; hence, these studies are likely transferable to the birds that scatterhoard acorns and disperse the oaks.

Importantly, there also appears to be some fundamental differences in the way birds and mammals process this spatial information. For mammals, it appears that distal cues may be more important than proximal cues (Lavenex et al. 1998), whereas the opposite may be true for birds (see review by Gould et al. 2010). Overall, mammals also may be more flexible in the use of various cues, relying on a combination of cues when available (Gibbs et al. 2007; Waisman and Jacobs 2008). Studies on captive flying squirrels (*Glaucomys volans*), for exam-

Table 11.1. Overview of use of spatial information by food-storing birds.

SPECIES	PRIMARY FINDING	REFERENCES
Eurasian jay (*Garrulus glandarius*)	Proximal (near) landmarks potentially more important than distal landmarks for cache recovery	Bennett 1993
Eurasian jay	Vertical edges (sticks in ground) are more important than horizontal edges (sticks across ground) for spatial memory of cache sites	Bossema 1979
Eurasian jay	Cache recovery highest when orientation of bird to cache site and when caching is more similar to orientation when recovering cache	Bossema and Pot 1974
Clark's nutcracker (*Nucifraga columbiana*)	Evidence of long-term spatial memory for cache sites; spatial memory began to decay only after 183 days	Balda and Kamil 1992
Clark's nutcracker	Spatial memory highly resistant to various forms of interference (e.g., retroactive and proactive interference)	Bednekoff et al. 1997
Eurasian jay, jackdaw (*Coloeus monedula*), marsh tit (*Poecile palustris*), and Eurasian blue tit (*Cyanistes caeruleus*)	Found that one species of corvid (European jays) and one species of parid (marsh tits) that scatterhoard use spatial information to relocate a food source, whereas another species of corvid and parid (jackdaw and blue tit) that do not store food did not use spatial information	Clayton and Krebs 1994 (see also Brodbeck 1994)
Clark's nutcracker	Evidence for the use of a cognitive map in some birds and "vector Integration" by others	Gibson and Kamil 2001a
Clark's nutcracker	Spatial memory for "cache site" (hidden food) is more accurate when site is located inside compared with outside landmark array	Gibson and Kamil 2001b
Several species	Review of strategies employed by scatter-hoarding birds and mammals; discusses contrasting use of spatial information	Gould et al. 2010
Clark's nutcracker	When proximal and spatial cues are both available, the proximal cues appear to be more important for remembering cache sites	Goodyear and Kamil 2004
Clark's nutcracker	Greater importance of global spatial information over local cues	Gould-Beierle and Kamil 1996
Clark's nutcracker	Use of local spatial cues vs. global spatial cues is flexible and dependent on previous experience	Gould-Beierle and Kamil 1999

SPECIES	PRIMARY FINDING	REFERENCES
Clark's nutcracker	Use of geometric relationship of landmarks used by birds over exact bearings (vectors)	Jones and Kamil 2001
Clark's nutcracker, jackdaws, and pigeons (*Columba livia*)	Nutcrackers superior at exploiting geometric information	Jones et al. 2002
Clark's nutcracker	Capable of accurate assessment of distance from global (distal) landmarks	Kamil and Cheng 2001
Clark's nutcracker	Preliminary support for the multiple-bearing hypothesis	Kamil et al. 2001
Clark's nutcracker	Definitive evidence of ability to learn geometric relationship between landmarks	Kamil and Jones 1997
Clark's nutcracker	Patterns of body movement prior to caching similar to movement during cache recovery, but not related to recovery success	Kamil et al. 1999
Clark's nutcracker	Evidence for use of exact bearings (vectors) over geometric relationship with landmarks	Kelly et al. 2008b
Clark's nutcracker	Birds vulnerable to proactive and retroactive interference during cache recover	Lewis and Kamil 2006

ple, show that this species chooses locations based on a strategy that averages the outcome of two of three spatial strategies (Gibbs et al. 2007). A similar flexibility in spatial cognition has been demonstrated for free-ranging fox squirrels (Waisman and Jacobs 2008). Yet, one should not discount the flexibility of birds in processing spatial information, which is clearly evident from the vast array of experimental procedures in which birds have successfully relocated stored resources (Gould et al. 2010). Cheng et al. (2007) further suggest the possibility that some scatter-hoarding birds and mammals may have the ability to integrate numerous bearings in a Bayesian framework from several locations by weighing differently the importance of specific cues.

CACHE PROTECTION STRATEGIES OF SCATTERHOARDERS

Scatter-hoarding birds and mammals are reported to engage in a variety of behavioral responses when faced with the risk of pilferage. Dally et al. (2006) provide a detailed review of the behavior and evolution of such responses, which I briefly summarize here. The reactions that food-storing animals show when they perceive an increase in pilferage risk include (1) an increase in the rate of food consumption compared with caching, (2) an increase, decrease, or cessation in the rate of caching, depending on the food-hoarding species in question, (3) increased aggression, (4) placement of caches out of sight of potential pilferers, (5) increased rate of re-caching in the presence of potential pilferers, and (6) increased rate of cache recovery followed by re-caching (Dally et al. 2006, and references therein). The studies reviewed by Dally et al. (2006) include observations on approximately 14 species, 11 of which are avian foodhoarders and only three of which are mammals. Six species of these avian scatterhoarders are corvids, five of which are likely dispersers of oaks (see Chapter 10). Among the three species of mammals, the white-footed mouse (*Permoyscus leucopus*) and the eastern chipmunk (*Tamias striatus*) contribute to oak dispersal in some situations.

Since the review and synthesis of Dally et al. (2006), several studies have reported on additional capabilities and strategies of scatter-hoarding birds and mammals to thwart pilferage (Table 11.2). Many of these more recent studies suggest both a high cognitive awareness and a more sophisticated response to such threats. Several studies out of the Clayton lab, for example, demonstrate that corvids can distinguish pilfering individuals from those that are less likely to pilfer and that they selectively respond to the former over the latter. Most recently, Kelley and Clayton (2017) showed that the California scrub jay, a highly social corvid, matches color of the food item with the caching substrate in the presence of a potential pilferer but does so only when acoustic cues (but not both acoustic and visual cues) of the potential pilferer are available. This suggests that the birds are highly discriminatory with respect to the activity of a potential pilferer and the future risk of pilferage. In the presence of a conspecific competitor, both the California scrub jay and the Eurasian jay selectively cache the item that potential pilferers find less desirable (Ostojic et al. 2017), which suggests that some corvids possess the ability to respond to a pilferer's knowledge, preferences, and likely intentions.

Compared to the innumerable studies on corvids, few studies on mammals have examined the specific responses to the presence of potential pilferers, especially under rigorously controlled laboratory conditions (but see Zhang et al. 2011). Nevertheless, several field and large enclosure studies illustrate the capability of rodents to manage the problem of pilferage or the future risk of pilferage (Table 11.2). Both the Korean field mouse and the Chinese white-bellied rat increase hoarding in the presence of a potential pilferer, and these two species along with Pére David's rock squirrel, striped field mouse, and ratlike hamster increase caching rates and disperse seeds farther when caches are robbed (Huang et al. 2011; Luo et al. 2014). In the event of a rapid food pulse, both free-ranging and captive Pére David's rock squirrels quickly sequester nuts to scatterhoards close to the food source, but then subsequently re-cache individual

Table 11.2. Cache protection strategies typically employed by food-storing mammals and birds.

SPECIES	PRIMARY FINDING	REFERENCE
Raven (*Corvus corvax*)	Pilfering ravens modify behavior and avoid pilfering in presence of conspecific cachers but not non-cachers	Bugnyar and Heinrich 2006
Clark's nutcracker (*Nucifraga columbiana*)	Nutcrackers engage in cache protection strategies in presence of potential pilferer	Clary and Kelly 2011
Clark's nutcracker	Social cues induce changes in caching, eating, and re-caching behavior	Clary and Kelly 2016
California scrub jay (*Aphelocoma californica*)	Jays use a combination of strategies (caching at greater distance, caching out of sight, and reaching) in presence of potential pilferers	Dally et al. 2005
California scrub jay	Jays alter caching behavior in presence of individual that watched them cache but not others that did not watch them	Dally et al. 2006
Corvids	Review of challenges of food-caching corvids and evolutionary responses to such challenges	Grodzinski and Clayton 2010
Pére David's rock squirrels (*Sciurotamias davidianus*), Korean field mice (*Apodemus peninsulae*), striped field mice (*A. agrarius*), Chinese white-bellied rats, (*Niviventer confucianus*), and ratlike hamsters (*Tscherskia triton*)	Complete pilferage of caches in field and captivity causes all 5 species to increase seed removal; the first 3 species which scatterhoard and larderhoard increase scatterhoarding and move seeds greater distances when caches are removed	Huang et al. 2011
Central American agouti (*Dasyprocta punctata*)	Agoutis repeatedly cache, recover, and re-cache palm seeds, and in the process increase dispersal distances to sites with low conspecific tree and seed density	Hirsch et al. 2012
California scrub jay	Jays match color of caching substrate with food color when only acoustic but not both acoustic and visual cues of potential pilferer present	Kelly and Clayton 2017
Florida scrub jay (*A. coerulescens*)	Re-caching behavior affected more by soil conditions (e.g., moisture content) than social context	Kulahci and Bowman 2011
Eastern gray squirrel (*Sciurus carolinensis*)	Squirrels trade-off predation risks for safer cache sites with lower pilferage risks via general rules of thumb than actual risks	Leaver et al. 2017
Eastern gray squirrel	Squirrels modify caching procedures in presence of conspecific but not heterospecific covid	Leaver et al. 2007

SPECIES	PRIMARY FINDING	REFERENCE
Eurasian jays (*Garrulus glandarius*)	Jays cached more food out of site in presence of conspecifics	Legg and Clayton 2014
Eurasian jay	Jays prefer more distant cache sites in presence of conspecific observer but not in absence of observer	Legg et al. 2016
Korean field mouse, Chinese white-bellied rat, Père David's rock squirrel	Complete cache removal of caches of a scatterhoarder, larderhoarder, and a species that engages in both hoarding strategies does not result in cessation of caching	Luo et al. 2014
Algerian mouse (*Mus spretus*) and wood mouse (*Apodemus sylvaticus*)	Acorns cached in open habitat to reduce pilferage by competitors	Muñoz and Bonal 2011
California scrub jay and Eurasian jay	Jays cache item that observers desire the least, demonstrating knowledge of potential pilferers' current preferences	Ostojic et al. 2017
Mountain chickadee (*Poecile gambeli*)	Chickadees choose hidden cache sites when observed by both conspecific and heterospecific pilferers	Pravosudov 2008
Eurasian jay	Jays altered caching, re-caching, and pilfering strategies depending on the presence of subordinate or dominant observer	Shaw and Clayton 2012
Eurasian jay	Jays cache less in "noisy" substrate in the presence of potential pilferers	Shaw and Clayton 2013
Eurasian jay	Pilfering jays use auditory cues in addition to visual cues when pilfering caches	Shaw and Clayton 2014
Eastern gray squirrel	Squirrels devalue cacheable food in response to vocalizations by blue jays (*Cyanocitta cristata*) when jays vocalize nearby but not at a distance	Schmidt and Ostfeld 2008a
Eastern gray squirrel	Squirrels engage in deceptive caching in presence of potential conspecific pilferers	Steele et al. 2008
Eastern gray squirrels	By selectively caching in open sites squirrels decrease probability of cache pilferage	Steele et al. 2011
Eastern gray squirrels	Squirrels selectively cached higher quality seeds farther from cover where predation risks are likely higher and pilferage rates are lower	Steele et al. 2014
Eastern gray squirrels	Measures of predation risks (via GUDs, giving-up densities) higher at cache sites than other sites	Steele et al. 2015

SPECIES	PRIMARY FINDING	REFERENCE
Wood mouse (*Apodemus sylvaticus*) and Algerian mouse (*Mus spretus*)	Rodents modify seed shadows by dispersing seeds from shrub habitat to open habitat	Sunyer et al. 2015
Florida scrub jay	Caching jays cache in hidden sights when observed by another jay; nonbreeding helpers were less likely to cache in presence of dominant jay	Toomey et al. 2007
Clark's nutcracker	Jays place caches farther from conspecific observers; light conditions do not influence caching decision in presence of potential pilferer	Tornick et al. 2016
Korean field mouse and Chinese white-bellied rat	Both species increase hoarding intensity in presence of conspecific observer	Zhang et al. 2011
Pére David's rock squirrel	In response to food pulse and risk of pilferage, both captive and free-ranging squirrels first rapidly scatterhoard nuts close to the source and then subsequently re- scatterhoard farther from the source and closer to the nest	Zhang et al. 2014a

nuts farther from the source and closer to the nest (Zhang et al. 2014a). Although this latter study was not designed to test whether these responses vary in the presence or absence of a potential pilferer, it logically follows that the most plausible reason for this behavior is to reduce pilferage risk and increase the probability of cache recovery.

A recent field study on the Central American agoutis found that this species, after caching palm seeds, will repeatedly recover and re-cache the same seed and, with each subsequent re-caching event, move the seed farther from its source, often across other agouti territories, to sites with decreasing conspecific tree and seed density. Whereas this behavior likely decreases the probability of pilferage by conspecifics, the agouti is also moving the seed to sites where it is less vulnerable to density dependent mortality and most likely to establish if not recovered from the cache. To accomplish this, however, the agouti must acquire considerable knowledge about the spatial distribution of seeds, trees, and competitors.

As I discuss in detail in Chapter 8, several field studies further demonstrate that rodents select sites with higher risk of predation for caching (Muñoz and Bonal 2011; Steele et al. 2015; Leaver et al. 2017) and that such cache sites are less likely to be visited and pilfered by competitors (Muñoz and Bonal 2011; Steele et al. 2014; but see Leaver et al. 2017), which may lead to a higher probability of establishment when cache owners are lost to predation (Steele et al. 2011; but see Sunyer et al. 2015) or fail to return to cache sites for other reasons. Although these studies do not directly address cognitive abilities, they certainly suggest that rodents are aware of the intent of pilferers, their perception of predation risks, and the probability of pilfering activity by conspecifics.

The one study that indeed suggests a higher cognitive ability employed by a scatter-hoarding rodent was conducted by my students and colleagues (Steele et al. 2008; see Chapter 9). Herein we established that gray squirrels engage in a form of behavioral deception—meeting all the criteria of tactical deception, previously considered the exclusive domain of the primates. Even at a conservative level, deception is perceived as a trait only exhibited by mammals with advanced cognitive abilities. In the presence of conspecifics, gray squirrels excavate a cache site, engage in the motions of planting a seed or nut in the cache site while holding the food in the mouth and appearing to vigorously push the food item into the cache, and then covering over the empty cache site, only to move off and cache the item in another location. We also show that it was possible to induce both deceptive caching and an array of other pilferage avoidance behaviors (e.g., moving caches greater distances to hidden or inaccessible locations) by experimentally robbing caches. My colleague, Dr. Lisa Leaver, and her associates (Leaver et al. 2007) report similar "audience" responses by gray squirrels in the presence of conspecifics but no evidence of deception as observed in our study sites. Interestingly, however, Leaver and her team never observed deceptive behavior in the introduced squirrels that they observed in the United Kingdom—an observation which has been recently confirmed by one of Leaver's former students, Dr. Pizza Kay Yee Chow. Dr. Chow has conducted recent behavioral studies at both the UK sites and one of our sites in the United States, but has only confirmed the deceptive caching behavior in the US site. Surely there may be additional insights to be gained by comparing the cognitive strategies of these native and introduced gray squirrels. Regardless, the behavior of deception in a rodent suggests that mammals possess some sophisticated strategies for dealing with cache pilferers. Perhaps we have just begun to understand the capabilities of the rodents and other mammals.

STRATEGIES OF PILFERERS

Among the strategies used by pilferers to raid caches, only a few seem common: random searches, olfactory searches, a combination of random and olfactory searches, and observational spatial memory. The latter is the ability to observe a conspecific scatterhoarding and then subsequently recall and raid the cache. Bednekoff and Balda (1996a) reported that the highly social pinyon jays (*Gymnorhinus cyanocephalus*) are capable of doing this under some circumstances. Similar studies with pinyon jays and Clark's nutcrackers reported evidence that both of these species are also capable of observational spatial memory, but the more solitary nutcracker appears less efficient at the task and less likely to remember locations a day or more after observations (Bednekoff and Balda 1996b). It has also been shown that when given the opportunity to observe conspecifics through a peep hole, western scrub jays were more likely to do so when those being observed were engaged in caching rather than any other activity (Grodzinski et al. 2012). Shaw and Clayton (2014) were the first to demonstrate that a scatterhoarder (Eurasian jay) could rely on acoustic information to locate and pilfer a conspecific's caches.

Other studies in mammals and food-storing birds other than corvids have failed to demonstrate the capability of some species to rely on spatial observations for pilfering (see review by Dally et al. 2006, and references therein), whereas at least one other study demonstrates that the eastern gray squirrel can use both social and nonsocial cues to locate hidden food and that they appear more efficient at using the former (Hopewell et al. 2010). In fact, the authors in this study were able to demonstrate that when using social cues, feature negative cues (e.g., failure of cache recovery) conveyed the strongest information when observing conspecifics.

As discussed by Dally et al. (2006; Bugnyar and Kotrschal 2002), both olfactory searches and observational spatial memory are likely to result in an evolutionary arms race, in which the pilferer and the scatterhoarder directly evolve behavioral and, potentially, cognitive strategies, in response to that of the other. Indeed, preliminary evidence for such an arms race follows from recent studies on corvids and squirrels. Some corvids use auditory cues (Shaw and Clayton 2014), specific visual cues, and even the prey color to locate and pilfer caches, whereas the hoarding subjects are capable of masking each of these cues in the presence of the pilferer (Table 11.2). Likewise, eastern gray squirrels rely on spatial observations to pilfer conspecific caches, and, in response, hoarding squirrels modify their behavior in the presence of potential conspecific pilferers (Leaver et al. 2007) and even engage in deceptive caching behavior to distract would-be pilferers (Steele et al. 2008).

EPISODIC MEMORY

The remarkable ability of food-hoarding birds and mammals to process complex spatial information when managing caches and guarding against potential pilferers raises the question of just how sophisticated the memory of food-storing animals can be. This very question led Clayton and Dickinson (1998) to test whether Florida scrub jays (*Aphelocoma coerulescens*) are capable of episodic memory—the ability to remember the where, what, when (and potentially who and why) of a particular event—a capability that was long assumed the exclusive domain of humans (Tulving 1984). Clayton and Dickinson showed that scrub jays will remember cache locations of both preferred, perishable insect larvae (wax-worm) and nonperishable, shelled peanuts but are also aware of the time they were cached. In training trials, birds were given the opportunity to learn that both peanuts and larvae were still intact when either was recovered four hours after caching. And, in contrast, birds learned that after 124 hours of caching, wax-worm larvae died, decayed, and became unpalatable. When birds were given the choice of recovering either item (peanut or worm) after four hours of caching, they selectively recovered the worms over the less preferred peanuts. However, when presented with the same opportunity 124 hours after caching, birds selectively recovered peanuts. Taken together, these results demonstrated that the birds were simultaneously processing the "where," "what," and "when" about their caches, and thus demonstrating episodic memory.

THE HIPPOCAMPUS

No discussion on spatial memory and cognitive capabilities of food-hoarding animals would be complete without a brief look at the hippocampus—that part of the vertebrate brain most important for short- and long-term memory, including the storage of spatial information. The hippocampus, named for its general morphological similarity to that of the sea horse (*Hippocampus* spp.), is part of the limbic system. In the mammalian brain, it is located under the cerebral cortex and is composed of the hippocampus proper and the dentate gyrus.

The importance of the hippocampus to scatter-hoarding birds and mammals was first brought to light by two seminal empirical studies by Sherry et al. (1989) and Krebs et al. (1989). Sherry et al. (1992) summarized how early comparative studies had clearly demonstrated how hippocampus size was significantly larger in food-hoarding versus non-hoarding passerine birds, polygynous male rodents versus females of the same species (see the breakthrough study by Jacobs et al. 1990), and in homing pigeons versus more sedentary domestic pigeons. In all cases, it appeared that selection had favored the species or sex that was most dependent on processing spatial information (Sherry et al. 1992). Soon thereafter, a review by Krebs et al. (1996) summarized the major avenues of inquiry that were emerging from research on the relationship between the hippocampus and food-storing behavior in birds and mammals. These included (1) comparisons of both the brain and behavior as it relates to both phylogeny and ecology of food hoarding, (2) the development of the brain (hippocampus) in relationship to food hoarding demands, and (3) seasonal variation in food hoarding and its relationship with the hippocampus. This overview synthesized the current knowledge on the subject and, at the same time, presented a road map for future inquiries.

The breakthrough studies by the Sherry and Krebs research teams, followed by their respective reviews on the hippocampus of food-hoarding animals, ignited an explosion of research on the subject that has continued to the present day. In Table 11.3, I provide a modest review of some of these studies in an attempt to characterize the direction this research has taken the field in the past three decades. To underscore both the breadth and historical perspectives of these investigations, the table is organized around major subjects and avenues of inquiry, and studies are presented in chronological order. I emphasize that this is by no means an exhaustive survey, but one intended to only equip you with a reasonable overview of this important area of investigation. The table also permits a brief narrative on the subject here. As this area of research is quite extensive, there are certainly some important papers I have overlooked. Hence, this should not be viewed as a definitive overview on the subject, as that work remains to be done. In a review of the hippocampal function in food-hoarding birds, Smulders (2006) outlined key approaches necessary to better understand the relationship between hippocampus function and the ecology and evolution of these species' behavior. He emphasizes the need to link corvid behavior with cognition and behavior in a natural setting, to compare these behavioral and cognitive traits, such as hippocampus function, with those of non-hoarding species, and to link physiological and morphological characteristics of the hippocampus with the behavior of food-hoarding species.

Comparative studies of the relative size of the hippocampus in hoarding and non-hoarding species (primarily birds) characterize much of the early work on attempts to understand the importance of hippocampal function in wild scatterhoarders (Table 11.3). For nearly 25 years, comparisons of the hippocampus of scatter-hoarding and non-hoarding avian species demonstrate that the relative size or volume of the hippocampus is consistently larger in the former. However, because size is only a proxy for the ability to process spatial information, this approach led to the criticisms that this was a weak, even adaptationist's approach to understanding hippocampus function in wild scatterhoarders. This comparative approach soon gave way to subsequent studies demonstrating that scatter-hoarding species with larger hippocampi also show higher neuron numbers, densities, and connectivity in the hippocampus. By 2010, a critical review on the subject underscored the efficacy and preference of the latter approach (Roth et al. 2010).

Another research direction that proved effective was comparative studies of the same species under different environmental conditions. Although still a comparative approach, these studies offered a more powerful way of showing how hippocampus structure and function may vary under different selective regimes. Studies on black-capped chickadees (Table 11.3), for example, were particularly critical for showing how hippocampus size and neuron number varied

Table 11.3. Overview of studies examining the relationship among hippocampus, spatial memory, and food-storing behavior. The table is organized by several topics of investigation: comparative studies, seasonal variation in hippocampus size, hippocampal manipulation, hippocampal development, neurogenesis, and reviews.

SPECIES STUDIED	FINDINGS	REFERENCES
Comparative Studies of the Hippocampus		
Red-billed blue magpie (*Urocissa erythrorhyncha*), European crow (*Corvus corone*), rook (*C. frugilegus*), jackdaw (*C. monedula*), Eurasian jay (*Garrulus glandarius*), magpie (*Pica pica*), and alpine chough (*Pyrrhocorax graculus*)	Correlation between relative hippocampus volume and amount of time each of the 7 species spends storing food	Healy and Krebs 1992
Merriam's kangaroo rat (*Dipodyms merriami*), banner-tailed kangaroo rat (*D. spectabilis*), and Ord's kangaroo rat (*D. ordii*)	Size of hippocampus (relative to total brain) is largest in species (*D. merriami*) that scatterhoards more; hippocampus also larger in males (vs. females), which range farther than females	Jacobs and Spencer 1994
Gray-breasted jay (*A. wollweberi*), scrub jay, Clark's nutcracker, and pinyon jay (*Gymnorhinus cyanocephalus*)	Clark's nutcracker possesses largest hippocampus (relative to body and brain size) compared to other 3 species of corvids (but see Pravosudov and de Kort 2006)	Basil et al. 1996
Review and synthesis of comparative studies of hippocampus of parids and corvids	Synthesis argues scatter-hoarding species (1) possess more accurate and long-term spatial memory compared to species that do not hoard or larderhoard, and (2) are more resistant to memory loss in part due to greater neuron production when scatterhoarding	Clayton 1998
Coal tit (*Parus major*) and blue tit	Food-hoarding coal tits with greater hippocampus volume performed better in tests of memory over time (but not tests of memory capacity and resolution) compared with non-hoarding blue tits	Biegler et al. 2001
Comparison of hoarding ability in numerous corvids and parids in relation to hippocampus size	By controlling for continent (because hippocampus size varies between NA and Europe), authors confirmed previous studies finding hippocampus size is related to caching ability	Lucas et al. 2004
Numerous species considered	Review countering criticism that neuro-ecological approach to understanding hippocampus function involves adaptationist's perspective	Healy et al. 2005
Black-capped chickadee	Hippocampus size and neuron number increase in relation to severity of environmental conditions across gradient of environmental severity (Kansas to Alaska)	Roth and Pravosudov 2009

SPECIES STUDIED	FINDINGS	REFERENCES
Review of several studies	Critical review of studies on hippocampus size (volume) relative to the need for spatial memory; neuron number, size, and connectivity may be better; measures of how brain structure relates to memory and behavior than to hippocampus size	Roth et al. 2010
Black-capped chickadee	Same results as Roth and Pravosudov (2009) after controlling for day length; harshness of environmental conditions but not day length lead to larger hippocampus and more neurons in hippocampus	Roth et al. 2012
Numerous species	Review concludes that selective pressures associated with scatter-hoarding behavior have shaped spatial memory and neural mechanisms controlling it	Pravosudov et al. 2013
Clark's nutcracker, pinyon jay, Western scrub jay, blue jay, and azure-winged magpie	Evidence of larger hippocampus and greater neuron number in scatterhoarders vs. species that do not store food	Gould et al. 2013

SPECIES STUDIED	FINDINGS	REFERENCES

Seasonal Variation in Hippocampus Size

SPECIES STUDIED	FINDINGS	REFERENCES
Black-capped chickadee	Hippocampus of chickadees largest in October when food-hoarding most intense	Smulders et al. 1995
Two South American cowbird species	Hippocampus of both species largest during breeding season	Clayton et al. 1997
Eastern gray squirrel	No seasonal variation in relative hippocampus size, but male hippocampus larger than female	Lavenex et al. 2000a
Song sparrow (*Melospiza melodia morphna*)	No seasonal change in size of hippocampus of males of this non-hoarding species	Lee et al. 2001
Primary focus on black-capped chickadee; other species discussed	Review of seasonal changes in hippocampus size in food-hoarding birds and its relationship to seasonal changes in neurogenesis	Sherry and Hoshooley 2010

SPECIES STUDIED	FINDINGS	REFERENCES
Hippocampal Manipulations		
Black-capped chickadee	Hippocampus damage via aspiration reduced memory for location of caches and "working memory"	Sherry and Vaccarino 1989
Mountain chickadee (*Poecile gambeli*)	Experimental variation in food supply does not result in variation in hippocampus size or neurogenesis	Pravosudov et al. 2002
Laboratory rodent	Review showing how new hippocampal lesion techniques (as well as focused pharmacological and molecular genetic treatments) identify how specific regions of the hippocampus store, encode, and retrieve specific spatialinformation	Martin and Clark 2007
Laboratory rodent	Lesions resulted in disruption of processing of self-movement cues in a foraging task, but rats could compensate with environmental cues	Martin and Wallace 2007

SPECIES STUDIED	FINDINGS	REFERENCES
Hippocampal Development		
Magpie and jackdaw	Relative volume of hippocampus not different in juveniles of these 2 species, but hippocampus of adult, food-storing magpie larger than that of jackdaw; neuron number greater in adult magpies than adult jackdaws	Healy and Krebs 1993
Marsh tit and blue tit	Hippocampus size was larger in adult marsh tits than blue tits but not different between nestlings of the 2 species; adult marsh tits had more neurons than that of adult blue tits; neuron density highest in nestlings of both species compared with adults	Healy et al. 1994
Marsh tit	Birds allowed to store and retrieve food had larger hippocampi, more neurons, and fewer apoptotic cells than birds not afforded the same experience	Clayton and Krebs 1995
Marsh tit and blue tit	Memory stimulates hippocampus growth even late in juvenile development	Clayton 1995
Marsh tit	Limited food storing and recovery experience is all that is needed to promote hippocampus development	Clayton 1996
Mountain chickadee	Birds provided with both experience to store and recover food had larger hippocampi than those with limited experience; a threshold level of experience may be needed to promote hippocampus development	Clayton et al. 2001

SPECIES STUDIED	FINDINGS	REFERENCES
Marsh tit and blue tit	Study shows the act of retrieving (but not necessarily storing) food triggers hippocampus growth in marsh tits but not in blue tits	Clayton 1995

SPECIES STUDIED	FINDINGS	REFERENCES
	Neurogenesis	
Black-capped chickadee	New neurons produced in hippocampus throughout the year, but production highest at peak of food-storing season (October)	Barnea and Nottebohm 1994
Black-capped chickadee	Neuron numbers higher in juvenile hippocampus than that of adults, possibly due to elevated environmental novelty for juveniles	Barnea and Nottebohm 1996
Marsh tit	Birds permitted to cache and recover food showed greater cell proliferation, cell number, and neuron number in hippocampus than birds not allowed to store and recover food	Patel et al. 1997
Eastern gray squirrel	No seasonal variation in neuron cell production or total neuron number in dentate gyrus of the hippocampus	Lavenex et al. 2000b
Review of studies on several species	Provides overview of 2 avian models for understanding neurogenesis and learning, with 1 based on food hoarding; also provides critical analysis of fundamental differences of neurogenesis in juvenile and adult birds and mammals	Lavenex et al. 2001
Black-capped chickadee	Number and relative number of immature neurons in the hippocampus increase across latitudinal gradient of environmental harshness	Chancellor et al. 2011
Several species	Includes review of evidence that food-hoarding birds possess larger hippocampi with more neurons and more active neurogenesis	Barnea and Pravosudov 2011
Siberian chipmunk (Tamias sibiricus)	Variation in degree of scatterhoarding by individual chipmunks correlated with neuron production and survival in the hippocampus	Pan et al. 2013

SPECIES STUDIED	FINDINGS	REFERENCES
	Reviews	
Chickadees and titmice	Food storing and recovery	Pravosudov 2007
Chickadees and titmice	Food storing and recovery	Sherry and Hoshooley 2007

across a latitudinal gradient of environmental severity (Roth and Pravosudov 2009), even when investigators controlled for photoperiod (Roth et al. 2011). Similarly, studies on seasonal changes in the hippocampus of scatterhoarders, but not non-hoarders, demonstrated that some scatter-hoarding bird species exhibit a larger hippocampus during the time of year when scatterhoarding is most intense. Here again the black-capped chickadee proved to be a model study organism, and by 2010, Sherry and Hoshooley were able to provide a review of several studies showing how both hippocampus size and patterns of neurogenesis vary with season.

Other approaches, including studies on hippocampus development and manipulation of the hippocampus, such as experimental damage to the structure, further strengthened the understanding of how the hippocampus is tied to the capability of scatterhoarding and cache recovery (Table 11.3). Several studies from the Clayton laboratory on marsh tits, for example, demonstrate that this species experiences greater hippocampus growth (compared with no-storing blue tits) soon after the nestling stage. In addition, limited food-storing experience is all that is needed to promote development of the hippocampus in marsh tits (Clayton et al. 2001). Finally, studies on neurogenesis have reported higher neuron numbers and higher rates of neurogenesis in scatter-hoarding birds, especially during the food-hoarding season (e.g., Barnea and Nottebohm 1994).

Although these studies on avian scatterhoarders demonstrate clear evidence of seasonal changes in hippocampus size and neurogenesis, at least two publications (Lavenex et al. 2000a, 2000b) in eastern gray squirrels suggest that the processes of development and neurogenesis in the hippocampus may differ in some mammals, or in mammals in general. Around the time research on structure and function of the hippocampal complex in food-hoarding animals was gaining momentum, I was fortunate enough to collaborate with Lucy Jacobs's team and, Pierre Lavenex, whose research on spatial memory in fox squirrels I discussed earlier. Working with eastern gray squirrels collected from some of my study areas, we conducted a detailed examination of the seasonal variation in the hippocampus complex of this long-lived mammal. In the course of this research, we observed a pattern of hippocampal growth and neurogenesis in eastern gray squirrels that was not consistent with many of these earlier studies (Table 11.3). We examined two hypotheses: (1) that the relative hippocampus size and neuron number in the hippocampus complex peak during food hoarding, and (2) that male squirrels, which typically range farther than females, show greater hippocampus size and neuron number. We collected 30 squirrels in and around two locations in northeastern Pennsylvania across three periods between October 1996 and June 1997: October (the peak in scatterhoarding), January (the primary breeding season), and June (when home ranges are generally smaller) (Thompson 1978).

Although we found evidence of seasonal changes in overall brain size, we found no seasonal changes in relative size in the hippocampal complex of adult squirrels, causing us to reject the first hypothesis. In contrast, however, we did observe a significantly larger volume of one part of the dentate gyrus in males compared with females. This supports previous findings that such differences in hippocampus structure between males and females of some mammal species may correspond with differences in home-range size between the sexes. In a second paper, we also reported no evidence of seasonal variation in neurogenesis or neuron number in adult gray squirrels, even during the season of peak scatterhoarding (Lavenex et al. 2000b). Consequently, these studies led us to conclude that there was no evidence of seasonal changes in size and neurogenesis of the adult hippocampus of this long-lived rodent and that perhaps such changes reported in other short-lived species may reflect developmental changes. Although producing largely negative results, these studies were important in cautioning about drawing conclusions about mammals from research on the corvids.

COGNITION AND SEED DISPERSAL

From this chapter it is evident that the typical scatter-hoarding mammal or bird possesses the cognitive facilities to recall many cache locations based on spatial memory, manage caches over time, disrupt, deceive, or avoid the activities of competitors/pilferers, and engage in effective pilfering themselves. With the emergence of many new techniques and breakthroughs in the study of social cognition, brain function and neurogenesis, the past two decades, have witnessed a growing array of new perspectives on the ability of scatter-hoarding animals to manage food stores. Hence, now is the time to integrate these findings with field studies in a way that places these impressive capabilities in the context of the natural world and, ultimately, the process of seed dispersal. The next step is to better understand how spatial memory and cognitive strategies ultimately relate to seed dispersal.

Looking ahead, several key questions regarding cognitive strategies of scatterhoarders seem most relevant to the process of seed dispersal, and, in particular, oak dispersal and oak regeneration. Here, I suggest just a few. First, we need to know the extent to which caches are recovered through the scatterhoarding season. Are caches recovered and re-cached as suggested by Bartlow et al. (2018b) and is memory recharged through the season as scatterhoarders manage food reserves or does intense pilferage predominate in many systems (Vander Wall and Jenkins 2003; Jansen et al. 2012)? Second, it will be important to better understand how cognitive strategies of rodents and corvids differ and how these differences relate to patterns of cache selection, cache recovery, patterns of pilferage, and seedling establishment. Building on the research of Delgado and Jacobs (2017), it also would be interesting to know how seeds or nuts of different utility (value) are stored and recovered over time by cache owners. Is greater energy invested in management, memory, and recovery of stores of greater value? Perhaps stores of greater value are cached at sites that provide better spatial information. I also suggest that more information is needed on scatterhoarders' cognitive awareness of both pilferers and predators, as awareness of each of these may influence seed fates.

Finally, given the knowledge to date on the impressive capability of scatterhoarders to remember and potentially manage their caches, I propose a conceptual scenario of scatterhoarding that requires further exploration. I suggest that with some notable exceptions (Vander Wall and Jenkins 2003), scatterhoarders generally maintain ownership over their own caches, placing caches where pilferage is less likely, remembering the precise location of many of their caches, and recovering and re-caching foods through the storage season. Moreover, as I have shown for eastern gray squirrels (Steele et al. 2011, 2014, 2015), scatterhoarders may trade off higher predation risks when caching to reduce pilferage rates. Whereas this behavior will consistently reduce pilferage rates, it will also result in the occasional loss of scatterhoarders to predators, and the likely establishment of seedlings from these caches. Predation happens, and if it happens slightly more often during the scatterhoarding season in a manner as described here, this may be a process by which seeds are successfully dispersed (see Chapter 15).

THE INSECTS: ACORN PREDATORS, PARASITES, OR COMMENSALS? 12

INTRODUCTION

I now turn to the extensive but incongruent information on the insects that feed on acorns and how they influence acorn survival, dispersal, and establishment. I will first review the general community of invertebrates (primarily insects) found in acorns in different oak ecosystems, discuss their collective impact on acorn survival, and highlight the relative frequencies of the insect community occurring in acorns of a few oak species in northeastern deciduous forests of North America. I also will compare insect damage in acorns from this long-term data set with observations from acorns of numerous oak species from central Mexico where damage is marked by extremely high infestation rates that most certainly limit acorn survival, dispersal, and forest regeneration.

After highlighting some of these generalities, I will then shift to a more detailed and rigorous evaluation of the impact of weevils (e.g., *Curculio*), arguably the most detrimental insect predator of acorns worldwide. I will review evidence that chemical gradients within acorns direct weevil damage away from the acorn embryo, and I will summarize several experiments demonstrating how acorns successfully survive partial damage by weevils. I also will use two long-term data sets collected in my lab to evaluate specific hypotheses regarding weevil damage. I will provide evidence that masting in oaks does little to satiate *Curculio* possibly because of the insect's ability to track acorn production in time and space within a forest. I will close this chapter with a review of the fascinating interactions documented by the author and others on the tri-trophic interactions between weevils, oaks, and the vertebrates that feed on acorns and/or weevils.

OVERVIEW OF INSECT ACORN PREDATORS

The insects that feed on acorns, although somewhat limited in diversity, are sometimes pervasive and devastating in their damage to acorns and assumed by many authors to be a major limiting factor in oak dispersal and regeneration. By far the most widely spread and populous seed predator are the nut weevils belonging to the genus *Curculio*, the family Curculionidae, and the order Coleoptera—the beetles. These weevils are generally found wherever there are oaks. And, while they can cause devastating consequences to acorn survival and establishment, and are thus regularly referred to as seed predators, the relationship between the oaks and the weevils and other insects is far more complicated. On the one hand, the oaks appear to have evolved a number of traits that provide resistance to weevil and other insect predators. Alternatively, the oaks show tremendous tolerance to seed and seedling damage by insects (see Chapter 6). In fact, I argue that there is plenty of evidence to suggest that the relationship between oak and insects may be better considered a host-parasite relationship or even one that may be evolving toward a commensal-like relationship in some cases.

First, I begin with a brief overview of the insects that feed on acorns. Although weevils are consistently the most common insect seed consumer, they are not the only insect to feed on the fruits of oak. The literature is in fact replete with reports of the relative abundance of weevils and other insect taxa found in acorns (Table 12.1). Despite this diversity of studies, few have considered the entire guild of insects that depend on acorns of various species nor the precise timing of their residence in acorns. Here I review two critical papers that together provide some of this information. Perhaps one of the most instructive, albeit qualitative, overviews of the insects that feed on acorns is that provided by Dr. Mark Moffett in National Geographic Magazine in June 1989. In this brief, but beautifully illustrated article, Moffett, a renowned entomologist and photographer and the curator at the Harvard's Museum of Comparative Zoology, provides an exquisite overview of the insects that either feed on, or develop in acorns of the oaks typically found in the northeastern United States. He begins with the *Curculio* spp. (see weevil life cycle below), the filbertworm moth (*Cydia latiferreana*), the acorn worm moth (*Velentina glandulella*), and the short-snouted weevil (*Conotrachelus naso*) (Fig. 12.1). The first of these, *Curculio*, drills a small hole in the acorn and deposits one or more eggs that hatch into larvae that feed on the cotyledon before cutting another hole to exit. The female acorn moth and the short-snouted weevil both deposit eggs through existing holes or other compromises in the acorn shell. In the case of the filbertworm, the egg is laid on the outside of the acorn, hatches, and the caterpillar excavates its own hole to enter the acorn and feed on the cotyledon. Two other species sometimes feed on the cotyledon of acorns in the eastern US. One is the gall wasp (*Callirhytis fructuosa*), which deposits eggs, around which a hard, spherical gall forms that later serves as a food source for the developing larva. The other is the sap beetle (*Stelidota octomaculata*) that also feeds on acorn cotyledon. Several additional species will invade damaged acorns to attack the acorn consumers reviewed above. These include several species of parasitic wasps and an entire sequence of other invaders that, according to Moffett (1989, p. 784), include "wireworms, springtails, fly maggots, cheese mites, minute fungus beetles, . . . beetle mites . . . millipedes, slugs and snails (which scavenge) . . . centipedes and some fungus gnat larvae (which prey on other species in the acorn), and ants." In short, an entire temporal community of insects and other invertebrates can occupy a single acorn from early maturation to its final fate on the forest floor. Yet, all but the first five species are secondary invaders, many of which depend on access and/or damage created by the primary seed predators (Moffett 1989). Thus, despite the diversity of insects and other invertebrates associated with the acorns, it is primarily just a few key species that appear as the primary acorn consumers of most oaks (Table 12.1).

The second study that provides some insight into the temporal guild structure of insect consumers of acorns was conducted by Fukumoto and Kajimura (2011). These researchers tracked the seasonal development of *Q. variabilis* and *Q. serrata* acorns and the corresponding emergence patterns of acorn-feeding insects. By placing mesh bags around tree branches containing

(a)

(b)

(c)

(d)

12.1. A few insects that commonly infest acorns. Shown are **(a)** larvae of *Blastobasis glandulella*, the acorn moth. *(Thomas Palmer)* **(b)** Adult of the filbertworm moth, *Cydia latiferreana*, the larvae of which infests acorns of several oak species and the fruits of hazelnut (*Corylus*). *(David D. Beadle)* **(c)** Adult weevil (*Curculio* spp.), the most significant insect predator of acorns. *(Andres DeKesel)* **(d)** Larvae of an unidentified gall wasp found in the acorn. *(Shealyn A. Marino)* SEE COLOR PLATE

Table 12.1. Sample of studies across the globe demonstrating the community of weevils and other insects that influence acorn fates.

LOCATION	OAK SPECIES	NUMBER OF TREES / ACORNS INSPECTED	WEEVIL SPECIES	OTHER SPECIES	REFERENCES
Korea	*Quercus* spp.	NA	*Mechoris ursulus* (Roelofs)	NA	Choi et al. 1993
Iberian Peninsula	*Quercus* spp.	7000 acorns	*Curculio elephas* (Gyllenhaal)	*Cydia fagiglandana* Zeller and *C. splendana* Hubner (Lepidoptera: Tortricidae)	Gallardo et al. 2011
Arkansas, US	*Q. alba*	NA	*Curculio* and *Conotrachelus* spp.	Filbertworm, *Melissopus latifereanus* (Walsingham) (Lepidoptera: Tortricidae); acorn moth, *Valentinia glandulella* Riley (Lepidoptera: Blastobasidae), cynipid gall wasps (stone galls) (Hymenoptera: Cynipidae), and midge larvae (Diptera)	Mangini and Perry 2004
Spain	*Q. suber, and Q. canariensis*	NA	NA	Dung beetle, *Thorectes lusitanicus* Jeckel (Coleoptera: Geotrupidae)	Pérez-Ramos et al. 2007; Verdú et al. 2007
Georgia, US	*Quercus* spp.	NA	NA	Dung beetle, *Mycotrupes lethroides* Westwood	Beucke and Choate 2009
Central France	*Q. robur* (L.) and *Q. petraea* (Mattuschka)	NA	*C. glandium* Marshall, *C. elephas*, and *C. venosus* (Gravenhorst)	NA	Rougon and Rougon 2001
Iran	*Quercus* spp.	NA	*C. glandium*	*Cydia fagiglandana* (Lepidoptera: Tortricidae)	Sadaghian et al. 2007
California, US	Coast live oak, *Q. agrifolia* (Nee), and Engelmann oak, *Q. engelmannii* (E. Greene)	62% of acorns collected on ground had insect damage; amount of damage to each acorn was <20%	*C. occidentis* (Casey)	*Cydia latiferreana* (Walsingham) (Lepidoptera: Tortricidae) and *Valentinia glandulella* Riley (Lepidoptera: Blastobasidae)	Dunning et al. 2002
Spain	Cork oak, *Q. suber* (L.)	NA	*C. elephas* (Gyllenhal)	*Cydia* spp.	Soria et al. 1999

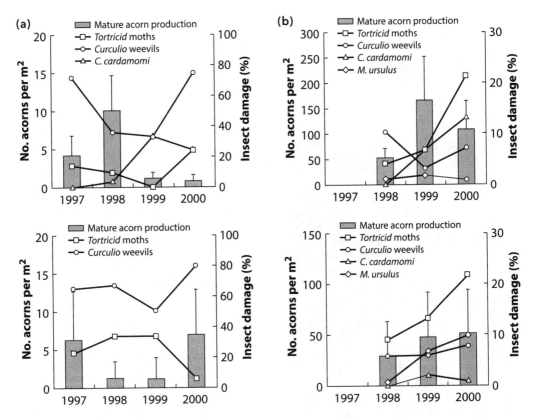

12.2. Shown are **(a)** annual fluctuations of mature acorn productions of *Quercus variabilis* and its insect damage over four years at site 1 (above) and site 2 (below). Bars on columns indicate standard deviations. **(b)** Annual fluctuations of mature acorn production of *Quercus serrata* and its insect damage over four years at site A (above) and site B (below). Bars on columns indicate standard deviation. *(From Fukumoto and Kajimura 2011)*

acorns at one of five times during acorn development between May and early October, these researchers could track the times at which different species oviposited in the acorns. By doing so, Fukumoto and Kajimura (2011) were able to identify two guilds of insects that sequentially invaded *Q. variablis* acorns and three guilds that attacked *Q. serrata* acorns (Fig. 12.2). The two guilds in *Q. variablis* acorns consisted of one guild of two *Curculionidae* spp. that were present during the immature development of the acorn, and a second guild during and after acorn maturation, which included two species of *Curculio* and one species of *Cydia* (Fig. 12.2). I highlight this study because it is one of the few to provide such a careful assessment of the temporal pattern of acorn-consuming insect assemblages occurring in oaks. It demonstrates how the sequence of events may influence one another and that interactions between species within the

(a)

(b)

(c)

(d)

(e)

acorn may be significant and influence both ovipositioning decisions by adult insects as well as competition of larvae once established in an acorn.

The Weevil Life Cycle

In nearly all surveys of insect damage in acorns (Table 12.1), *Curculio* larvae are consistently the most prevalent seed consumer. Although other insect taxa are consistently associated with acorns, they almost always occur in considerably lower frequency. For this reason, much of the remainder of this chapter focuses on the weevils.

Adult acorn weevils, typically less than 1 cm in length, are most easily recognized by a long rostrum (snout). In some species, this rostrum exceeds the species' body length (Fig. 12.3a). Once gravid, the female searches immature acorns in the oak canopy, excavating a long hole in the acorn by moving its head back and forth and chewing through the shell and the cotyledon (Moffett 1989). The rostrum is equipped with small cutting "teeth" at the distal end, which allows it to cut through the shell and to feed on the cotyledon and/or other oak tissue in the case of some species. A gravid female, while feeding on these immature acorns, is also searching for a suitable site in which to lay an egg (Fig. 12.3b).

12.3. Shown are **(a)** adult *Curculio* spp. showing typical body form, long rostrum. *(Janet Graham)* **(b)** Adult female *Curculio* feeding on acorn prior to ovipositing. *(Joe M. Devereaux)* **(c)** A larva of *Curculio* spp. in acorn showing exit hole of another individual. *(Shealyn A. Marino)* **(d)** Larva of *Curculio* spp. exiting acorn. *(Shealyn A. Marino)* **(e)** Pupa of *Curculio* spp. in soil chamber. *(Jerry A. Payne)* SEE COLOR PLATE

Immediately after the canal is drilled into the immature acorn, the adult female weevil may choose to oviposit an egg, after which the surface of the acorn heals, and the acorn continues to mature with the egg inside the cotyledon. The hole closes as the acorn matures, but in many oak species a scar on the surface of the acorn shell can serve as an indication of previous weevil activity. Although some authors have used these scars as a record of ovipositioning, it is not clear to what extent scars correspond directly to the presence of larvae. According to Moffett (1989), the weevil egg then hatches within two weeks into a developing larva (Fig. 12.3c), which will begin to feed on the cotyledon as it burrows its way through the acorn. The weevil larva remains in the acorn until the acorn matures and drops to the ground, at which time it is thought that the larva is signaled to exit the acorn. This is accomplished as the larva chews a small hole, significantly smaller than its body diameter, then slowly rotates its flexible body until free of the acorn (Fig. 12.3d).

Once outside the acorn, the weevil then burrows into the soil and remains dormant for one to five years depending on both the weevil species and environmental conditions for each species. In the soil, the larva will pupate (Fig. 12.3e) and then later emerge as an adult which moves to a nearby oak tree (often the same tree from which it developed), climbs the tree, and mates. While still in the tree, the female will then oviposit before dying. Although adult weevils can fly, they generally exhibit limited mobility and usually complete the life cycle in the same or nearby trees from which they started their life cycle (but see Bogdziewicz et al. 2018b).

Although little is known about the timing of the underground stage of weevil development, there is considerable evidence in one system that rainfall, which in part drives masting, can also influence the emergence of weevils (Espelta et al. 2017). These authors found that weevil numbers in the autumn were related primarily to autumn rainfall and only secondarily to weevil densities the previous year.

Because of the cryptic nature and complexity of the *Curculio* life cycle, it has been difficult to untangle the weevil's interactions with the oaks and other hardwood species. Yet, early taxonomic and basic ecological research by Gibson in North America (Gibson 1964, 1972, 1982, 1985), the breakthrough research by Menu and colleagues on the life cycle and diapause in the chestnut weevil (*C. elephas*), which also infests acorns of some oaks (Menu 1993; Menu and Debouzie 1993; Menu et al. 2000; Menu and Desouhant 2002), the extensive research by Bonal and colleagues on the ecology of weevils infesting oaks of the Iberian Peninsula and elsewhere (Espelta 2009ab; Bonal et al. 2010, 2011, 2015; Espelta et al. 2017; Bogdziewicz et al. 2018b, 2018c; Arias-Leclaire et al. 2018), all coupled with new advances in DNA barcoding (Peguero et al. 2017) have provided a rich foundation for a growing area of investigation. New students in this area should consider the literature above before delving into studies on *Curculio* life cycles.

LONG-TERM PATTERNS OF WEEVIL PREVALENCE IN PENNSYLVANIA

Monitoring acorn predation by weevils has been an ongoing component of the research activities in my lab since I joined the faculty at Wilkes University in 1989. While still in graduate school, I had analyzed the data amassed from a sophomore laboratory exercise in ecology at Wake Forest University in which students systematically collected acorns from individual trees of red and white oak on the campus and then systematically evaluated the prevalence of weevil infestation in acorns, individual trees, and the two species of oaks (Steele et al. 1993). The overall goal of the exercise, adopted from a lab originally developed at Duke University and brought to Wake Forest by Dr. Ellen Simms, was to introduce students to some basic questions in ecology by assessing spatial variation in weevil infestation.

Soon after joining the faculty at Wilkes, I adopted a slightly modified version of the lab for my newly developed sophomore requirement in population ecology and evolution. Each year, in preparation of the exercise, students and other personnel from my research team and I systematically collected acorns beneath the canopy of individual red and white oak trees and on occasion other species, such as chestnut oak (*Q. prinus*) and black oak (*Q. velutina*) when available. Acorns are collected from a small sample of individual trees (typically 5–10) by raking acorns under the canopy of each tree and ensuring that acorns from all areas of the canopy were sampled and that all acorns collected are unlikely to have fallen from another tree. Acorns are then bagged, labeled, and stored at 4°C until the time of a lab, usually just a few weeks after collection.

Each student in the class takes a random sample of approximately 20 acorns, a few from each bag (tree). The student then measures the diameter, length, and mass of the acorn, and records the oak species, tree number, and acorn number. The student then inspects the acorn for weevil exit holes and records whether they are located in the top (basal) half or bottom (apical) half of the acorn. Each exit hole in the acorn is assumed to represent one weevil and is recorded separately. The student then systematically dissects the acorn by first severing the acorn in two equal halves and then carefully dicing the cotyledon from each half, looking for insect larvae and recording their position in the acorn. The class data from three to four labs of approximately 20 students each is then compiled and redistributed to the students for analyses and a final laboratory report. This means the students analyze as many as 1500 acorns each year, which provides them an adequate sample for testing hypotheses of how weevils are distributed

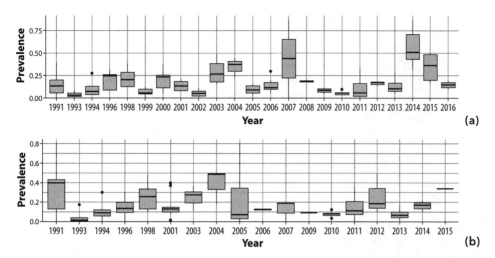

12.4. Long-term patterns of insect infestation in acorns of two oak species. Shown is prevalence of all insect larvae (% acorns infested per tree) in individual trees of **(a)** red oak (*Q. rubra*) and **(b)** white oak (*Q. alba*) from 1991 to 2016. Sample sizes range from 1 to 10 trees per year. *(Michael A. Steele, unpublished data)*

within an acorn, between acorns of the same tree, between different trees of the same species, and between oak species. In early versions of the lab, students dissected many more acorns and their responsibility for analyses were limited. Over the course of 26 years (omitting a few years because of mast failure), these students analyzed more than 52,000 acorns.

In recent years, my colleagues Jeffery Stratford and Debra Chapman, who assist in the team-teaching of this course, reduced the acorn number analyzed by students to allow us to add two additional components to the exercise. We now include a separate set of instructions for extracting weevil DNA (from a small number of weevils) that is then stored at −80°C. Research students then subsequently amplify the cytochrome oxidase I gene and send the DNA to a private company for sequencing. When results of the barcoding analyses are returned, each student in the class then searches GEN-Bank to identify her/his weevil species. Although acorns are not necessarily collected from the same trees and we do not have estimates of acorn abundance during each year of the study, these long-term data provide an excellent opportunity for assessing overall patterns of weevil abundance in one location (41° 14′ 44″ N/ 75° 52′ 54″ W; Fig. 12.4). A few key points follow from this long-term data set. First, the mean prevalence (percentage of acorn infested with insects) ranged from < 10.0% (1993) to 50.0% (2014) in red oaks and from 17.6% (1993) to 50.0% (2004) in white oaks (Fig. 12.4). The lowest prevalence in both species was observed in 1993. Across all 26 years of the study, prevalence in both species was consistently below 25%

12.5. Adult weevil (*Curculio* spp.) trap. Trap modified from that developed by Mulder et al. (2007) for the pecan weevil. (*Shealyn A. Marino*)

and the highest prevalence occurred in years when relatively few trees were sampled, suggesting that even in those years these prevalence measures may be somewhat inflated. These observations thus dispute two assumptions that frequently appear in the literature: that weevil infestations are higher in white oaks than red oaks because of differences in tannin levels, and that, in some years, weevils decimate acorn crops by attacking nearly all the acorns in individual trees (Weckerly et al. 1989a; Govindan and Swihart 2014, and references therein). Although I do not doubt the rare events in which a single tree may experience extensive weevil damage, it does not appear to be common, at least within northern areas of the range of *Q. rubra* and *Q. alba*.

Although we are only beginning to barcode weevils in large numbers under the guidance of my colleague Dr. Raul Bonal of Spain, clearly one of the leading authorities on weevils, we have to date identified weevils of three genera in the oaks in our region: *Curculio* spp., *Conotrachelus* spp., and *Cyrtepistomus* spp.

We have done this by attaching homemade weevil traps to oak trees (Fig. 12.5) and then regularly monitoring these traps during mid- to late summer to recover adult weevils headed to the canopy to mate and oviposit. Barcoding of larvae potentially will reveal additional species. The other insects frequently found in acorns in northeastern Pennsylvania are the Valentina moth larva (*Valentina* spp.), fly larva (*Diptera* spp.), and the galls containing wasp larvae (*Callirhytis* spp.). The relative abundance of these species is consistently a small fraction of that represented by weevils.

PATTERNS OF WEEVIL PREVALENCE IN INDIVIDUAL TREES IN PENNSYLVANIA

Despite obvious limitations, these class data shown above provide a nice long-term perspective on weevil infestation at one location. Yet, the lack of information on tree identity and acorn crop size constrain some of the questions that can be addressed with this data set. This is in part why we chose to separately measure patterns of acorn abundance in 135 forest trees across three study areas beginning in 2000. We tracked acorn production under 15 individual trees of each of three species at three study areas for a total of 19 years before we conducted formal analyses of these data (Bogdziewicz et al. 2017c, 2018b; see Chapter 4). Previously I discussed some of these data in the context of small mammal abundance (see Chapter 4). However, our primary intention was to use these data to assess the relationship between acorn abundance (at both the individual tree and population levels) and patterns of weevil and other insect infestation. For every acorn collected from these seed collectors, my research team carefully inspected each acorn for evidence of insect infestation, as described above.

Over the 19-year period, the mean percentage of white oak acorns infested with weevils (mean prevalence ± [SD]) at individual trees at the three sites was 13.2% ± 10.7, 14.6% ± 15.5, and 20.4% ± 27.3. The corresponding measures of infestation for red oaks at the same sites were 24.08% ± 18.28, 25.9% ± 22.0, and 37.2% ± 23.7, respectively. Patterns of prevalence in this more controlled setting were similar to those observed in the 26-year class data. Moreover, in years when acorn abundance increased, so did the weevil abundance. Generally, weevil abundance never seemed to reach the devastating effects reported in some systems. In fact, at all sites and for all oak species, weevil numbers were significantly correlated with acorn numbers (Fig. 12.6). Across the three sites, for example, correlation coefficients (r) between weevil number and acorn number ranged between 0.69 and 0.80 for red oaks and between 0.62 and 0.79 for white oaks—all were statistically significant. And as we discuss in a bit more detail later in this chapter (see also Chapter 4), it appears that weevils in this system seemed to somehow track acorn abundance by both increasing reproduction when acorn crops are higher and possibly moving within the forests to individual trees that produce more acorns than their neighbors (Bogdziewicz et al. 2018b).

PATTERNS OF WEEVIL PREVALENCE
IN CENTRAL MEXICO

Now I return to Central Mexico, where the impact of weevils appears significantly more devastating. As described previously, I spent a year in Mexico in the town of Puebla working out of the laboratory of Maricella Rodriguez, a leading expert on Mexican oaks. With the help of Juan Radillo and several other students, I was able to assess patterns of insect infestation in numerous oaks. The rich diversity of oaks in Mexico, especially in the state of Puebla and neighboring states, meant I had a remarkable number of species in which to study acorn-insect interactions. The oaks here included numerous species of both red oaks (section *Lobatae*) and white oaks (section *Quercus*) with many, but not all,

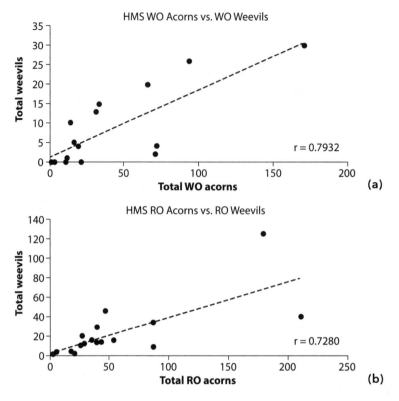

12.6. Shown is **(a)** the relationship between number of acorns of white oak (*Q. alba*) produced by 15 trees and number of weevil larvae over a 19-year period, and **(b)** the same relationship between number of acorns of red oak (*Q. rubra*) produced by 15 trees and weevil numbers at the same site (HMS). Years in which oaks did not produce acorns are not shown. Pearson r values are both significant at $p < 0.01$.
(Michael A. Steele, unpublished data)

of the characteristics of the oaks I had studied in eastern deciduous forests (Table 12.2). In the autumn of 1997, students and I systematically collected 5075 acorns from beneath 83 trees of eight oak species: three red oak and five white oak species. Table 12.3 summarizes the prevalence of insect infestation per tree for each species. For red oaks, mean prevalence per tree ranged from only 11.5% (*Q. crassifolia*) to 40.3% (*Q. cassipes*) and were generally comparable to patterns of weevil infestation observed elsewhere. However, the patterns of infestation in the five white oak species was remarkably high. The lowest mean prevalence occurred in *Q. glaucoides* (36.1%) and *Q. obtusata* (42.5%), whereas the other three species (*Q. laeta*, *Q. liebmannii*, and *Q. microphylla*) all exhibited prevalence measures near or above 60%. Couple this rather impressive weevil damage with considerable additional damage from other insects (6.8% to 20.0%) and fungi,

Table 12.2. Acorn and germination characteristics of nine species of oaks studied by the author in central Mexico.

SUBGENUS	SPECIES	ACORN MASS (g)	PERCENTAGE GERMINATION X ± SE	GERMINATION ON TREE	DOUBLE-SEEDED ACORNS (HIGHEST % OBSERVED)	SEED FALL
Quercus	Q. glaucoides	2.14 ± 0.10	31.0 ± 7.7	No	No	Rapid/Early
	Q. laeta	3.42 ± 0.22	51.9 ± 13.3	Yes	Yes	Rapid/Early
	Q. liebmannii	8.51 ± 0.64	33.8 ± 17.8	No	No	Rapid/Early
	Q. microphylla	2.00 ± 0.18	64.0*	Frequently	No	Mod/Early
	Q. obtusata	3.61 ± 0.20	76.4 ± 5.7	Yes	Yes	Mod/Early
Lobatae	Q. acutifolia	2.12 ± 0.31	0.90 ± 0.90	No	Yes	Delayed/Late
	Q. crassipes	1.80 ± 0.09	4.88 ± 1.82	No	Yes	Delayed/Late
	Q. crassifolia	1.75 ± 0.07	11.5 ± 5.25	Possibly	No	Delayed/Late
	Q. mexicana	1.20 ± 0.06	1.50 ± 1.20	No	No	Delayed/Late

Source: Unpublished data from Michael A. Steele.

* Represents a composite sample for approximately 25 trees.

including mold (5.8% to 28.2%), and relatively few sound acorns of most of these species ever made it past the initial stage of seed fall. As I describe below, acorns from many of these same species of oaks are faced with additional predation by insects and extensive desiccation once they disperse to the ground before they ever have a chance of dispersal by birds or mammals.

Although this represented only one year of observation, I was convinced that oaks at these latitudes were dealing with a different set of selective pressures imparted by the weevils and their kin. I recall one particular day, in mid-December 1997, when I was examining the acorns on a scrub oak, *Q. microphylla*, and I observed an adult *Curculio* working its way along the branches of an adult tree presumably looking for an acorn in which to deposit its eggs. I was awestruck that acorns of some of the oaks had matured as much as three months earlier and weevils were still doing their thing. Maybe the weevils have an edge at these latitudes. The long season means weevils are afforded ample time for their larvae to develop and feed on acorn tissue before they exit the acorn and head into the soil prior to diapause. But, as I discuss below, the oaks may exhibit a slightly different set of defenses against the insects in Mexico and similar climates.

ARE MEXICAN OAKS DISPERSAL LIMITED?

In the autumn of 1997, while in Mexico, Juan Radillo, my other student assistants, and I designed a "dispersal" experiment similar to those we conducted at other sites in Pennsylvania. Our goal was to determine how rodents responded to acorns of native oaks. At both sites, we presented tagged acorns of three red oaks and three white oak species. We established six feeding plots at each of two study areas in the state of Puebla: one located on the edge of the village of La Preciosita approximately 90 km northwest of the city of Puebla and the other at Africam Zoological Park, 17 km south of Puebla. The acorns presented in the experiment were native to both study sites (Table 12.4). At each feeding plot, we placed 175 sound acorns of one red oak species and another 175 acorns of one white oak species on a 2 m × 2 m plot of ground, over which we staked chicken wire about 30 cm above the ground and the acorns to prevent access by larger vertebrate and avian seed predators. These exclosures still allowed access by rodents. Across the six plots at each site, we presented identical pairs of red and white oak species as shown in Table 12.4. Individual plots were separated by approximately 150 m.

Every few days, we initiated visits to all plots at each of the two study areas to evaluate acorn fates. To our surprise, evidence of acorn dispersal and predation by rodents was extremely limited. By day 17, we ceased following seed fates at both sites due to alternative acorn fates. Shown in Table 12.4 is the summary of the fates of the 2100 acorns followed at Africam Zoological Park. By day 17,

Table 12.3. Overview of infestation of acorns by weevils (*Curculio* spp.) and other insects in eight species of oaks studied by the author in central Mexico.

SECTION	SPECIES	n (TREES, TOTAL ACORNS)	WEEVILS (% INFESTED) MEAN ± SE	OTHER INSECTS (% INFESTED) MEAN ± SE	OTHER FACTORS (% DAMAGED) MEAN ± SE	VIABLE ACORNS (%) MEAN ± SE
Quercus	Q. glaucoides	(7388)	36.1 ± 3.0	6.8 ± 4.2	17.9 ± 7.4	38.8 ± 6.8
	Q. laeta	(7456)	58.3 ± 5.5	10.7 ± 2.3	13.8 ± 3.9	21.9 ± 8.7
	Q. liebmannii	(6153)	57.6 ± 5.6	8.3 ± 3.3	27.4 ± 2.0	8.2 ± 5.7
	Q. microphylla[1]	(NA, 100)*	68.0	20.0	5.0	15.0
	Q. obtusata	(10666)	42.5 ± 7.1	10.3 ± 3.5	13.9 ± 4.3	33.4 ± 7.1
Lobatae	Q. acutifolia	(11700)	23.6 ± 3.4	6.0 ± 1.6	13.4 ± 2.5	56.4 ± 4.2
	Q. crassipes	(6443)	40.3 ± 1.8	4.0 ± 1.7	28.2 ± 7.6	28.4 ± 10.8
	Q. crassifolia	(3236)	11.5 ± 5.3	13.0 ± 2.5	6.9 ± 3.8	37.1 ± 18.4

Source: Unpublished data from Michael A. Steele.

* Represents a composite sample for approximately 25 trees.

[1] Casual inspection of 1333 additional acorns of *Q. microphylla* revealed complete damage of 47.5% (633) of these acorns.

Table 12.4 Fate of acorns after 17 days on the ground at the African study site. Initial sample sizes were 175 acorns of each species in each experimental plot.

| PLOT | ACORN SPECIES | VIABLE (%) | DESICCATED (%) | FUNGI (%) | INSECT/SPLIT (%) | ANT PREDATION (%) | MAMMAL | |
							DISPERSED (%)	EATEN (%)
2A	*Q. laeta*	54 (30.9)	23 (13.1)	86 (49.1)	9	7 (4.0)	1 (0.6)	2 (1.1)
	Q. acutifolia	107 (61.1)	18 (10.3)	26 (14.9)	67	2 (1.1)	2 (1.1)	20 (11.4)
2B	*Q. obtusata*	117 (66.9)	2 (1.1)	46 (26.3)	19	5 (2.9)	3 (1.7)	2 (1.1)
	Q. mexicana	141 (80.6)	0 (0.0)	29 (16.6)	4	0 (0.0)	5 (2.9)	0 (0.0)
2C	*Q. microphylla*	87 (49.7)	0 (0.0)	88 (50.3)	16	0 (0.0)	2 (1.1)	0 (0.0)
	Q. mexicana	140 (87.0)	3 (1.7)	30 (17.1)	0	0 (0.0)	0 (0.0)	0 (0.0)
2D	*Q. obtusata*	63 (36.0)	17 (9.7)	78 (44.6)	48	9 (5.1)	5 (2.9)	3 (1.7)
	Q. acutifolia	125 (71.4)	5 (2.9)	29 (16.6)	17	12 (6.9)	4 (2.3)	0 (0.0)
2E	*Q. microphylla*	66 (37.7)	11 (6.3)	95 (54.3)	22	2 (1.1)	1 (0.6)	0 (0.0)
	Q. acutifolia	122 (69.7)	13 (7.4)	32 (18.3)	25	2 (1.1)	4 (2.3)	1 (0.6)
2F	*Q. laeta*	88 (50.3)	2 (1.1)	82 (46.9)	6	2 (1.1)	1 (0.6)	0 (0.0)
	Q. mexicana	145 (82.9)	2 (1.1)	26 (14.9)	3	1 (0.6)	1 (0.6)	0 (0.0)

Source: Unpublished data from Michael A. Steele.

a sizeable portion of acorns of both red oak (17.1% to 30%) and white oak (26.3% to 54.3%) received significant damage to insects and fungi (Table 12.4). Many of the other acorns had received damage by ants or had desiccated and split; only 29 and 28 of all 2100 acorns were either dispersed or eaten by rodents, respectively. By day 17, fewer than 30.9% to 66.9% of white oak acorns and 61.1% to 80.6% of red oak acorns appeared viable. Although not presented here, results from the replicate experiment at La Preciosita were similar to those collected at Africam, leading my team to close the experiment at 17 days.

Given that we had inspected each acorn for evidence of pre-dispersal damage and attempted to present only sound acorns in this experiment, it was clear that acorns suffered significant additional predation and other damage well before any opportunity for dispersal.

Although live-trapping and other observations demonstrated that rodents (e.g., rock squirrel [*Otospermophilus variegatus*] and the painted spiny pocket mouse [*Heteromys pictus*]) were common at both sites, at least one species, the ground squirrel, appeared to function almost exclusively as a seed predator. In separate experiments, we observed free-ranging ground squirrels consistently larderhoarding in underground burrows when directly presented with acorns. Although I have observed young oak seedlings establishing just outside the burrows of this ground squirrel at La Preciosita, Puebla, this appears to be a limited opportunity for dispersal by this species. Further evidence that this ground squirrel is an intense seed predator followed from our observations that this species frequently entered greenhouses to excavate germinating acorns and prune the radicles and epicotyls before consuming the acorns.

Collectively, evidence of intense pre-dispersal weevil predation followed by significant, post-dispersal predation by insects and fungi and limited opportunities for dispersal following seed drop, suggest the distinct possibility that oaks in central Mexico may be seriously dispersal-limited. This does not account for the possibility that avian agents of dispersal may be important in moving and caching acorns prior to seed fall. This indeed should receive important attention in future research. One way to approach this question is to examine patterns of pre-dispersal predation across latitudinal or elevational gradients, an approach my lab has begun to address in North America, north of Mexico.

THE OAK'S RESPONSE TO WEEVIL AND INSECT SEED PREDATION

As I discuss in Chapters 4 and 5, the oaks display several characteristics that are best interpreted as defenses against seed predators. These include large mast crops that likely satiate seed predators and chemical gradients in acorns that likely promote cotyledon damage in primarily the basal half of the acorn. Both such adaptations offer resistance and/or tolerance to a suite of both vertebrate and invertebrate acorn predators and are likely the result of diffuse coevolution. In addition, however, several other specific characteristics of oaks may represent defensive adaptations in part, or exclusively, to the insects, especially the weevils.

Polyembryonic Acorns

As shown by McEuen and Steele (2005), potential adaptations by some oaks to both weevil predation and embryo excision by squirrels may include rapid germination, multiple-seeded acorns, and atypical seed morphology (i.e., in which the embryo [seed] is not located in the apical end of the acorn (see Chapters 3 and 6). Although McEuen and Steele (2005) experimentally demonstrate how these latter two characteristics specifically enable acorns to thwart mortality as a result of embryo excision by squirrels, they may also be involved in defense against weevils. McEuen and Steele (2005) summarize observations on polyembryonic acorns in temperate, tropical, and subtropical forests across the globe. The 14 species cited include six white oak and eight red oak species. The frequency with which this occurs in both subgenera suggests that this characteristic is not solely an adaptation to escape embryo excision, since this behavior is performed primarily on acorns of white oak species. The most direct evidence that this trait is involved in defense against insect seed predators comes from Garrison and Augspurger (1983). They observed greater seedling survival immediately following partial predation by insects in multi-seeded acorns of *Q. macrocarpa* than in those with only a single seed. Elsewhere multi-seeded fruits of a tropical palm (Janzen 1971) have been observed to escape partial damage by both rodents and insects and are thus interpreted as an adaptation to deal with both selective pressures (Bradford and Smith 1977). I argue that while there is sufficient evidence to conclude that the polyembryonic condition in acorns provides defense again both insect and rodent seed predators, more information is needed to understand the genetic basis of this trait and the specific conditions under which it facilitates escape from seed predators.

Germination Patterns

In addition to contrasting germination schedules in red oaks (section *Lobatae*) that typically require a period of dormancy prior to germination, and the white oaks (section *Quercus*) that usually germinate soon after seed fall, there are additional interspecific differences in germination patterns within each section of oaks, especially in lower latitudes. Some white oaks in central Mexico (Table 12.2) and at least one white oak species in the southern US (*Q. geminata*), for example, germinate while still on the tree. They are, in a matter of speaking, ready to go as soon as they hit the ground. Moreover, even a small percentage of some of

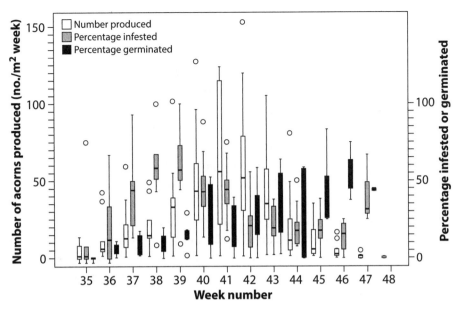

12.7. Pattern of acorn production of *Quercus schottkyana*, percent of acorns infested, and percent of acorns germinated by week of the year, combined across all years (2006-2014). *(From Xia et al. 2016)*

these red oaks in Mexico appear to break dormancy in early to mid-autumn, such as *Q. crassifolia* for which I found a mean of 11% of the acorns from three trees had germinated (Table 12.2). Even 1–5% of the acorns of the other three red oak species I examined in Mexico had germinated at seed fall. In northern latitudes, this behavior is not observed in red oak species. Although this may, in part, be due to climatic factors that exert a significant selective pressure on all oaks to germinate before the impending winter drought in Mexico and similar latitudes, I suggest this rapid germination may involve a strategy for escaping weevil damage as well, which is exceptionally high even in the red oak species (Table 12.2).

One of the few studies to consider the detailed pattern of germination in relation to acorn maturation and weevil infestation was conducted by Xia et al. (2016). Over a nine-year period, they determined detailed patterns of weevil infestation and development and germination of *Quercus schottkyana* acorns on a weekly basis from late August to mid-November. They found that this oak produced smaller, poor quality acorns (of a lower cotyledon mass) early in the season and that these acorns were more heavily infested by weevils, whereas later in the autumn, it produced larger, more viable acorns that were more likely to germinate. In effect, individual oaks were providing a "food source" to satiate weevils early in the season and then directing more energy per seed when infestation rates dropped later in the season (Fig. 12.7).

Acorn Size, Acorn Crop Size, and Acorn Abscission

Research by Bonal and colleagues (Bonal and Muñoz 2008) demonstrates that *Q. ilex* is able to satiate its primary insect seed predator, the chestnut weevil (*C. elephas*), by both acorn number (masting) and acorn size, and early acorn abscission during the latter stages of acorn development. Adult female beetles oviposit in acorns throughout autumn (the last few months of acorn development); however, weevils that establish in early autumn when acorns are still growing are limited in body size because of smaller cotyledon size. Moreover, these smaller acorns are generally shed earlier, thereby constraining survival of the weevils. Weevils developing from eggs laid later in the season grow larger due to larger acorn size and are likely to experience greater survival.

The work by Bonal and Muñoz (2008) confirms previous predictions by Boucher and Sork (1979; Yu et al. 2003) that early fruit abscission may be a common strategy for oaks and other nut-producing species to defend against insect seed predation. However, observations by Yi and Yang (2010b) on weevil infestation of Oriental white oak (*Q. aliena*) acorns suggests an opposite response in which larger individual acorns are consistently super-parasitized due to their preference by adult females for ovipositing. As a result, weevil larvae that emerge from these larger acorns are smaller than those that developed individually in smaller acorns. Clearly, additional research is necessary to understand better how acorn size and early abscission of acorns may influence interactions between the oak and the weevil. Perhaps in some systems the oak has the upper hand, whereas in others, it is the weevil who has the edge.

Bonal et al. (2007) explored further the satiation resulting from seed size and seed number and demonstrated a satiation effect due to both the acorn crop in individual trees and acorn size. Larger acorns produce larger seedlings and are likely to survive even when infested with weevils. However, Bonal et al. (2007) also showed a negative correlation between crop size and seed size in individual trees of *Q. ilex*, which resulted in less of a satiation effect at the crop level, as well as the individual acorn level in trees that produced the smallest crops and the largest acorns. In southeast Asia, Wang et al. (2008), Yi and Yang (2010b), and Cao et al. (2013) each reported a preference of weevils for larger acorns of *Q. mongolica*, *Q. aliena*, and *Q. variabilis*, respectively.

Acorn Traits and Weevil Infestation

My lab's research on chemical gradients in acorns suggests a plausible defense of individual acorns against both vertebrate and insect predators (Steele et al. 1993). Higher frequencies of weevils in the basal half of acorns, where tannin levels are lower and lipid and some nutrient levels are higher, provide strong circumstantial evidence of an adaptation in the oaks for defense against predation in the apical end of the acorn closer to the embryo. These observations are further strengthened by those of Hou et al. (2010) reporting that infested acorns are more likely to germinate and establish when weevil damage is limited to the basal half of the acorn (Bartlow et al. 2018a). Yi and Yang (2010b) found similar patterns of higher seedling success for *Q. variabilis* acorns infested in the basal half compared to those in which weevils fed closer to the seed (embryo) in the apical half.

Equally compelling, though, is how pericarp thickness varies in response to weevil activity. Yi and Yang (2010b) found that pericarp thickness in the apical, middle, and basal portion of *Q. variabilis* acorns was negatively associated with the percentage of oviposition scars, eggs/larvae, and weevil exit holes across the same location of the acorns. Pericarp thickness thus varies in a way comparable to chemical gradients such that these characteristics should synergistically direct weevil predation toward the basal half of the acorn. Couple these characteristics with evidence that acorns of numerous species are large enough to sustain considerable damage (Bartlow et al. 2018a) and even acorn removal soon after seedling establishment (Yi et al. 2019), and it appears that a suite of acorn characteristics function to tolerate seed predation by weevils.

Masting and Insect Acorn Predation

Masting as a strategy of oaks to defend against seed predators is covered in Chapter 4. Here, I briefly discuss how masting specifically relates to weevil infestation. Whereas large-seed crops will almost always satiate avian and rodent seed predators, it does not always follow that this will occur for weevils. Moreover, the culling of seed-predator populations during low-mast events or mast failures may apply primarily to vertebrates. The weevils may have evolved some strategies to overcome the effects of seed-crop failure and masting.

My attention was drawn to this subject when we first evaluated the relationship between acorn numbers and weevil numbers in our seed collection data from individual oaks in Pennsylvania. A preliminary comparison of seed number to weevil number across years resulted in a surprisingly strong positive correlation between acorn numbers and weevil abundance (Fig. 12.6). When acorn numbers increased so, too, did the weevil populations. Although a very preliminary analysis, this suggested to me that weevils were somehow tracking seed production, possibly emerging from diapause in years of higher seed production. If squirrels, as suggested by Boutin et al. (2006), can anticipate and respond to future seed crops long before they occur, why not insects? With a shorter generation time, it seems insects might be able to keep up with acorn production. In our first analyses of our long-term data of individual trees, we produced evidence that some masting oaks appear not to satiate the weevils (Bogdziewicz et al. 2018b). Instead, we found evidence of a rapid bottom-up effect of acorn masting on weevil populations in two white oak species (the white oak and chestnut oak [*Q. montana*]) but not in the red oak. The two oaks were not able to satiate these insect seed predators by masting. We attributed these results to both increased reproduction and the ability of weevils to aggregate in seed-abundant trees during years of moderate mast production. Increases of weevils in seed-rich trees were more likely to occur when other trees in the population had failed at mast production (Bogdziewicz et al. 2018b).

To date, several other studies suggest that satiation of weevils is a bit more complicated, because satiation may occur at the seed level and/or the tree level (but see Bogdziewicz et al. 2018b) even if it does not occur at the population level. As Bonal et al. (2007) further demonstrate, if larger acorns serve to satiate individual weevils, which appears to be the case in a number of studies (see above), the trade-off between acorn size and crop size can complicate patterns of satiation.

Some studies have found little evidence of weevil satiation at the level of the population, while others have reported that it is more likely to occur at the tree level. Xiao et al. (2017) investigated patterns of seed predation of *Q. serrata* acorns by two weevils' species (*Curculio* spp. and *Cyllorynchites* spp.) and one moth species (*Cydia* spp.) while acorns were still attached to the tree. Their goal was to determine whether pre-dispersal seed predation was consistent with the Janzen-Connell model, which predicts higher rates of predation when seed densities are high (positive density dependence) or the predator satiation hypothesis, which predicts lower rates of seed predation when seed densities are high (negative density dependence). Overall, they found that total insect infestation rates of the three species were twice as high in the low-mast year at the population level, supporting the predator satiation hypothesis. However, the opposite was true within individual trees where infestation rates were correlated with seed number. Furthermore, different patterns were observed for the three different species, which may suggest that individual life-history traits may influence the relationship between seed number and patterns of insect infestation.

Maeto and Ozaki (2003) found no significant negative relationship, and therefore no evidence of predator satiation at the population level, between seed crops of *Q. crispula* and *Curculio* seed predators in Japan. The authors attributed this result to the extended diapause of the weevils. Similar results were reported by Fukumoto and Kajimura (2011) for *Q. serrata* and *Q. variabilis*, and by Perez-Izquierdo and Pulido (2013) for *Q. ilex*; both studies reported that infestation rates were independent of acorn crop size, which they too attributed to diapause.

In contrast, Higaki (2016) found a strong negative relationship between acorn production by *Q. acutissima* and the percent infestation of acorns, as well as the number of *C. robustus* infesting acorns, demonstrating strong evidence for predator satiation. Higaki (2016) also reported that two successive years of low acorn production appeared to significantly reduce the population of weevils, which may be one of the few studies to demonstrate this. Higaki further argues that extended diapause in *C. robustus* allows this weevil to survive multiple mast failures and is likely an adaptation to counter the effects of oak mast failures. Finally, Espelta et al. (2017) also reported evidence of predator satiation of *Curculio* spp. in *Q. ilex* based on two studies, one of which was conducted over an 11-year period. However, these researchers also examined how a specific environmental factor (rainfall) in the Mediterranean can drive the relationship between acorn crops and weevil populations and further complicate the interactions between weevils and acorns. They found that *Curculio* populations in a given year were directly related to rainfall during acorn maturation and levels of weevil infestation the previous year. Both weevil emergence and acorn production were influenced by rainfall, but these factors also varied between individual trees and with the density of trees (Espelta et al. 2017).

The above studies indicate a range of evidence that both supports and refutes the concept of predator satiation in insect seed predation in oaks. Despite the vast number of studies to address the subject, little common ground appears evident. Satiation can occur at the level of acorn, individual tree, or population. However, the extent to which satiation does or does not occur is likely to depend on a number of factors, many of which may be related to the approaches used by researchers. Many studies, for example, are limited to three or fewer years of data and relatively few study areas. Such data provides little insight. What is sorely needed is a standardization in approaches so studies around the world can be compared. The ideal studies needed most are (1) those conducted over a decade or more (Espelta et al. 2017), (2) those replicating forest sites, (3) those that include the community of nut-weevil species in a forest, as well as the community of tree species affected (Govindan 2013; Govindan and Swihart 2014), (4) those that control for the effects of forest fragmentation and dispersal abilities of weevils, and (5) those that seek to better understand the proximate and ultimate factors influencing diapause in the weevils. Finally, moving forward, patterns of acorn production in oaks (masting) must be better evaluated in the context of all insect and vertebrate seed predators and seed dispersers.

WEEVIL COMMUNITY STRUCTURE

Until recently, there has been limited research on community structure, patterns of emergence, diapause, and population dynamics of acorn weevils. However, the last few years has, indeed, seen some significant breakthroughs in this area. From those that have explored these phenomena, it appears that there is a rich diversity of questions regarding the structure of oak weevil communities and the underlying ecological and evolutionary processes structuring these communities.

Among some of the first to carefully investigate the mechanisms structuring communities of nut weevils were my colleagues, Byju Govindan and Robert Swihart (2014), who tracked use of the fruits of two oak species (red oak and white oak) and the shagbark hickory (*Carya ovata*) by nine weevil species. This three-year study focused on the effects of the spatiotemporal variation in nut abundance on weevil community structure in these three hardwood species across forest patches in the central hardwood region of Indiana. They found that species richness of weevils was highest in a year when mast production was highest in the three tree species. Species of less specialized weevils that could rely on nuts of multiple tree species were more abundant in less preferred nuts when primary hosts experienced mast failure. The authors argued that such spatial storage effects, coupled with long-term storage effects (extended diapause), especially by specialist weevils, maintained species coexistence in this system.

Venner et al. (2011) examined communities of four species of closely related acorn weevils (*C. glandium*, *C. elephas*, *C. pellitus*, and *C. venosus*) in two oak species (*Q. robur* and *Q. petraea*) near Lyon, France. In this five-year study, the authors first demonstrated that acorns were a limiting resource for this community of weevils. The number of larvae that developed in acorns in a given tree per year was related to acorn abundance in that tree. In addition, acorns were attacked at a lower rate in years in which the acorn crop within a tree was higher. Given this resource limitation, it was perhaps not surprising to find that there was evidence that the weevils partitioned acorn resources both within a year and across years. Within each annual acorn crop, three species relied on the acorns for a brief period. Two species (*C. pellitus* and *C. venosus*) were found in acorns early in the season and the third used the acorns late in the season (*C. elephas*). The fourth species (*C. glandium*) relied on acorns across the entire season of acorn maturation. Patterns of emergence across years also suggested that acorn resources were partitioned by all but two species (*C. pellitus* and *C. venosus*). Thus, patterns of weevil activity across species are likely to be highly variable within oak stands, making assessment of the relationship between acorn abundance and weevil populations quite complex.

In another first, Peguero et al. (2017) compared the structure of acorn weevil communities in temperate forests of California with those of tropical forests in Nicaragua. They specifically tested the host-specialization hypothesis that has long been forwarded to explain diversity of some insect communities in the tropics. By controlling for the effects of host diversity (oak species diversity) on levels of host specialization, these authors were able to demonstrate that weevil species in Nicaragua are significantly more specialized with respect to host range and interaction strength (number of larvae emerging from a primary host relative to all emergences). Evidence of host specialization was also supported by observations that head width (and body size) of the larvae of tropical weevils was correlated with acorn size. No such relationship was observed in California weevils. Whereas this study was a major step forward in understanding coevolution of weevil-oak interactions, it was equally important for its contributions to ecological theory and its test of the host-specialization hypothesis in understanding community structure of a particular insect guild in the tropics.

Similar measures of host specificity in four species of nut weevils in northeastern Spain showed that the one highly specialized species (*C. nuncum*) that feeds exclusively on patchily distributed hazelnuts (*Corylus avellana*) exhibited a markedly different genetic structure than that of three oak weevils (*C. venosus*, *C. glandium*, and *C. elephas*), which show far less niche specialization in the two oaks (*Quercus ilex* and *Q. pubescens*) in which they reside (Arias-Leclaire et al. 2018). Other studies suggest that niche specialization by *Curculio* species may be balanced with trade-offs between dormancy and dispersal, in which species with short dormancy periods often exhibit greater dispersal ability (Pélisson et al. 2012). Although there is much yet to learn about how guilds of *Curculio* partition the ephemeral nature of an acorn crop, it seems certain the answers, in part, follow from a diversity of these beetles' complex and cryptic traits (e.g., host preferences, diapause, dormancy, dispersal, genetics).

OTHER INDIRECT EFFECTS OF INSECTS ON ACORN PRODUCTION / DISPERSAL

Insect Herbivory, Acorn Production, and Seedling Survival

The many herbivorous insects that attack the leaves of oaks also exert a significant effect on acorn production. One needs only to witness the defoliation of an oak stand by the invasive gypsy moth (*Lymantria dispar*) in North America—followed by a second leaf-out in the same season and then a complete mast failure later that year—to recognize the devastating impact such herbivory can have on acorn production. But, as demonstrated experimentally by Crawley (1985), only light defoliation (<15% of leaf area) by herbivorous insects is all that is required to significantly reduce acorn production in *Q. robur*, suggesting that insects play an equally critical role in early acorn development as they do in acorn survival once present and mature on the tree. It has also been shown that young seedlings of a shade tolerant oak (*Q. crispula*) near the parent tree can also succumb to herbivory by lepidopteran larvae that fall from canopy trees after leaf quality declines in these adult trees (Wada et al. 2000).

Although oak leaf herbivory has been studied in great detail, this subject is well beyond the scope of this book. I mention it briefly to emphasize how important insect fauna can be on oak ecology. This is certainly evident by classic studies by Feeny (1970) and Schultz and Baldwin (1982) demonstrating the relationship between leaf-feeding lepidopteran larvae and phenolic (tannin) levels in oak leaves. Related to this was evidence that simulated leaf damage of only 7% causes increased tannin levels in host trees, as well as neighboring undamaged trees, which also launched the important discussion of chemical communication between trees (Baldwin and Schultz 1983).

The Special Case of Oaks, Ants, and Aphids

I now turn to two examples in which insects may benefit the oak and its potential for acorn production or dispersal. In the first, Ito and Higashi (1991) studied the effect of a colony of the red wood ant, *Formica yessensis* (Forel), on leaf herbivory, acorn production, and acorn predation in *Q. dentata* in northern Japan. By experimentally excluding ants, they found evidence of a strong mutualism between the ant, a myrmecophilous aphid, *Tuberculatus quercicol*, and the oak. The presence of ants resulted in a significant increase in aphids and a reduction in leaf damage by several genera of Lepidopteran and weevil herbivores. The ants contributed little difference in the overall acorn number maturing, but accounted for a significant decrease in acorn infestation by Lepidopteran seed predators such as *Pammene nemorosa* and *Cydia glandicolana* and a nearly 100% increase in oak fecundity. One of the few of its kind, the study by Ito and Higashi (1991) suggests a fertile area for additional research.

And Along Came the Dung Beetle

The second example of a potentially strong insect-oak mutualism involves the dung beetle (*Thorectes lusitanicus*) and at least two oaks (*Q. suber* and *Q. canariensis*) in southern Spain. First reported by Perez-Ramos et al. (2007), these authors demonstrated that the dung beetle regularly feeds on, disperses, and buries acorns in sites suitable for germination when not recovered or only partially damaged by beetles. Dung beetles were more common where large grazers and their dung were abundant and the activity of rodents less frequent. Dung beetles rolled acorns in the same manner as dung and then buried acorns in a few centimeters of soil, generally in sites suitable for germination (Perez-Ramos et al. 2007). Most significantly, this study demonstrated that dung beetles depended on acorns as a food source, contrary to Vander Wall and Longland (2004) who suggested that secondary dispersal by dung beetles is likely coincidental because dung beetles generally do not consume seeds.

Subsequent comparisons of the effectiveness of rodents and the dung beetle (*T. lusitanicus*) on the potential dispersal of these two oak species revealed that rodents dispersed and cached a significantly greater portion of acorns but also recovered and consumed most of these acorns. In contrast, the "quality" of cache sites selected by dung beetles were generally superior to that of rodents, suggesting that they may have a meaningful impact on oak regeneration in the southern Iberian Peninsula. Moreover, this dung beetle shows a preference for acorn cotyledon over dung (Verdú et al. 2013), and its relationship with the oaks may be strengthened by advantages the beetle procures in thermal tolerance and enhanced ovarian development during autumn seed fall (Verdú et al. 2010). When given a choice between dung and acorns the beetle prefers acorns and, in response, shows a bolstered immune response and lower incidence of mortality (Verdú et al. 2013). Collectively, these observations suggest a potentially strong mutualism between this dung beetle and the oaks on which it feeds. Other reports of dung beetle-oak interactions are limited, perhaps only because researchers have not fully explored this relationship. Beucke and Choate (2009) recently reported that *Mycotrupes lethroides*, a dung beetle found in the southeastern US (Georgia), also feeds on acorns, although nothing was reported regarding acorn dispersal and caching.

TRI-TROPHIC INTERACTIONS BETWEEN OAKS, WEEVILS, AND VERTEBRATE ACORN PREDATORS AND DISPERSERS

Weevils introduce an interesting dimension in the interaction of oaks and vertebrate acorn seed predators and dispersers. Weevils provide a potential food source for both rodents and jays, but the response of vertebrates to infested acorns is far from consistent across studies (Table 12.5). A few studies indicate that shrews and rodents can be important predators of weevils significantly reducing larvae numbers in the soil (Andersen and Folk 1993) and, in the case of gray squirrels, selectively eating infested acorns, along with the weevil (Steele et al. 1996). In this latter study, we demonstrated that the keen ability of squirrels to detect infested acorns meant that sound acorns were selectively dispersed and cached significantly more often than those which were infested. Similar studies by Perea et al. (2012) demonstrated that the wood mouse (*Apodemus sylvaticus*) can also distinguish weevil-infested acorns from noninfested acorns and eat infested acorns more often and invest far less in dispersing and caching infested acorns.

In Steele et al. (1996), we also posited that consumption of weevils by squirrels and other rodents may counter the toxic effects of tannins. Interestingly, both squirrels (see Chapter 9) and jays (see Chapter 10) consistently lose body mass when fed only acorns, further suggesting that weevils provide a key nutrient supplement when consuming acorns. Johnson et al. (1993) found that it was necessary to supplement a diet of acorns with 5 g or more of weevil tissue to prevent weight loss in blue jays. Together these three studies (Johnson

Table 12.5. Overview of the impact of acorn weevils (*Curculio* spp.) on acorn predation and dispersal by vertebrates.

VERTEBRATE(S)	OAK SPECIES	LOCATION	INTERACTION/ OUTCOME	REFERENCES
White-footed mouse (*Peromyscus leucopus*) and Short-tailed shrew (*Blarina brevicauda*)	*Q. rubra* and *Q. velutina*	Central Indiana, US	Significant predation of weevils in soil by small mammals	Anderson and Folk 1993
Rodent spp.	*Q. aliena*	Qinling Mountains in Shaanxi Province, China	Infested acorns avoided in autumn when caching and eating; no avoidance in spring	Bo et al. 2014
Edward's long-tailed rats (*Leopoldamys edwardsi*)	*Q. variabilis*	Dujiangyan, Sichuan province, China	Cached and ate fewer larvae-emerged seeds than sound seeds; did not distinguish between sound seeds and larvae-concealed acorns	Cheng and Zhang 2011
Blue jay (*Cyanocitta cristata*)	*Q. palustris*	Southern Virginia, US	Jays selectively consumed non-infested acorns over infested nuts, even when consuming high tannins	Dixon et al. 1997b
Mexican jays (*Aphelocoma wollweberi*)	*Q. emoryi*	Southern Arizona, US	Jays selectively consumed non-infested acorns	Hubbard and McPherson 1997
Blue jay (*Cyanocitta cristata*)	*Q. rubra* and *Q. alba*	Southern Virginia, US	Jays lost mass on diet of acorns only or acorns plus 1.5 g of weevils, but maintained mass with diet of acorns plus 5.0 g of weevils	Johnson et al. 1993
Algerian mouse (*Mus spretus*)	*Q. ilex*	Central Spain	Inexperienced (naïve) rodents showed no preference of infested vs. sound acorns, but wild and captive-experienced rodents avoided infested acorns	Muñoz and Bonal 2008b
Wood mouse (*Apodemus sylvaticus*)	*Q. pyrenaica* and *Q. petraea*	Central Spain	Rodents distinguished between infested and non-infested acorns and consumed infested acorns more, dispersed them later, shorter distances, and less often than sound acorns	Pérea et al. 2012
Eastern gray squirrel (*Sciurus carolinensis*)	*Q. rubra* and *Q. alba*	Ohio, US	Limited evidence to support selective caching of red oaks over white oaks due to reduced losses to weevil infestation	Smallwood et al. 2001

VERTEBRATE(S)	OAK SPECIES	LOCATION	INTERACTION/ OUTCOME	REFERENCES
Eastern gray squirrel (*Sciurus carolinensis*)	*Q. rubra* and *Q. alba*	Eastern Pennsylvania, US	Squirrels distinguished between infested and non-infested acorns, selectively cached non-infested acorns, and ate infested acorns, along with weevils	Steele et al. 1996
Eastern gray squirrel (*Sciurus carolinensis*)	*Q. nigra*	Tennessee, US	Squirrels consumed and dispersed infested and non-infested acorns at indistinguishable rates	Weckerly et al. 1989a

et al. 1993; Steele et al. 1996; Perea et al. 2012) suggest an important interaction in which vertebrates may selectively consume weevil-infested acorns and disperse and cache sound acorns. The evidence for such an important tri-trophic interaction, however, is undermined by other studies demonstrating that some rodents consistently avoid infested acorns or cannot distinguish infested from noninfested acorns and that jays will often selectively consume noninfested acorns (Table 12.5). These inconsistencies may be due to specific experimental conditions (e.g., captivity) or environmental situations (e.g., food availability); thus, more information is clearly needed to understand the full nature of this multi-trophic interaction and its impact on oak dispersal.

Finally, I touch on another question that has been lurking in my mind for some time (see Chapter 15): the potential for dispersal of *Curculio* by both rodents and/or avian dispersers of acorns. As emphasized throughout the literature, an adult weevil is limited in its mobility, often not dispersing beyond the tree that produced the acorn from which it emerged. Nevertheless, dispersal— in fact, significant dispersal—may occur when acorns are moved by vertebrate dispersal agents. As I propose in Chapter 15, despite the evidence that jays and rodents selectively disperse and cache sound acorns, the occasional dispersal of infested acorns could result in infrequent but significant dispersal of *Curculio* and other insect species.

In the next two chapters, I shift to discussions on the human impact on oaks and oak dispersal (see Chapter 13) and then focus on the threatened oak forests (see Chapter 14), where dispersal limitations often prevail. I close with a brief statement of future directions and questions that require additional research efforts (see Chapter 15).

ANTHROPOGENIC FACTORS INFLUENCING OAK DISPERSAL, ESTABLISHMENT, AND REGENERATION

13

INTRODUCTION

In this chapter, I briefly highlight some of the many factors related to human activities that influence oak regeneration. Forest fragmentation, silviculture practices, nutrient deposition, modifications in fire regimes, excessive herbivory by deer, and acid deposition represent only a few important examples. I cite studies from across the globe, but also focus on the arguably dire situation in the Central Hardwoods Region of North America and the eastern deciduous forests, where oak regeneration is seriously compromised due to a perfect storm of anthropogenic factors (Abrams 1992). Current conditions in this region are considered in the context of both the presence and absence of American chestnut, which was virtually eliminated from the landscape by the middle of the last century due to an introduced fungal blight. I will also offer some predictions regarding the proposed introduction of a blight-resistant, hybrid American chestnut soon planned for much of the deciduous forests of the eastern US. I will consider the potential increase in seed consumers and how changes in forest context could influence the dispersal process.

I also touch on ongoing research in my lab that is now directed at understanding how some aspects of climate change may decouple or, alternatively, strengthen the disperser-oak mutualisms, and in turn, dramatically influence dispersal dynamics. Two such examples include the increase in frequency of late spring frosts across the Central Hardwoods Region, and the greater frequency of drought conditions. The first of these will most likely result in both greater frequency of mast failure and increased amplification in mast (bumper) years, changing ecosystem dynamics in many unpredictable ways. The latter, I suggest, may influence patterns of cache placement, cache recovery, and seedling establishment, as has already happened in other oak forests.

FACTORS LEADING TO THE LOSS OF OAK FORESTS

The human fondness for the oak, exemplified best by the age-old oaks that are frequently associated with historic sites and other locations of importance to people (Fig. 13.1), has long led to the over exploitation of oaks for a diversity of human needs. The high density of the wood, the resistance to insect damage because of high tannin levels, and the attractive finish of oak wood render oak an important cash crop. From lumber for building material to fine furniture, oak wood barrels for the aging of whiskey, scotch, and finer wines, cork, and many other prized products, oak forests have long been cut for human use. In addition to the exploitation of oak for important products, however, oak forests have also been removed for other needs, such as agriculture and the growth of the human population across the landscape. In some parts of the world (Mexico and Central America), oak is intensively collected for charcoal and other fuel. The high density of oak makes it slow burning and thus ideal fuel for cooking and heating. Although forests are generally not cut for these purposes, the collection of down timber likely leads to significant forest degradation due to the loss of woody debris on the forest floor (Chidumayo and Gumbo 2013). Despite the long history of human exploitation of the oaks, many forests still stand, and many others have been intensively managed for regeneration, although there is much yet to learn about this process. Here I provide just a brief overview of some of the ways in which anthropogenic activity influence oak dispersal and regeneration of oaks in these managed systems.

LOSS OF OAK FROM THE ECOSYSTEM

Deforestation at the hand of humans has accounted for major losses of oak in some regions of the world. Here I cite just two examples where massive stands of oak of an entire region have been lost to human deforestation. My account is based simply on personal observation and anecdotal knowledge and should not be considered an authoritative or judgmental assessment of oak deforestation. In my travels to four of the five continents where oaks occur, I have been perhaps most impressed by the impact early humans had on oak forests in the United Kingdom, especially Ireland. In my two short visits to the enchanted country of Ireland, I was quite impressed with the lack of native forests, especially those of oak. Most accounts indicate that following the last ice age (12,000–15,000 years ago), Ireland was recolonized by plants and by the time

13.1. Often associated with many aspects of human history and culture, towering oaks symbolized strength and success due to their long life. Although adult oaks take many forms, a sample of the giants included here are **(a)** the white oak (*Q. alba*) from West Hartford Connecticut. *(Msact at English Wikipedia)* **(b)** This famous Angel Oak, a live oak (*Q. virginiana*) estimated at nearly 400 to 500 years old in Angel Oak Park, near Charleston, South Carolina. *(Stephen Yelverton)* **(c)** The blue oak (*Q. douglasii*) found only in California. *(Stuart T. Wilson)* **(d)** This sessile oak (*Q. petraea*) from western Ireland, a species found throughout Europe, often in association with the English or pedunculate oak (*Q. robur*). *(Michael A. Steele)* **(e)** the pin oak (*Q. palustris*) found in the eastern deciduous forests. This pin oak, estimated at more than 100 years of age, is located on the eastern shore of the Chesapeake Bay. *(Michael A. Steele)* SEE COLOR PLATE

humans first arrived (9000 years ago) most of Ireland was covered with forests of oak or pine. However, within 3000 years of their arrival, humans had contributed to a significant loss in forest cover, which continued to the middle of the 19th century when most of the native forests were gone, replaced in many places by extensive stretches of acidic bogs (Fig. 13.2), the oaks are now far and few between with less than 1% of native forests remaining.

In North America, a more recent significant decline of oak is seen in the Midwestern and eastern US. Much of the loss here, which followed European settlement, resulted from the combined effects of extensive harvesting of oaks for timber and the clearing of land for agriculture. Today, many of the oak forests of this Central Hardwoods Region still exist but as documented by Abrams (2003), many of the oaks, such as the white oak (*Q. alba*), are at risk of a slower process of degradation due to combined effects of fire suppression, invasive species, and intensive herbivory (Abrams 2003).

13.2. The acidic pete bogs (blanket bogs) of western Ireland, formed by the massive elimination of oaks and other trees by the Neolithic settlers (~3900 BC–3000 BC), now serve as an important fuel source. (Michael A. Steele)

OAK FOREST DEGRADATION

As reviewed by Abrams (2003), the deciduous forests of the eastern US were dominated by the oaks prior to European settlement. This was particularly true of the longer-lived and more widely distributed white oak, compared with red oak (*Q. rubra*) and chestnut oak (*Q. montana*). In addition to clear-cutting and catastrophic fires, much of this loss of white oak is attributed to fire suppression and increasing herbivory by white-tailed deer (*Odocoileus virginianus*) (Abrams 2003), which is assumed to have undermined oak regeneration throughout eastern deciduous forests (but see Abrams and Johnson 2012). Disturbance due to natural fires and those set by native Americans appear to have dramatically favored oak establishment and recruitment and its dominance in many ecosystems of North America prior to extensive European settlement in the mid-1800s (Dey 2002).

Shumway et al. (2001) documented historic patterns of oak recruitment in relation to non-catastrophic fire over a 400-year period in an old-growth forest in western Maryland. During the 17th, 18th, and 19th centuries, major fires occurred on average every 7.6 years until 1930 after which fire suppression policies prevented any major fires. During these first three centuries, oak was regularly recruited into the forest until fires were no longer evident (1930), after which recruitment increased for red maple (*Acer rubrum*) and black birch (*Bentula lenta*) and declined for oak (Shumway et al. 2001). Hence, it is argued that the most common factor likely contributing to the dominance of oak in these presettlement forests was drier conditions and the frequent occurrence of fire (Abrams 1992; Abrams and Seischab 1997; but see Clark and Royall 1996). Even today, at sites in eastern deciduous forests where fire is frequent and occasionally severe enough to create some canopy openings and thinner tree densities, oaks appear to achieve successful regeneration (Signell et al. 2005).

According to Nowacki and Abrams (2008), fire suppression by humans, formalized by law in the early 1900s, favored shade-tolerant species that grew best in cool, moist shaded conditions over the fire-tolerant oaks that required drier, open areas with sufficient light. These shade-tolerant species, in turn, created conditions that further favor their establishment and growth, a process that Nowacki and Abrams (2008) refer to as "mesophication." The authors argue that this process makes it increasingly difficult to reverse the conditions necessary for oak establishment, as increased mesophication further modifies forest conditions.

As Abrams (2003) describes, this transformation of the landscape and decline in forest health was further complicated by invasive species and heavy browsing by white-tailed deer. Although not contributing to the direct loss of oak, the chestnut blight (*Cryphonectria parasitica*), a fungal pathogen introduced near the beginning of the 20th century, which had contributed to the near extinction of the American chestnut (*Castanea dentata*), likely dramatically changed the dynamics of oak dispersal and regeneration in this region (see Chapter 7). Although our studies suggest that white oak has a clear dispersal disadvantage in the presence of American chestnut (Lichti et al. 2014), the complete elimination of the chestnut would have also dramatically lowered the overall abundance of food for rodents and thus the likelihood of any oak dispersal in low-mast years.

The slow degradation of oak forests in the eastern US appears therefore to be the result of numerous causes that have collectively reduced the probability that an acorn is dispersed, a seedling established, and a sapling recruited into the population. Indeed, there are many questions that remain and certainly no simple solution to this problem.

TIMBERING OF THE OAKS: SILVICULTURE PRACTICES AND THE PROMOTION OF DISPERSAL AND REGENERATION OF THE OAKS IN MANAGED SYSTEMS

Silviculture is the refined practice of managing forests for a variety of human needs. Although silviculture is borne out of the need to manage forests for utilitarian purposes, it also emphasizes the importance of maintaining forests for other critical functions, such as habitat management, maintenance of biodiversity, and carbon-sequestration (Johnson et al. 2009). Silviculture is thus an evolving practice of forest management that falls at the interface of the ecological sciences, social sciences, humanities, and environmental law and policy (Johnson et al. 2009). Central to its mission is the goal of managing forests for these multiple objectives in a way that also maximizes forest regeneration following harvest. However, whereas maximum regeneration is a central goal in silviculture practices, only recently have studies focused on the importance of understanding acorn dispersal, seedling establishment, and seedling growth in the context of these managed environments.

New studies, involving careful experimentation in the context of advanced silviculture practices on oak regeneration out of the Department of Forestry and Natural Resources of Purdue University, have begun to investigate patterns of oak infestation, acorn dispersal, seedling establishment, and the early stages of regeneration in the context of large-scale silviculture disturbance. Oak regeneration in the Central Hardwoods Region of the United States has been faced with significant challenges due to a combination of factors, including light competition with more shade-tolerant species, and the large scale suppression of fire since the early 20th century, which otherwise helps to control competitors of oak that are less fire tolerant (Dey 2002). As described by Dey (2002), the silviculture methodology that best facilitates oak regeneration is gradual shelterwood harvesting coupled with low temperature, prescribed burns. Shelterwood harvesting involves the removal of canopy trees in a mature forest in a way that offers more light for seedlings. Progressive or expanding group shelterwood removal (also known as Femelschlag) rather than that done all at one time provides the light requirements that best favor oak. In the case of the oaks, such a method is complicated because excessive light on the forest floor can also favor competitors (Dey 2002; Kellner et al. 2014).

In one of the first studies of its kind, Kellner et al. (2014) tracked rates of pre-dispersal predation of acorns and patterns of acorn removal by rodents (eastern gray squirrel, *Sciurus carolinensis*, and white-footed mice, *Permomyscus leucopus*) of both black oaks (*Q. velutina*) and white oaks (*Q. alba*) three years prior to silviculture treatment and three years following treatment. Shelterwood harvest involving midstory removal resulted in a reduction in weevil infestation but no changes in patterns of acorn removal by mammals. In a related study, Kellner et al. (2016) tracked tagged acorns dispersed by gray squirrels, white-footed mice, and eastern chipmunks (*Tamias striatus*) on progressive shelterwood sites and undisturbed control sites. Significantly more acorns were removed from sources on forested sites than those presented on undisturbed sites. Acorns on shelterwood sites were also less likely to survive the winter. The authors attributed higher acorn predation rates and an overall reduction in "dispersal effectiveness" on the increased ground cover and woody debris that likely increases the abundance and activity of small mammals (Kellner et al. 2014, 2016). Based on some of the empirical data from these studies, Kellner and Swihart (2017) developed an individual, spatially explicit model (Simulate Oak Early Life history [SOEL]) that can predict patterns of oak regeneration under a variety of conditions. This model, in effect, connects the effects of forest disturbance on acorn dispersal, acorn germination, seedling emergence, and seedling and sapling growth and, ultimately, patterns of regeneration.

Working in this same system, under the direction of Michael Sanders and Robert Swihart of Purdue University, Skye Greenler's recent thesis (Greenler 2018) focused on the specific effects of expanding shelterwood harvests and controlled burns on patterns of acorn dispersal and oak regeneration. By monitoring and mapping woody tree regeneration, Greenler found that the expanding shelterwood harvests, which increase edge-interior ratios on the perimeter of harvest gaps, was successful at producing the variable and intermediate light levels that favor oak regeneration in all but the most mesic sites. She also observed that the survival of both dispersed acorns and those placed in artificial caches were higher in harvested stands than control stands and in burned versus unburned stands. She argues that prescribed burns following acorn dispersal decreases ground cover, which in turn, increases rodent predation risks, decreases pilferage rates, and facilitates patterns of oak establishment. As Greenler aptly notes, "these results suggest that the environmental conditions following timber harvesting and prescribed fire shifts the oak-granivore mutualism in a direction beneficial for oak regeneration" (Greenler 2018, p. 26).

FOREST FRAGMENTATION

Often, when oak forests are cut, the result is not complete elimination of entire stands but removal of forest blocks that results in a network of smaller patches (Fig. 13.3). As any habitat like an oak forest becomes increasingly fragmented, the ecological properties of the system begin to change. As stand size decreases, for example, the relative importance of edge habitat (ecotone between the core forest and surrounding matrix) becomes increasingly significant for the patch. This, in turn, can result in a change in the relative abundance of edge species compared to larger stands where species that prefer interior forests may dominate. Just different shapes of forest fragments of the same size could result in different species assemblages and therefore different ecosystem processes. Also, as habitat patches become increasingly isolated, the effectiveness of corridors between fragments also determines how species within the patch may be able to move between habitat fragments (Smith and Smith 2012). Habitat fragmentation also changes ecological processes, such as patterns of seed dispersal, but can be overcome with corridors between fragments, which promote dispersal, establishment, and greater species richness in the face of habitat loss (Levey et al. 2005; Damschen et al. 2006).

Our understanding of the effects of fragmentation on various ecological processes is grounded in the well-established field of landscape ecology (Harris 1984; Wiens 1992), largely derived from the basic applications of the classic

13.3. Images of **(a)** continuous closed-canopy forest, and **(b)** fragmented forest patches as a result of agriculture. *(Shealyn A. Marino)*

theory of island biogeography (MacArthur and Wilson 1963). Marked by the introduction of its own important journal in the early 1990s, *Landscape Ecology*, this field of investigation remains at the center of conservation of biodiversity.

Studies on the effects of forest fragmentation on oak biology are somewhat limited to subject and location, and focus extensively on how fragmentation patterns and isolation of forest patches influence genetic variation and gene flow, especially as it relates to patterns of pollination (Table 13.1). Many of these studies, conducted predominantly in the Midwestern US (on Bur oak [*Q. macrocarpa*]) and the Iberian Peninsula (on holm oak [*Q. ilex*]), suggest that genetic variation is high across fragmented stands and that there appears to be limited effects of fragmentation on genetic structure primarily because of pollination success (Table 13.1). One exception to this generalization is reported for *Q. oleoides* in Costa Rica, where Deacon and Cavender-Bares (2015) observed distinct genetic differences in isolated stands of this species. The numerous studies on holm oak in the Iberian Peninsula suggest that although gene flow between forest fragments may be reasonably high due to pollination, there is evidence of reduced genetic variation and evidence of clonal events in younger trees in response to fragmentation (Table 13.1).

Beyond the genetic studies employed to assess the effects of fragmentation on oaks, there has been limited research on how fragmentation influences other aspects of oak ecology, especially acorn dispersal. One such example is a study by Moore and Swihart (2008) in which the authors examined patterns of seed dispersal of several hardwood species, including two oaks (Table 13.1), across highly fragmented stands of hardwoods that characterizes the Midwestern US. The authors reported that one critical dispersal agent, the eastern gray squirrel, is lost from isolated fragments across this landscape due to its inability to disperse across the open matrix of agricultural fields (Zollner 2000). However,

Table 13.1. Overview of studies on the effects of forest fragmentation on oaks.

SPECIES	LOCATION	PRIMARY FINDINGS	REFERENCES
Bur oak (*Q. macrocarpa*)	Illinois, US	Genetic variation across highly fragmented stands is extremely high (97%) compared with variation within forests (3%); fragmentation has limited effect on genetic structure across the landscape	Craft and Ashley 2007
Bur oak	Illinois, US	Gene flow due to pollination in extremely isolated stands is considerably high (~50%), indicating that such stands are not reproductively isolated	Craft and Ashley 2010
Q. deserticola	Michoacán, Mexico	Infection by mistletoe parasite (*Psittacanthus calyculatus*) negatively related to forest fragment size; insect herbivory higher in oaks in larger fragments possibly as a result of physiological interactions of mistletoe and oaks	Cuevas-Reyes et al. 2017
Q. oleoides	NW Costa Rica	Highly fragmented populations of this species show distinct genetic differences but no evidence of local adaptation	Deacon and Cavender-Bares 2015
Bur oak	Illinois, US	Pollination success not related to distance or direction from maternal tree, genetic relatedness to maternal tree, or size of pollen source tree	Dow and Ashley 1998
Tanoak (*Notholithocarpus densiflorus*) and potentially other California oaks	California, US	Connectivity between forest stands important for spread of sudden oak death but not as critical as other environmental factors	Ellis et al. 2010
Holm oak (*Q. ilex*)	Spain and US	Genetic studies suggest considerable gene flow between populations, but protected glacial populations have not been a critical source of genetic diversity	Guzmán et al. 2015
Mexican red oak (*Q. castanea*)	Central Mexico	Examined genetic diversity in seedlings and adult trees across 33 forest fragments and found that genetic diversity of seedlings, but not adults, was correlated with fragment size, indicating that forest fragmentation is limiting genetic connectivity via seed dispersal	Herrera-Arroyo et al. 2013
Review of several studies on oaks	North America	Pollen movement in oaks may be limited by forest fragmentation	Koenig and Ashley 2003
Pin oak (*Q. palustris*) and Northern red oak (*Q. rubra*) as well as shagbark hickory (*Carya ovata*) and black walnut (*Juglans nigra*)	Indiana, US	Examined dispersal of 5 hardwood species in 18 forest fragments in Midwestern US, including some isolated fragments where gray squirrels were no longer present; in the absence of gray squirrels, dispersal was successfully achieved by larger bodied fox squirrels (*S. niger*), which tolerated significant fragmentation	Moore et al. 2007

SPECIES	LOCATION	PRIMARY FINDINGS	REFERENCES
Holm oak (*Q. ilex*)	Central Spain	Dispersal effectiveness higher in larger rather than smaller fragments where greater stand density favored dispersal; delayed breeding in southern sites resulted in lower rates of acorn dispersal	Morán-López et al. 2015
Holm oak	Central Spain	Caching rates lower in forest fragments; seed size drives dispersal where mouse (*Apodemus sylvaticus*) densities and competition is higher; dispersal more effective in larger stands	Morán-López et al. 2016b
Holm oak	Central Spain	Pollen flow generally <100 m but predominantly sourced from larger fragments; gene flow constrained by limited seed dispersal	Morán-López et al. 2016b
Holm oak	Central Spain	Fragmentation reduced genetic diversity (and heterozygosity) of younger trees established after fragmentation events; also found evidence of some clonal events following fragmentation	Ortego et al. 2010
Holm oak		Limited evidence of the effect of fragmentation on mating patterns, genetic diversity, and seed production	Ortego et al. 2014
Q. laeta	Central Mexico	Buried acorns escaped seed predation by seed predators (*Peromyscus* and *Aphelocoma* spp.) in abandoned agricultural fields and mixed habitats surrounding forest patches, but were recovered at the same rates as unburied acorns in forest fragments and fragment edges	Ramos-Palacios and Badano 2014
Holm oak	NE, Spain	Weevil (*Curculio elephans* and *C. glandium*) prevalence higher in old forests compared with connected new forests and isolated new forests, whereas moth larvae with greater vagility did not differ in prevalence between forest stands; the net effect was no difference in overall acorn predation by insects	Ruiz-Carbayo et al. 2018
Numerous oaks (e.g., Bear oak [*Quercus ilicifolia*], black oak [*Q. velutina*], and white oak [*Q. alba*])	New Jersey Pine Barrens, US	Conducted simulation of current conditions of fragmentation and limited prescribed burning to determine changes in forest structure since precolonial times; found that landscape is shifting from pitch pine-dominated to oak-dominated landscape, especially where fire is limited; fragmentation had limited effect on successional dynamics compared with fire	Scheller et al. 2008
Castanopsis sclerophylla	Central China	Examined predation by *Curculio* spp. in forests of tanoak across islands and peninsulas of Thousand-Island Lakes; predation rates and relationship between crop size and insect predation was significantly influenced by island (forest) size and isolation from mainland	Tong et al. 2017

SPECIES	LOCATION	PRIMARY FINDINGS	REFERENCES
Black oak	Michigan, US	Mean number of egg masses of gypsy moth larvae (*Lymantria dispar*) decreases with stand isolation; distribution of egg masses also differs with matrix type (old field vs. swamp vegetation)	Vandermeer et al. 2001
Pedunculate oak (*Q. robur*)	Northern Belgium	Measured genetic diversity in adult trees and seedlings across forest stands of different tree density and found higher diversity among adults when all samples were pooled across forest patches; no differences were observed at the stand level	Vranckx et al. 2014

they also observed that the larger bodied fox squirrel, capable of such dispersal, occurs in all fragments and therefore adequately compensated for the lack of dispersal by gray squirrels in these more isolated stands.

In fragmented forests of holm oak, Morán-López et al. (2016a) found that rate of acorn caching by its primary dispersal agent, the wood mouse (*Apodemus sylvaticus*), is significantly reduced in smaller forest fragments. The authors attribute this primarily to lower mouse densities and a reduced necessity for caching, compared with conditions in larger stands. They conclude that dispersal is less effective in smaller stands, and in a related study suggest that gene flow may be affected by these constraints on dispersal (Morán-López et al. 2016b; see also Ramos-Palacios and Badano (2014) for similar results for fragmented forests in central Mexico).

THE LOSS OF THE AMERICAN CHESTNUT AND ITS IMPACT ON OAK ECOLOGY

The American chestnut, once widely distributed across much of eastern North America (Fig. 13.4), frequently accounted for 25% to 50% of forest canopy where it occurred (Russel 1987; Foster et al. 2002; Dalgleish and Swihart 2012). This impressive tree, often associated with oak stands, was nearly eliminated by the middle of the 20th century due to the chestnut blight, *Cryphonectria parasitica*. Introduced from SE Asia in 1904, this fungal blight rapidly spread and killed adult trees across its entire range. As a critical food source for many wildlife, the loss of American chestnut drastically altered ecosystem dynamics, including patterns of masting, dispersal, and the population dynamics of several rodent species critical for nut dispersal.

As reviewed by Dalgleish and Swihart (2012), several characteristics of chestnut likely contributed to tremendous alterations in forest dynamics. First,

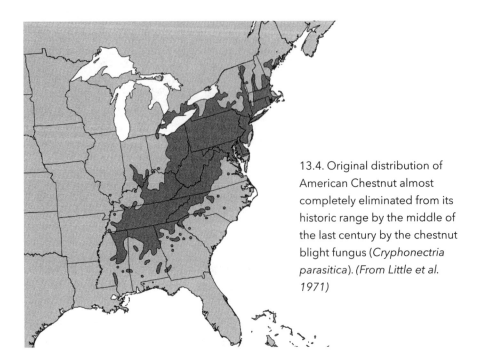

13.4. Original distribution of American Chestnut almost completely eliminated from its historic range by the middle of the last century by the chestnut blight fungus (*Cryphonectria parasitica*). (*From Little et al. 1971*)

chestnut produced prolific nut crops on an annual basis that were well protected from seed predators on the tree, but once shed, were made readily available to rodent seed consumers and dispersers on the ground. This meant that when other species, such as oaks, experienced mast failures there was alternative nuts to store and consume. In addition, the nuts of chestnut were highly palatable (low in tannin) and exhibited a short dormancy. So, in the presence of oaks, chestnuts were also dispersed and cached. And as I discussed in detail in Chapter 7, American chestnut's increased palatability and short dormancy provided this species with a dispersal advantage in the presence of white oaks but not red oaks because of their size and extended dormancy (Lichti et al. 2014). This likely resulted in markedly different patterns of oak and chestnut dispersal in forests dominated with red oak species versus those with white oak species.

Today, in the absence of the American chestnut, overall mast abundance is lower and far more episodic. Based on historic data at two sites, Dalgleish and Swihart (2012) modeled potential patterns of mast production and the population dynamics of rodents (eastern gray squirrels [*Sciurus carolinensis*], eastern chipmunks [*Tamias striatus*], the white-footed mouse [*Peromyscus leucopus*]), and the white-tailed deer (*Odocoileus virginianus*), all of which depended heavily on oak and chestnut. Their models indicated that overall mast abundance decreased by 80% and 35% at the two sites, and that variation in mast (measured by CV-coefficient of variation) increased by 60% and 76%, respectively, because of the absence of chestnuts. The authors' simulations of mammal populations estimated a decline in all four species following chestnut loss, but a most pronounced drop (48%) in the population of the white-footed mouse. This estimate was coupled with a 57% increase in annual variation in mouse numbers. Overall, these results provide some quantitative estimates of just how significantly American chestnut contributed to forest dynamics, and further forecast the potential alterations that will likely follow from the planned reintroduction of the blight-resistant chestnut.

CLIMATE CHANGE AND OAK REPRODUCTIVE ECOLOGY

Forest ecologists are particularly interested in human-induced climate change because it has the potential to significantly impact trees across the globe. Global warming trends, of course, follow directly from increases in carbon dioxide (CO_2), which in turn influence carbon sequestration by trees, as well as seedling and tree development and mortality. However, most studies that investigate the effects of CO_2 on tree physiology and growth do so at an ecosystem scale. Few studies have considered how increases in CO_2 likely influence the reproductive biology of individual trees. In one of the first studies to explore such a question experimentally, Stiling et al. (2004), working on the property of the Kennedy Space Center in southern Florida, grew three scrub oak species (*Q. geminata*, *Q. chapmanii*, and *Q. myrtifolia*) in chambers in which they could expose one group of the three oak species to elevated CO_2 levels and the other control group to normal CO_2 levels. Elevated CO_2 resulted in a significant increase in acorn production in the latter two species but not the first. The authors observed no effect of CO_2 on acorn mass, germination rate, or levels of weevil (*Curculio* spp.) infestation in any of the three species. The authors suggested that the increase in reproductive output is likely to provide a recruitment advantage in two of these trees at the cost of the other, *Q. geminata*, a less competitive species. They predicted that over time, forest structure would likely change as a result.

Other aspects of human-induced climate change that are likely to influence oak ecology is the frequency of both drought and late spring frosts. Periodic drought, and to a lesser extent late spring frosts, have been shown to cause abrupt decreases in growth, followed by mortality in nine tree species, including oaks (Vanoni et al. 2016). In addition, increasing warming trends often resulted in premature flowering in oaks in the spring, for example, if followed by late spring frosts. This could result in death of flowers and then failure of acorn crops in white oaks for the current year and red oak species in the subsequent year. We are currently investigating the effects of late spring frosts on masting patterns in oaks of northeastern Pennsylvania. Our analyses of these data have already demonstrated support for the veto hypothesis to explain masting patterns in oaks (Bogdziewicz et al. 2018a). This veto hypothesis predicts that external environmental factors (e.g., late spring frosts) result in correlated reproductive failure and, in turn, coordination of the resource budget of the tree population. Analyses of our data revealed that such a veto of reproduction followed by a coordinated resource gain adequately explained masting without invoking other explanations of masting patterns (i.e., pollen coupling) at our study areas.

We are now exploring the extent to which important weather events (e.g., late spring frosts and droughts) are the vetoes that inhibit reproduction and coordinate mast cycles in oaks. Severe droughts, especially in late summer, frequently result in large-scale abortion of immature acorns, followed by mast failure. Thus, the combination of these extreme weather events may not only contribute to masting patterns but also increase the variation in masting cycles.

EFFECTS OF INCREASED SOIL NITROGEN ON OAK REPRODUCTION AND REGENERATION

Another significant impact of humans on their environment involves their modification to the nitrogen cycle as a result of the burning of fossil fuels for transportation, agriculture, and other industrial uses (Galloway et al. 2008), as well as significant deposition of nitrogen in both the aquatic and terrestrial environment (Galloway et al. 2008; Bogdziewicz et al. 2017a, and references therein). In the terrestrial world, nitrogen (e.g., fertilizer) often enhances plant growth, but its impact on many other aspects of individual plant biology, reproduction, and seed dispersal are less well understood.

As reviewed by Bogdziewicz et al. (2017a), studies of the impact of nitrogen deposition on tree growth and reproduction are mixed, with some research—even in oaks—showing an increase in both growth and acorn production (Callahan et al. 2008). In a similar system, Yi et al. (2016) showed that even short-term supplementation of soil nitrogen increased seed quality (the ratio of edible seed to seed coat) of the Korean pine (*Pinus koraiensis*) and, in turn, enhanced dispersal and caching by both the Siberian chipmunk (*Tamias sibiricus*) and the Korean field mouse (*Apodemus peninsulae*). However, nitrogen deposition can also have significant negative effects by depressing seedling survival, increasing herbivory, and decreasing regeneration in many temperate forest species (Bogdziewicz et al. 2017a).

Recognizing the complexity of these interactions and the many ways in which nitrogen could both enhance or depress dispersal and regeneration in oaks, my colleague, Dr. Michal Bogdziewicz, seized a perfect opportunity to explore this further. Michal wisely chose to conduct part of his dissertation research on the Long Term Ecological Research (LTER) site at Harvard Forest Research Station in western Massachusetts, where, among other long-term studies, researchers there have been supplementing forest patches with nitrogen since 1988. As I was supported by a fellowship at Harvard Forest when Michal began this research, I was afforded the wonderful opportunity to assist in this endeavor. The Chronic Nitrogen Amendment Study at Harvard Forest involved monthly applications of fertilizer (NH4NO3) during the growing season to study plots in middle-age forest dominated by northern red oaks (*Q. rubra*), first planted in the middle of the 20th century.

Working under the guidance of Dr. Rafal Zwolak, his major professor in Poland (University), and Dr. Elizabeth Crone, now a professor at Tuft's University, Michal's goals were well focused and aimed at assessing the impact of nitrogen on all aspects of the oak dispersal process, from acorn production through seedling establishment. Over the three-year study, he collected acorns from individual trees on high-nitrogen and low-nitrogen control sites, estimated acorn production for individual trees, measured acorn size, conducted dispersal experiments with acorns from contrasting treatment sites, and prepared artificial caches of acorns to determine patterns of germination and establishment under conditions in which rodents were both excluded or allowed access to these "cached" acorns. My laboratory also assisted with measurement of tannin gradients and overall tannin levels in acorns.

Whereas some of the results of Michal's experiments were as expected, the overall story was quite compelling. As previously demonstrated by Callahan et al. (2008) at the same site, acorn numbers were indeed higher, in fact as much as 2–9 times higher, in trees receiving high nitrogen treatments compared with control trees. Acorns in these trees, however, were smaller in overall size and were less likely to be eaten at the source or dispersed by rodents. And, although acorns at the source experienced less predation, establishment is less likely under a parent tree because of light competition and other density-dependent factors reducing seed and seedling survival. Even more interesting, however, was a notable increase (10–20%) in infestation by weevils, a similar increase in the damage to acorn cotyledon (20–30%), and a marked decrease (200–300%) in acorn germination irrespective of insect and rodent damage. Hence, the overall effect was a decrease (3–29%) in the rate of seedling establishment. From this study, it appears that several indirect effects of nitrogen supplementation can contribute to unexpected results in the multiple oak-animal interactions that collectively contribute to the dispersal process. In Bogdziewicz et al. (2017a), for example, we argue that the tremendous response of the insects was likely the result of disruption in the masting process, which interfered with normal process of predator satiation of the weevils. The nitrogen supplement likely favored weevil growth and fecundity and dampened the impact of mast failure, thereby increasing overall weevil damage to acorns, despite the increase in acorn production. It seems to follow that this unexpected result would most certainly constrain oak recruitment and regeneration.

INVASIVE SPECIES AND THE SPECIAL CASE OF INVASIVE OAKS

The invasion of alien (non-native) species that colonize and often proliferate in areas outside their historic (native) ranges, often as result of release from native predators, competitors, or parasites (Wolfe 2002; Torchin et al. 2003), represents a growing problem worldwide. This increasing challenge is often the result of human movement across the global landscape for the purposes of travel and trade; but now, climate change facilitates species' movement as well. As detailed by the Center for Invasive Species Prevention, the oaks in North America, just as an example, are susceptible to a diversity of invasive species. Included among the most significant threats are the European gypsy moth (*Lymantria dispar*) and winter moth (*Operophtera brumata*), both responsible for severe defoliation and tree death, and the fungus *Ceratocystis fagacearum* that causes the condition known as oak wilt. One of the most significant outbreaks in California, only first discovered near the end of the last century, is the disease known as sudden oak death caused by *Phytophthora ramorum*, a fungal pathogen that is responsible now for the death of a million tanoaks (*Notholithocarpus densiflorus*), thousands of coastal live oaks (*Q. agrifolia*), and two other oak species as well. Several other species of wood-boring insects can also be a problem. I limit my discussion on this topic, because a detailed treatment of these alien species is well beyond the scope of this book. In most cases, these invasive species are attacking adult trees and generally not directly influencing acorn dispersal, seedling establishment, and recruitment.

However, a particularly interesting case of invasion involves the establishment and spread of oaks themselves in non-native forests. Here I discuss two such examples: the northern red oak (*Q. rubra*) of North America, which is established and spreading in Europe, and the sawtooth oak (*Q. acutissima*), an oak native to southeast Asia, which is now well established as an ornamental and is invading isolated sites in the eastern half of the United States. In yet another study led by Michal Bogdziewicz, several of us investigated the effects of larval insect seed predation (both weevils [*Curculio*] and moths [*Cydia*]) in the invasive red oak in Poland. Introduced in the 17th century, the northern red oak is now recognized as a leading invasive tree species in Europe with a significant recruitment advantage over native oaks, as well as other tree species (Major et al. 2013). It also has a negative impact on other understory flora (see references in Bogdziewicz et al. 2018d).

Over a three-year period, we compared patterns of insect infestation in acorns of native sessile oak (*Q. petrea*) and invasive northern red oak in Poland, as well as northern red oak in its native range in North America (Bogdziewicz et al. 2018d). Compared with red oaks in its native range and sessile oaks in Poland, red oaks in their invasive range harbored 10 times fewer weevil larvae. The prevalence of moth larvae did not vary between these three types of oaks, but their prevalence was generally quite low, <5%. Of particular interest was the probability of weevils surviving to emergence from the acorn. By selecting acorns with oviposition scars and no emergence holes and then following subsequent larval emergence in the laboratory, we were also able to determine rates of larval survival and emergence in acorns of these three tree types. The results were compelling. In acorns of the invasive red oaks, once infested, weevil emergence rates were 20 times less likely than those of sessile oak and red oaks from their native range. We also found that tannin levels in acorns of native red oaks were nearly two times higher than that of their invasive counterpart.

The results of Bogdziewicz et al. (2018d) collectively suggest that in its invasive range, the red oak may be released from many of the selective pressures typically exerted by insect seed predators. Acorns of this invasive red oak may also, in effect, trap and kill insect larvae and, in turn, reduce numbers of insect seed predators, thereby enhancing the species ability to invade.

Myczko et al. (2014) present preliminary evidence that, in Poland, the Eurasian jay (*Garrulus glandarius*) will remove acorns of the pedunculate oak (*Q. robur*) twice as often as those of the non-native northern red oak, although nothing was reported on the final fate of these removed acorns. The authors also reported, however, that seedling establishment of the non-native red oak was significantly greater than that of the native species. Moving forward, it will be important to better understand differences in dispersal of these oaks and the patterns of establishment and recruitment that follow directly from these dispersal patterns.

A second example of an invasive oak is the sawtooth oak (Section *Cerris*; *Q. acutissima*), commonly planted throughout the eastern US as an ornamental tree (Fig. 13.5). Native to southeast Asia (from the Himalayas to China, south through Indochina, Korea, and Japan), this oak is now found in more than 15 states in the eastern US and some isolated sites in Europe. Typically planted in residential and government properties, as well as a food supplement for wildlife, this species appears to sometimes establish and spread throughout neighboring forests. Although considered naturalized in some areas, it is also now firmly recognized as an invasive species.

My interest in the sawtooth oak was first aroused by my discovery of a mature individual of this species on my daughter's college campus just outside one of the university dining halls in Philadelphia. After saying my goodbyes to my daughter one evening, I decided I needed a sample of acorns that I could exam-

13.5. Shown are **(a)** the adult tree. (Ruud De Block) **(b)** The acorn and leaves of *Quercus acutissima*, arguably the most invasive oak in North America. *(Andrey Zharkikh)* SEE COLOR PLATE

ine more closely. To this day, my daughter reminds me of her disbelief, as she sat for dinner with her roommates only to witness her father on his hands and knees collecting acorns just outside the dining hall window. Two years later, however, I was comforted to know she had recovered from her devastation when, as a senior, she asked if I needed her to collect some acorns from the same tree. Soon after learning of this lone tree, I discovered a number of mature trees at a rest stop on a Maryland freeway, a line of immature trees in an urban neighborhood in Philadelphia, and several patches of this tree species at two sites in Pennsylvania Game Commission forests in Schuylkill County, Pennsylvania. These latter populations of the invasive tree species, brought to my attention by my research associate, Shea Marino, led us to investigate further the status of the sawtooth oak in the eastern US. After a thorough literature search and several discussions, we decided to assign an undergraduate research team to determine if and how these acorns are dispersed by native scatterhoarders.

Because acorns of sawtooth oak (STO) are large, typically as large or larger than acorns of northern red oak (RO), and appear to have a short dormancy period, we hypothesized that rodents may, under some circumstances, selectively disperse acorns of STO over native species, such as RO and white oak (*Q. alba*; WO). We also assumed that because of the diameter of these acorns, corvids, such as the blue jay (*Cyanocitta cristata*), were unlikely to disperse these acorns. To test the response of rodents to STO, we worked with this group of undergraduate students to conduct a preliminary experiment much like that reported by Lichti et al. (2014), in which we performed pair-wise dispersal experiments between American chestnut (*Castanea dentata*), RO, and WO (see Chapter 8). In this STO study, students tagged 6000 nuts of each of the three species (STO, RO, and WO) and presented these metal-tagged acorns in exclosures that permitted access by only rodents. In each seed box, rodents were presented with 200 acorns of each of two species: either STO and RO, STO and WO, or RO and WO. Each of these three treatments were replicated with five exclosures at each of three study areas: two where STO had established and regularly produced acorns and one where STO was absent. Students presented these 400 tagged acorns in each of the 45 exclosures (18,000 acorns) in November and then followed removal rates and acorn fates over the next few months.

This study required considerable effort and was a tall order for three students who were simply completing a required senior research project for their degree. Moreover, the results were somewhat preliminary because the study was conducted in a low-mast year. As a result, most acorns were more likely to be eaten, not cached—and if cached, they were larderhoarded below ground. This, of course, limited the conclusions that could be drawn. In addition, as with most metal tag studies, only a small percentage of tagged acorns were recovered. In this case, it was only 1230 recovered tags and tagged acorns, <7% of those pre-

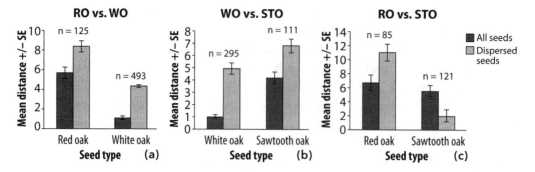

13.6. Shown is the dispersal of acorns in three experimental treatments in which each represented a unique pair of acorns **(a)** RO/WO, **(b)** RO/STO, and **(c)** WO/STO (RO = red oak [*Q. rubra*], WO = white oak [*Q. alba*], and STO = sawtooth oak [*Q. acutissima*]). Shown are the dispersal distances of all acorns and those that were only dispersed and cached (± SE). Sample sizes represent the number of each species recovered in each treatment. *(Michael A. Steele, unpublished data)*

sented. Nevertheless, we logged some interesting observations. First, nearly all acorns were removed from the 45 exclosures in approximately 16 days of presentation. In addition, removal rates differed little between species, with only a slight delay in the removal rate of STO compared with the other two species. However, even at the site where STO trees were absent, the rodents eventually removed nearly all the tagged STO acorns within the 16 days. Thus, it was apparent that the naïve rodents had sampled the acorns and were assessing their potential for future use. For each tag or tagged acorn recovered, the research team also determined dispersal from the seed source. As expected, RO acorns were dispersed and cached more often and farther from the source than those of WO acorns (Fig. 13.6). A similar pattern was observed between STO and WO, with the former being dispersed farther and cached more often than the latter (Fig. 13.6c). When RO and STO were paired, RO acorns were carried significantly farther than those of STO before being cached (Fig. 13.6b). Although the total number of recovered cached acorns was quite low (n = 41), the frequency of acorns germinating was highest for STO (Fig. 13.7). Hence, the conclusions that can be drawn from this study are indeed preliminary; yet, they strongly suggest that STO may have a dispersal advantage over WO. Our plans to repeat this experiment during a mast year and to carefully investigate the effects of insect seed predators on STO should provide a more complete picture of the advantages STO is afforded in ecosystems where it is considered invasive. As with the northern red oak in its invasive range in Europe, it will be necessary to study the acorn fate, from maturation and survival of insect attack, to dispersal, ger-

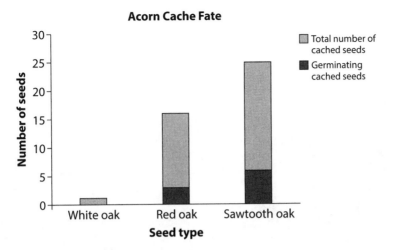

13.7. Frequency of white oak (*Q. alba*), red oak (*Q. rubra*), and sawtooth oak (*Q. acutissima*) acorns dispersed, cached, and germinated in caches. *(Michael A. Steele, unpublished data)*

mination, establishment, and recruitment to fully understand the STO's ability to establish and invade in its non-native range.

In Chapter 14, I now extend my discussion to oak ecosystems across the globe that are considered highly vulnerable. I highlight specific systems with which I am most familiar and those where limits to oak dispersal, oak seedling establishment, and forest regeneration are likely the primary factors contributing to conservation concerns.

THREATENED
OAK FORESTS AND
OAK FOREST
CONSERVATION

14

INTRODUCTION

From eastern North America, where I have concentrated much of my discussion on oak dispersal ecology and where we arguably understand the human impact on oak forest dynamics the best, I now move to some of the endangered oak ecosystems across the globe. These include the Garry oak (*Quercus garryana*) ecosystem of the Pacific Northwest of the United States and southwest Canada, the cork oak (*Q. suber*) forests and other oaks of the Mediterranean, the bear oak (*Q. ilicifolia*) found in the pine barrens of the northeastern US, the California oaks, and the dry forests of *Quercus oleoides* in Costa Rica. Across these studies, I will discuss how fire suppression and various seed predators, pathogens, and diseases threaten these systems, and I will consider a common problem that hinders management and protection in *all* of these systems—the paucity of information on oak dispersal and its underlying role in regeneration. I also note periodically how issues related to climate warming may restrict regeneration even in the face of suitable dispersal and acorn germination. I then close this chapter with a brief glimpse at the Red List of Oaks (Oldfield and Eastwood 2007), highlighting the oak species of greatest conservation concern. My coverage here is only a glimpse at threatened systems and is not intended as a comprehensive review, as I am well aware that many such systems are not included.

14.1. Garry oak (*Q. garryana*) of northwestern North America. Shown are **(a)** Garry oaks on the southeastern shore of Vancouver Island, Canada, **(b)** a mixed stand of Garry oak and Douglas fir (*Pseudotsuga menziesii*) at Piper's Cove, Vancouver Island near Nanaimo, British Columbia, and **(c)** leaves of Garry oak. (*Michael A. Steele*) SEE COLOR PLATE

GARRY OAK FORESTS OF THE PACIFIC NORTHWEST

Garry oak is a white oak species found in the Pacific Northwest coast of the United States and Canada (Fig. 14.1). The species usually grows up to 30 m but can also occur as a scrub oak, especially in the southern part of its range.

The species' limited distribution extends from southeastern British Columbia to southern California. In Canada, the species' range is limited to the southeastern coast of Vancouver Island, adjacent coastal islands, and three relatively isolated stands on the mainland of southwestern British Columbia (Garry Oak Ecosystem Recovery Team 2003). It is estimated that the current range of Garry oak in Canada consists of <5% of its original range, prior to European settlement (Jones et al. 2011). From southwestern British Columbia, the distribution of Garry oak extends southward primarily on the west coast of the Cascade Range into southern California.

The Garry oak ecosystem can vary tremendously in structure as it can consist of closed-canopy oak forests, mixed stands of Garry oak and other species (e.g., Douglas fir [*Pseudotsuga menziesii*]), open meadows with scattered Garry oak, or even scattered trees on rocky coastal outcrops (Garry Oak Ecosystem Recovery

Team 2003). And, although the ecosystems associated with Garry oak comprise a relatively small geographic area, they tend to be associated with a tremendous biodiversity, which includes a high number of species of conservation concern. Clements (2013) makes this point best. Occupying only 20 km^2 in Canada, this habitat is home to 10% of all species at *At Risk* in Canada. Approximately 30 species of plants (Clements [2013]) and another 70 species of mammals, birds, reptiles, insects, and one earthworm are associated with the Garry oak ecosystem (Garry Oak Ecosystem Recovery Team 2003).

My interest in the dispersal of Garry oak was first brought to my attention by my close friend and colleague, Dr. Timothy Goater, professor of biology at Vancouver Island University. As one of my closest friends from graduate school, Tim knew well of my research on oaks and insisted that I consider research in this system, which I plan to pursue in the near future.

Despite the tremendous concerns and fragility of the Garry oak ecosystem and the general lack of regeneration in Garry oak stands (Erickson 1996), little attention to date has been devoted to understanding patterns of either acorn predation or dispersal in this oak. There is, however, one study (Fuchs et al. 2000) based on the thesis of Fuchs (1998) that examined predation patterns of cached (buried) acorns that provides some limited insight into both of these processes. Conducted at two sites both of varying vegetative composition in the southern end of Vancouver Island, Fuchs et al. (2000) buried acorns in varying habitats and at one of three "depths" (on the ground surface, beneath the leaf litter, and in a few centimeters of soil) and then followed their fate through seedling establishment and one year of seedling growth. Habitat type varied considerably but generally corresponded with habitat types selected by Steller's jays Cyanocitta stelleri) for caching (Fuchs 1998). Not surprisingly, acorns survived longer if stored in the ground and >65% of seedlings survived two years following initial establishment. The majority of buried acorns that were not pilfered by vertebrates (83%) survived to establishment.

Fuchs et al. (2000) observed that more than 50% of acorns on the ground surface were removed by vertebrates across all habitat types. But, in sharp contrast to my own experience and many observations in other oak ecosystems, in nearly half of the study sites, predation (removal) by vertebrates only ranged between 7% and 48%, suggesting low densities of acorn consumers/dispersers. Although this may be due to an unusually high seed crop during the study, even under such conditions the reported rate of predation would be unlikely in eastern deciduous forests. Another interesting finding from Fuchs et al. (2000) that was consistent with my own observations (Steele et al. 2011, 2014, 2015) regarding vegetation density and acorn dispersal was that acorn survival was lowest in open areas, where shrub density and canopy cover was lowest. This is consistent with my lab's prediction that dispersal to open sites, compared to those with denser vegetation, is more likely to translate into both seed survival and seed-ling establishment.

Historically, the primary dispersal agent of Garry oak on Vancouver Island is reported to be Steller's jay, although it is now also dispersed by the introduced and invasive eastern gray squirrel (*Sciurus carolinensis*) (Fuchs et al. 2000). On mainland Canada and in the remainder of its range, a greater diversity of ro-dents and corvids are likely to consume and disperse the acorns of Garry oak. However, studies are few. Glendenning (1944) described an inland population of Garry oak and speculated that the species may be dispersed by the band-tailed pigeon (*Patagioenas fasciata*), although no detailed observations were provided.

As a white oak, but the lone oak species in most ecosystems where it is found, the Garry oak may not face some of the typical challenges of dispersal and es-tablishment of other white oaks. Because it is not found in the presence of red oak species in particular, the white oak species may experience successful dis-persal and establishment in years of high acorn production. It would be of par-ticular interest to know if the introduced gray squirrel or other rodents in this system perform embryo excision on the Garry oak, to what extent it occurs, and whether the relative frequency of the behavior (Xiao et al. 2010) is related to successful establishment of the oak.

Although not directly focused on acorn dispersal, Frank et al. (2009) exam-ined the important tri-trophic relationship between Garry oaks, ectomycorrhi-zal fungi, and the rodents that disperse these fungal spores. Elsewhere, I have only briefly touched on this relationship. However, Frank et al. (2009) raise the important question, that in addition to dispersing acorns, rodents also may per-form the important task of inoculating young seedlings with the fungal spores that they accumulate in their digestive tract from feeding on the fruits of hy-pogeous fungi. To the extent that these fungi form important mutualistic rela-tionships with Garry oak and other oaks species, it may follow from Frank et al. (2009) that dispersal and failure of cache recovery may not be enough in some systems to promote regeneration. If seedling growth is dependent on inocula-tion of ectomycorrhizal fungi, then dispersal and regeneration of the oak may be in some systems more exclusively dependent on the activity of rodents. This indeed is an interesting subject for future research.

Marsico et al. (2009) conducted genetic studies across the range of Garry oak to assess historic patterns of gene flow. By collecting tissue samples from nearly 400 trees from 22 populations across the species' range and assessing genetic diversity at both nuclear loci (genes) and chloroplast loci, they were able to infer patterns of both pollen flow and acorn dispersal. Because chloro-plast DNA is inherited only from the mother in most higher plants, measure-ments of the genetic diversity in the chloroplast provide a measurement of the dispersal of the maternally produced acorns. Marsico et al. (2009) found high diversity in nuclear markers within and even across populations, suggesting pollen flow was not limited. However, chloroplast DNA told a different picture. Maternally inherited diversity showed little variation within populations on the mainland and no variation on isolated island populations, indicating that seed dispersal was limited between sites and especially between the mainland and island populations.

Current studies in the remaining Garry oak stands of Canada indicate that today these ecosystems are particularly vulnerable to multiple invasive species, such as Scotch broom (*Cytisus scoparius*) (Ussery and Krannitz 1998) and a number of perennial grasses (MacDougall 2001). Maintenance of today's forests critically depend on strategies focused on how to eradicate these invasive species (Mikaïlou et al. 2009). Susceptibility of Garry oak to gypsy moth larvae (*Lymantria dispar* [L.]), which has colonized in the Pacific Northwest, is also a looming threat in the future (Miller et al. 1991).

A comparison of land surveys prior to European settlement in the Pacific Northwest (southern Vancouver Island) with current vegetation and habitat surveys strongly suggest that fire suppression over the past 150 years has contributed significantly to changes in forest structure during this period (Bjorkman and Vellend 2010). These authors document an increase in fire-susceptible species (e.g., western red cedar, [*Thuja plicata*]), and a decrease in fire-resistant species (e.g., Douglas fir and hemlock [*Tsuga heterophylla*]), as well as a marked increase in tree density and a decrease in open areas. Although oak dominance did not change significantly over that period, the relative density of oak stands did. Bjorkman and Vellend (2010) found that in the past, open stands were occupied primarily by Douglas fir, whereas today, open stands are typically dominated by Garry oak. This historic perspective suggests that current management of Garry oak stands may be misguided, and that fire management practices today may result in some unexpected changes in Garry oak stands in the future (McCune et al. 2013, 2015). Based on pollen and charcoal records, Pellatt and Gedalof (2014) report strong evidence of frequent fires and periodic severe fires every few decades in southwestern British Columbia over the past few centuries, although disturbance fires are rare since European settlement. Finally, it would be important to consider the effect of fire on successful establishment of Garry oak, as it has been shown to facilitate regeneration in other systems (see Chapter 13).

THE CORK OAK AND OTHER MEDITERRANEAN OAKS

Mediterranean oak forests represent a diverse range of habitat types associated with the western Mediterranean Basin. Although reference to "Mediterranean forests" often refers to all forest types associated with a Mediterranean climate, which includes coastal areas of California, Chile, and other small regions of sim-

14.2. Shown are **(a)** the cutting of cork from the cork oak (*Q. suber*). *(Fernando Moital)* **(b)** The cork shown in cross section of a cork oak. *(Nicholas Service)* **(c)** An adult tree with cork recently harvested. *(Andrey Bogdanov)* SEE COLOR PLATE

ilar climate, here I refer specifically to oaks associated with the Mediterranean Basin, primarily the western Mediterranean coasts and many of the forest ecosystems of Portugal and western Spain. Vegetation types in this region include forests of several types, shrubland, and open agro-forestry systems, all of which include oaks, which in many cases are the dominant species in the ecosystem. Perhaps the most prominent species of this region is the cork oak (*Q. suber*), an evergreen oak that reaches approximately 20 m in height. This particular species (section *Cerris* of the genus *Quercus*) (Fig. 14.2a) is most common in western Europe (Portugal and western Spain) and is also found in northwestern Africa (Morocco, Algeria, and Tunisia), southern France, and the western coast of Italy. Many fragmented populations also occur throughout the species' range and on some western Mediterranean islands (Gil and Varela 2008).

14.3. Cork oak (*Q. suber*) occurs in **(a)** shrublands and pine-oak forests *(Richard Orr)* and **(b)** Open agro-savannahs where livestock often feed on acorns. *(Thomas Wustefeld)*
SEE COLOR PLATE

Cork oak is of particular interest to many because of its importance to the cork industry (Fig. 14.2b). *Q. suber* is the primary source of cork in the world and as such is of particular concern for economic reasons; the only other source of cork comes from *Q. variabilis*, found in southeast Asia, also referred to as *the* cork oak. The cork industry relies on the periodic harvesting of the very thick cambium (bark) that regrows and can be re-harvested from the same tree in less than 10 years (Fig. 14.2c). Thus, the industry itself is not a major threat to cork forests, as it is a sustainable resource (Gil and Varela 2008).

Throughout its range, the Mediterranean cork oak occurs in several types of ecosystems, including thick forest stands where it is frequently the dominant species and other forest communities, such as pine-oak forests, shrubland, and agro-forestry systems that appear as open savannas (Fig. 14.3). This latter system, common throughout Spain and Portugal, are referred to, respectively, as dehasas or montados. These systems are heavily managed for cork, a number of forest products other than timber, and for grazing by livestock, including the Iberian pig, which feeds extensively on acorns. Two tree species dominant in these agro-systems are the cork oak and the holm oak (*Q. ilex*). However, across its range, the cork oak is also associated with at least six other oak species (Table 14.1).

Table 14.1. Overview of the many factors influencing establishment and regeneration in Mediterranean oak forests, where these questions have been studied extensively.

OAK SPECIES	DISPERSERS AND/OR SEED PREDATORS	PRIMARY FINDINGS	REFERENCES
Cork oak (*Q. suber*)	Numerous species of seed predators and dispersers	Four mechanisms of recruitment limitation (acorn source, dispersal, germination, and establishment) are more critical in oak scrublands than in oak forests or "savannas"	Acácio et al. 2007
Holm oak	Red deer (*Cervus elaphus*) and wild boar (*Sus scrofa*)	Ungulates consume large numbers of both infested and non-infested acorns but do not exert a benefit to oaks by consuming larvae	Bonal and Muñoz 2007
Cork oak, downy oak (*Q. pubescens* also *Q. humilis*), and holm oak (*Q. ilex*)	NA	Recruitment of holm oak and downy oak in shrublands is low due to frequent fire, but higher in mixed oak or pine-oak forest stands where fire is less frequent; cork oak less affected by fire frequency	Curt et al. 2009
Holm oak and downy oak	*Curculio elephas* and *C. glandium*	Predation in both oaks was lower in higher mast years; holm oak showed greater masting patterns and greater patterns of infestation, but satiation at the seed level (~germination of infested acorns); high mast crops were more important than seed size for predicting patterns of oak establishment	Espelta et al. 2009a
Holm oak and downy oak	*C. elephas* and *C. glandium*	Seed phenology, seed size, and size of *Curculio* spp. interact to drive patterns of acorn predation; ,smaller weevils have an advantage in smaller acorns	Espelta et al. 2009b
Holm oak and downy oak	*C. elephas* and *C. glandium*	Rainfall influences both acorn production (masting patterns) and patterns of weevil emergence and infestation	Espelta et al. 2017
Cork oak, downy oak, Portuguese oak (*Q. faginea*), and Pyrenean oak (*Q. pyrenaica*)	All large vertebrate seed consumers	Removal of acorns by large herbivores not affected by microhabitat structure; seeds of higher seed mass (cork and Pyrenean oak) preferred. When large herbivores were excluded, habitat structure influenced acorn removal	González-Rodríguez and Villar 2012
Cork oak and Algerian oak (*Q. canariensis*)	Primarily rodents	Experimental placement of acorns revealed that acorns were more likely to be consumed if not buried even in heavy scrub vegetation when large herbivores were excluded; cork oak also preferred by seed predators	Pérez-Ramos and Maranon 2008
Cork oak	Large herbivores (cattle, deer, boar, and rabbits) and likely rodents	Seeds spaced on ground in scrub-dominated habitat suffered high rates of predation due to large vertebrates. Single buried acorns experienced highest germination success	Herrera 1995

OAK SPECIES	DISPERSERS AND/OR SEED PREDATORS	PRIMARY FINDINGS	REFERENCES
Cork oak, holm oak, and Kermes oak (*Q. coccifera*)	Avian dispersers and possibly humans	Examined genetic variation in oaks of Balearic Islands of western Mediterranean and found complex genetic structure that suggests a refuge effect during Pleistocene, and both hybridization and numerous long-distance dispersal events	López-de-Heredia et al. 2005
Cork oak	Partial predation of acorns by common rabbit (*Oryctolagus cuniculus*)	Partial acorn consumption by common rabbit may frequently result in seedling establishment	Mancilla-Leyton et al. 2012
Cork oak and Algerian oak	Insect, avian, and mammalian acorn seed predators	Both specialist and generalist seed predators show a Type II functional response to acorn availability in which proportion of acorns attacked declines with acorn availability	Moreira et al. 2017
Holm oak	Algerian mouse (*Mus spetrus*) and wood mouse (*Apodemus sylvaticus*)	Proportion of acorns dispersed, cached, and not recovered were lower in the presence of red deer (*Cervus elaphus*) and wild boar; dispersal distances not affected by ungulates although spatial distribution differed in presence of ungulates	Muñoz and Bonal 2007
Holm oak	Algerian mouse and wood mouse	Rodents selectively disperse acorns to open sites where pilferage was reduced	Muñoz and Bonal 2011
Cork oak, holm oak, Portuguese oak, and Pyrenean oak	*Curculio* spp.	Limited specificity for 4 species of *Curculio*; evidence of satiation at the acorn level for large-seeded species	Muñoz et al. 2014, and references therein
Cork oak, holm oak, and pedunculate oak (*Q. robur*)	Rodents most likely dispersal agent	Oak seedlings spatially aggregated with same species as a result of short-distance seed dispersal	Muhamed et al. 2015
Cork oak	NA	Seedling establishment similar in cork forests and pine forests but absent in shrublands; long-term regeneration generally absent in all forests, suggesting habitat manipulation needed for regeneration for cork oaks	Pausas et al. 2006
Cork oak and Algerian oak	Primarily rodents	Probability of acorn removal higher in heavier cover; cork oak acorns preferred over those of Algerian oak	Pérez-Ramos et al. 2008
Cork oak	Both avian and rodent acorn dispersers and predators	Cork oak recruitment in fragmented populations of eastern Iberian Peninsula is nearly absent in shrubland	Pons and Pausas 2006

OAK SPECIES	DISPERSERS AND/OR SEED PREDATORS	PRIMARY FINDINGS	REFERENCES
Cork oak, holm oak, Kermes oak, and Portuguese oak	Eurasian jay	Holm oak acorns preferred over all others for dispersal and Kermes oak least preferred; within oak species, larger acorns preferred	Pons and Pausas 2007b
Cork oak	*Curculio* spp.	Individual trees produce small acorns that become more heavily infested and larger acorns that are sound	Ramos et al. 2013
Cork oak	Community of dispersers and seed predators	Regeneration in savannas lower than in forests not due to pilferage but to lower seedling survival as a result of fewer safe sites for scatterhoarding	Smit et al. 2009
Holm oak and cork oak	NA	Examination of genetic relatedness (based on microsatellite analyses) reveals limited dispersal in cork oak and higher risk of inbreeding compared with holm oak	Soto et al. 2007
Cork oak	Dung beetle (*Thorectes baraudi*)	Strong preference by this dung beetle for acorns supporting evidence of secondary effect on dispersal	Verdú et al. 2011, and references therein

Many of the ecosystems where cork oak is found support high biodiversity. This is certainly one important reason for the focus on conservation and preservation of these systems.

And, whereas regeneration appears to be less of an issue in many of these forests, the highly fragmented nature of some of these oak ecosystems, illustrated by the distribution of cork oak, suggests significant risk of local extinction in the face of regeneration failure. Another potential threat to oaks of this region is disease. At least 100 fungi and water molds that occur on cork oak, for example, are pathogenic (Gil and Varela 2008, and references therein). One such species appears to be a growing threat to *Q. suber* in particular: the endophytic fungus, *Biscogniauxia mediterranea*, responsible for charcoal canker. The fungus' ascospores appear to be first released by rain and then dispersed through the air (Henriques et al. 2014). Although spores are found on acorns, it appears that they are not passed to developing seedlings (Henriques et al. 2014).

The oak ecosystems of the Iberian Peninsula and western Mediterranean Basin are subject to a diverse array of natural factors that interact in complex ways to influence patterns of acorn predation, acorn dispersal, seedling establishment, and, ultimately, the regeneration of these oaks (Table 14.1). Despite this complexity, one could argue that more is known about these oak ecosystems than many of the other threatened oak forests across the globe.

I provide Table 14.1 as an attempt to highlight some of the ecological interactions that ultimately contribute to, or prevent, successful dispersal and establishment in these systems. Thanks to the novel studies of R. Bonal, A. Muñoz, and J. Espelta, as well as I. M. Pérez-Ramos, J. Pons, and J. G. Pausas, and several others, more is known about pre-dispersal predation (by insects), post-dispersal predation (by mammals), and successful dispersal and establishment in these oaks than in nearly all other systems. For example, depending on acorn size, pre-dispersal seed predators (e.g., weevils) in this system have been shown to be sometimes satiated at the seed level, especially for larger-seeded oaks (Muñoz et al. 2014, and references therein), although masting patterns appear more important than seed size for predicting establishment success (Espelta et al. 2009a). Espelta et al. 2009b also showed that acorn size interacts with both weevil size and acorn phenology in different oaks to determine patterns of acorn predation. Moreira et al. (2017) found that all acorn predators (insects, birds, and mammals) of both cork oak and Algerian oak (*Q. canariensis*) show a Type II functional response to acorn availability; that is, the proportion of acorns predated declines with availability. Beyond the insects, important vertebrate acorn predators, which seem to exert a more significant direct or indirect effect on oak regeneration than in other systems, include the red deer (*Cervus elaphus*), the wild boar (*Sus scrofa*), and domestic livestock. Although frequent and significant consumers of acorns, these larger-seed predators also consume significant numbers of insect predators of acorns, and their presence can influence patterns of acorn predation and dispersal by rodents (Table 14.1).

Acorn dispersal can be accomplished by an entire guild of seed consumers that often include the Eurasian jay, rodents, predominantly the Algerian mouse (*Mus spretus*), and wood mouse (*Apodemus sylvaticus*) that scatterhoard acorns, the common rabbit (*Oryctolagus cuniculus*) that may frequently contribute to seedling establishment by its partial consumption of acorns (Mancilla-Leyton et al. 2012), and the dung beetle (*Thorectes baraudi*), which has been shown to engage in secondary dispersal and establishment of these oaks (Table 14.1). Despite the many species that contribute to the dispersal of the Mediterranean oaks, their overall impact varies considerably with respect to the forest type, oak assemblages, or the community of other acorn consumers. Hence, one salient lesson to follow from these well-studied systems in the Mediterranean Basin is that we should use care in drawing generalities from single studies about the dispersal process.

Finally, I note, that despite the presence of a diverse community of acorn dispersers in many of these oak forests, and evidence of frequent dispersal and establishment of seedlings, there is also evidence of a lack of long-term regeneration for some oak species in this region (e.g., cork oak) (Pausas et al. 2006). This suggests the need for a better understanding of the genetic structure and diversity across the fragmented distribution of these oaks (but see Soto et al. 2007; Canelo et al. 2018). In addition, there is evidence that regeneration of cork oak may be limited by the loss of ectomycorrhiza fungi, potentially critical for establishment and growth of seedlings. There is now evidence that these important fungi may be declining in abundance due to effects of climate change and increased desertification in parts of the oaks' range (Maghnia et al. 2017).

QUERCUS OLEOIDES FORESTS OF COSTA RICA

Quercus oleoides of Costa Rica (Fig. 14.4), much like Garry oak, is another isolated white oak species that comprises ecosystems of considerable concern. The species' range extends from the state of Tamaulipas, Mexico, adjacent to the US border along the Gulf coast, to the province of Guanacaste, Costa Rica. As emphasized by Boucher (1983), *Q. oleoides* is the only oak that is found in the lowland tropics of the western hemisphere. *Q. oleoides* is one of two sister species belonging to the monophyletic lineage of *Quercus* (Nixon 1985) that extends from North Carolina to Costa Rica (Cavender-Bares et al. 2011). The other species in this complex is the live oak, *Q. virginiana*, which extends from southeastern Virginia along a narrow band along the southeastern coast of the United States down through Florida along the Gulf Coast to the most southeastern edge of coastal Texas.

The *Q. oleoides* forests of Costa Rica are of particular interest and concern because they represent the terminal distribution of this oak, are further isolated by a notable gap in distribution between forests to the north in Honduras, and are biologically unique in that they are located in a lowland tropical forest. There is also limited information on the basic ecology of this species. Throughout its range, the distribution of *Q. oleoides* is significantly fragmented as it is in Costa Rica, where the distribution is represented by three isolated populations (Boucher 1983). In all of these sites, *Q. oleoides* is the most dominant tree species in the forest (Boucher 1981).

Several characteristics of *Q. oleoides* in Costa Rica are particularly relevant to understanding the nature of dispersal in this species. First, because of its isolation at the terminal end of its distribution, *Q. oleoides* is free of defoliating insects that occur elsewhere in its range, as well as *Curculio* larvae that prey on acorns, although at least one species of lepidopteran larvae are known to feed on the acorns (Boucher 1983). Second, like Garry oak, *Q. oleoides* is a white oak (section *Quercus*) that generally occurs in the absence of other oak species, which may free it from some of the challenges of dispersal when faced with the dispersal advantages of red oak species.

Third, as noted by Boucher (1983) the human transformation of the landscape to savannas and pastures for grazing has introduced a number of new

(a) (b)

14.4. Shown are **(a)** leaves and acorns and **(b)** trees and saplings of *Quercus oleoides* in the dry forests of Guanacaste, Costa Rica. *(Michael A. Steele)* SEE COLOR PLATE

factors influencing the potential for dispersal and regeneration of *Q. oleoides*. In these human-modified landscapes, the oak is faced with frequent fires, interaction with grasses, including several introduced, invasive species, and increased predation by livestock. Although predation of acorns by native rodents (see below) is reduced at these sites, so too is the probability of dispersal (Boucher 1983). Fourth, in Costa Rica alone, the time of flowering, and thus acorn production, varies considerably (up to seven months) depending on the location (Boucher 1983). And, although individuals of *Q. oleoides* can produce thousands of acorns, individual trees are often isolated or occur in densities as low as 30% of the forest trees. In these forests, there appears a limited benefit of masting. In contrast, there are pure forest stands of *Q. oleoides* in Costa Rica where synchronized masting may aid dispersal, seed survival, and seedling establishment.

Collectively, the characteristics above render *Q. oleoides* unique in comparison to many other oaks. From its location in lowland tropics to variable stand densities and a relationship with seed predators and dispersers that differ from many other oaks, *Q. oleoides* presents some interesting challenges when it comes to characterizing patterns of seed predation and dispersal in this species.

I was introduced to the unique ecology of *Quercus oleoides* by Dr. Salvatore Agosta, a former student, who at the time was a doctoral student under the codirection of Dr. Daniel Janzen at the University of Pennsylvania. Dr. Daniel Janzen, one of the world's most renowned evolutionary ecologists and conservation biologist, is also broadly known for the establishment and coordination of Area de Conservación Guanacaste, one of the world's largest conservation and habitat restoration projects (>1400 k2), and Instituto Nacional de Biodiversidad (INBio), a private foundation targeted, among other goals, at documenting the complete biodiversity of Costa Rica. Given the estimated species richness of this country, the immensity of such a goal cannot be overstated. Some estimates put the biodiversity of Costa Rica at approximately 2% of that of the entire globe, which is not too shabby for a country the size of the US state of West Virginia.

Although Sal eventually settled on a different set of ecological problems for his dissertation, before he began his doctoral research, Sal and Janzen had discussed an array of different research problems, one of which was the mystery of dispersal and regeneration of *Q. oleoides*. Sal was quick to share the problem, and he vigorously recruited me to investigate. On Sal's advice, I visited the park in Santa Rosa on four occasions with my son, Tyler, and three Wilkes University students to explore the situation and to attempt some preliminary research. Although none of this work has been published to date, I fully expect to see it through to completion, as soon as I have the time and resources to return to Santa Rosa. In the few short months that I spent in the beautiful country of Costa Rica, I was afforded the opportunity to conduct some preliminary experiments. Although I must admit, that from the wet and dry tropical forests, to the montane cloud forests and volcanoes, the diverse coast lines and, above all else—the hummingbird diversity—my interest in *Q. oleoides* was frequently interrupted. Yet, my message home was always that we were fast at work. And, *most* of the time we were.

During my first visit, I was introduced to an array of potential acorn dispersers, which to my surprise, was not as extensive as I expected. After a review of the limited literature on the subject, a focused, question-and-answer session with Dan Janzen in his dwelling in Santa Rosa National Park and my own limited observations in several *Q. oleoides* forests, I started to acquire some perspective on the species likely to feed on acorns of this oak. Among the mammals, the spiny pocket mouse (*Liomys salvini*), collared peccary (*Pecari tajacu*), white-lipped peccary (*Tayassu pecari*), variegated squirrel (*Sciurus variegatoides*), the Central American agouti (*Dasyprocta punctata*), and, in some areas, the white-faced capuchin (*Cebus capucinus*), all can be considered acorn seed predators (Boucher 1981, 1983). According to Hallwachs (1994), the peccaries consume large numbers of acorns and frequently raid scatterhoards as well. Several additional mammals, including other species of pocket mice (Lyomys spp. and Heteromys; Fleming 1983ab), the red-tailed squirrel (*S. granatroensis*) (Heaney 1983), and the white-tailed deer (*Odocoileus virginanus*) (Boucher 1982; Janzen 1983) are also likely to be opportunistic predators of *Q. oleoides* acorns. In addition, in areas where *Q. oleoides* is associated with open grasslands, free-ranging cows and horses are also common acorn predators. Although, Janzen and Martin (1982) have argued that these domestic herbivores fill important ecological roles vacated by the extinction of large Pleistocene herbivores, the dispersal of acorns is not likely to be one of them. The domestic replacements of these Pleistocene relics are likely to masticate and kill acorns when consuming them, rather than passing them through the gut intact.

Among the acorn consumers discussed above, the literature at the time suggested that *only* the agouti actively scatterhoards and disperses acorns (Hallwachs 1995). In this dissertation, Dr. Winnie Hallwachs explored in detail the potential for the agouti to disperse oaks and other native species (Hallwachs 1995) via their scatter-hoarding behavior (see below). And, to the present day, this dissertation still stands as the leading authority on the subject.

One group of mammals in Costa Rica that has been overlooked as potential agents of oak dispersal are the carnivores. Of the 21 species of carnivores in the country, 19 of them are known to feed extensively on various fruits (Janzen and Wilson 1983). Although their consumption of acorns is likely to occur only opportunistically, perhaps only during heavy acorn crops, it is also likely that one or more of these species passes whole acorns through the gut allowing them to germinate and establish following defecation. Although my research assistant, Shea Marino, and I have observed whole, intact acorns in the feces of several foxes (either the gray fox [*Urocyon cinereoargenteus*] or red fox [*Vulpes vulpes*]) during a mast year, we regrettably did not test the viability of these acorns. We are, however, waiting for our next mast crop so we can conduct such a study. It is also noteworthy that many carnivores exhibit extensive home ranges, which means as potential frugivores they could have a significant effect on dispersal. Indeed, there is evidence that some carnivores aid in seed dispersal, either directly as frugivores, such as the red fox (Cancio et al. 2017; Farris et al. 2017), black bear (*Ursus americanus*) (Enders and Vander Wall 2012), and the palm civet (*Paradoxurus hermaphroditus*) (Nakashima et al. 2010) or through other indirect activities as in the case of some members of the weasel family (Mustelidae), which has been shown to influence the establishment of the Liaodong oak (*Q. wutaishanica*) by burying acorns when digging for prey (Gao and Sun 2005). To date, the importance of carnivores in the dispersal of oaks is understudied and definitely requires a look, both in Costa Rica and in other oak forests around the globe.

My limited experiments in Costa Rica focused on the potential dispersal of acorns by two mammal species: the agouti and the pocket mice. Although based on her input from Dan Janzen, Hallwachs (1994) suggested that pocket mice were physiologically constrained by the tannins in acorns and were therefore unlikely to regularly prey on acorns, I was not convinced that these rodents would not contribute to their dispersal, especially if they were regularly exposed to them. We, therefore, started with the mice and other rodents with which they are likely associated. We attempted our standard seed-tagging ex-

14.5. Shown are **(a)** Wilkes students, equipped with snake guards on their legs, prepare **(b)** tagged acorns for placement in **(c)** exclosures that allowed access by rodents when placed along **(d)** forest edges. *(Michael A. Steele)*

periments, first collecting whole intact acorns beneath mature *Q. oleoides* from a diversity of sites across Guanacasta, metal-tagging the seeds and then building 10 differential exclosures to provision the mice (Fig. 14.5). We placed the exclosures approximately 100 m apart at two sites: one near Santa Rosa and the other in a montane site in the Cordillera de Guanacaste. At both sites, we placed exclosures on forest edges and hoped to determine if acorns were dispersed back into the forest or out into more open grasslands. Equipped with metal detectors, we were determined to document patterns of dispersal. However, to our surprise, few acorns were removed from the exclosures, and all were consumed directly inside or just outside the boxes. The mice, at least in our experiments, contributed little to the dispersal process and, although they consumed acorns, this was done at a very slow pace, nothing like what we typically observe in eastern deciduous forests of North America. Our observations in these experiments, although limited, were consistent with the early studies of Boucher (1980, 1981, 1983), who measured rates of acorn predation by all mammals (pocket mice, squirrels, agouti, peccaries, and deer) at different acorn densities. He found that acorn predation was clearly density dependent, and from that concluded that some acorns in dense oak stands will likely survive to germinate and establish because of higher acorn availability, but where oaks are isolated, acorns are likely to experience complete predation and little chance of successful establishment. Although Boucher concluded that there was little, if any, evidence of dispersal of acorns of *Q. oleoides*, I was still skeptical especially after reviewing Hallwachs's work with the Central American agouti.

Based on this logic, in one of our later visits, my crew deliberately sought out a location with *Q. oleoides* stands where agoutis were also regularly cited. We selected a site along a forest edge, placed acorns on the ground, and then followed their fate using a modified version of the tin-tagging method frequently employed by my colleagues from China (Xiao and Zhang 2006; Xiao et al. 2006). We modified the technique with material available from local hardware stores. Instead of using wire to attach plastic tags to the acorn, we used heavy monofilament line attached to the acorn through a small hole drilled into the basal end of the acorn. At the other end of the monofilament line (~30 cm), we attached a 5 cm length of fluorescent flagging, which was uniquely numbered. We placed a sample of approximately 100 acorns in a circle with the tags out to prevent them from getting tangled (Fig. 14.6). Our goal was specifically to see if this method would have some potential for tracking acorn dispersal by the agouti. We established the seed station at a site where agouti had recently been seen. The next morning, I arose early, prepared some coffee, and wandered the few hundred meters to the seed station. To my surprise the seeds had all been handled, a small percentage eaten, and the others all dispersed and cached. How did I know? With the tagging method used, it was possible to quickly scan the area and see the widely dispersed tags within 10 m of the seed station. I rushed back to our quarters, recruited a student and two colleagues, Dr. Sal Agosta and Dr. Jeffery Klemens, to assist in determining seed fates and dispersal distances for the cached acorns. Although we had not used camera traps to document any visitors, it was evident that these acorns had been moved by agouti. We also observed that several acorns had the bottom fifth of the cotyledon removed. The bottom tip of the acorns were severed off. Yes, it appeared that the acorn was manipulated in a way that would prevent it from germinating, just as rodents elsewhere excise the embryo of the acorn (see Chapter 5). This, to me, seemed further evidence that we were observing agouti behavior. The larger incisors of an agouti would make it impossible to excise the embryo. Although only a pre-

14.6. Student assistant and Dr. Jeffrey Klemens preparing a version of the tin-tag method for tracking acorn dispersal (see Chapter 7) in which Central American agouti (*Dasyprocta punctata*) were presented with acorns attached to a heavy mono-filament line and a piece of orange flagging. (*Michael A. Steele*) SEE COLOR PLATE

liminary experiment, this experience spoke volumes about how best to further explore dispersal of this oak by the agouti.

Although I have not yet returned to this system to conduct my anticipated experiments with agouti, I have on a number of occasions returned to the Hallwachs's dissertation for insight on my research in other oak systems. In her experiments with relatively habituated agoutis, Hallwachs (1994) was able to demonstrate several important aspects of acorn dispersal by this rodent. She showed that acorn crops of larger size, as well as individual acorns of larger size, were more likely to be cached than smaller ones. Larger acorns were also more likely to be dispersed farther when cached, and more likely to be moved to the owner's defended territory. Smaller acorns were more likely to be consumed. She also found that a number of environmental factors, such as the hardness of the soil and vegetation density (and therefore predation risks), influenced cache site selection by the agoutis. When predation risk is generally high, agoutis are also less likely to cache acorns. Although pilfering was also common, acorns cached in the agouti's territory were far less likely to be pilfered. Hallwachs also reported that in the rare cases of an agouti's death, cached acorns in the owner's territory were likely to establish. Hallwachs also cleverly characterized the factors driving acorn dispersal in the *Q. oleoides* system as contributing to either the "rhythm" or the "noise" in the system. Rhythm, for example, would include characteristics of the seed that are under genetic control (e.g., seed size) and, the noise would include the many unpredictable environmental factors that also influence seed fate (e.g., climate and soil hardness and predation risks). Hallwachs concluded that the extensive noise in this system weakened the argument for a tight coevolutionary relationship between the oak and the agouti as was evident in other systems (e.g., the Eurasian jays and oaks) (Bossema 1979). Hence, her dissertation title, "The clumsy dance between agoutis and plants: scatterhoarding by Costa Rican dry forest agoutis (*Dasyprocta punctata*: dasyproctidae: rodentia)" was quite ap-

14.7. Two likely candidates for dispersal of *Quercus oeloides* acorns are **(a)** the white-throated magpie jay *(Calocitta formosa)* and **(b)** the variegated squirrel *(Sciurus variegatoides)*. Despite our many observations and behavioral experiments in the range of these two species, we were not able to document regular dispersal of acorns by either of these species. *(Michael A. Steele)*
SEE COLOR PLATE

propriate. Importantly, Hallwachs's observations regarding predation risks contradict several of my own in which I argue that rodents in temperate forests may select cache sites with higher predation risk to purposely reduce the risk of pilferage. This marked difference may follow from contrasting forest structure and/or differences in the intensity of predation risk in these systems.

Finally, among the birds in Costa Rica that are likely dispersal agents of oaks, there are few. The jays, for example, include only the brown jay *(Cyanocorax morio)* (Lawton 1983) and the white-throated magpie jay *(Calocitta formosa)* (Fig. 14.7), although much of the range of the former species is not coincident with that of *Q. oleoides*. Despite speculation throughout the literature that the jays may contribute to dispersal of this oak, definitive research on this subject does not exist. My own countless attempts to present ripe acorns to white-throated magpie jays met with little success, although colleagues have shared reports of dispersal by this species.

Except for *Q. oleoides*, the other dozen or so oaks in Costa Rica are located in evergreen forests in higher elevations > 600 m (Burger 1983). *Q. costaricensis*, for example, is found at elevations of 2700–3300 m. This red oak species produces exceptionally large acorns (2–3 cm in diameter) and is sometimes found in single species stands in montane forests of Costa Rica.

BEAR OAK AND PINE BARRENS OF THE
NORTHEASTERN UNITED STATES

The bear oak, also referred to as the scrub oak where it occurs, is a small decid-uous shrubby oak (usually less than 6 m in height) with a limited range that ex-tends from southeastern Maine and southern Ontario to North Carolina (Stein et al. 2003). Common on the coast in New England, New York, and New Jersey, the species also occurs at higher elevations across the northern Appalachian Mountains to southeastern Virginia and northwestern North Carolina.

Although the species reproduces sexually and therefore produces acorns, across much of its range it often reproduces vegetatively (asexually) from root sprouts. When acorns are dispersed, this small-seeded red oak (section *Lobate*) species is likely moved considerable distances by jays from isolated stands to other sites. Short-distance rodent dispersal is also likely within stands, although acorn dispersal in general has not been studied in this species.

In many coastal regions, bear oak is frequently associated with pitch pine (*Pinus rigida*) in stands often referred to as pine barrens. Many of these stands support avian species of conservation concern (Grand and Cushman 2003; Akresh et al. 2015), as well as the pine barrens tree frog (*Hyla andersonii*) of the New Jersey Pine Barrens, once considered endangered.

Throughout much of their restricted range, such bear oak-pitch pine stands are threatened by long periods of fire suppression and occasional severe canopy fires. As a result, many of these barrens are changing in forest structure and de-mography (Bried et al. 2014) as well as forest composition (Howard et al. 2011). In New Hampshire, for example, an extended period of fire suppression in the 20th century has led to the gradual succession of pine barrens to stands now dominated by white pine (*Pinus strobus*) and red maple (*Acer rubrum*).

THE CALIFORNIA OAKS

I mention briefly the California oaks, which collectively represent about 23 spe-cies and approximately five additional hybrids all exclusive to western North America (Pavlik et al. 2002). This unique assemblage of *Quercus* includes nine tree species and 14 species of shrub oaks and accounts for many varied Califor-nia landscapes. This includes at least 15 types of oak communities, such as five types of forest communities (e.g., Northern Mixed Evergreen Forests and Coast Live Oak Riparian Forests), four chaparral communities (e.g., Scrub Oak Chap-arral), and six woodland and/or savanna communities (e.g., Valley Oak Wood-lands). This diversity of oak ecosystems is, in turn, associated with a diversity of federally or state threatened or endangered plants and animals (see detailed list by Pavlik et al. 2002). Threats to the oaks of California include an exponential growth in the human population, which now far exceeds 33 million, and accompanied with that, a tremendous rate of deforestation for housing and agriculture and grazing by livestock (Pavlik et al. 2002).

Just as significant as deforestation is the lack of regeneration in both the oak forests and savannas of California. Studied most intensively by Griffin (1971), regeneration appears not to be limited as much by seed production as it is by periodic drought and vertebrate seed predation. Griffin's (1971) experiments focused on the establishment of the blue oak (*Q. douglasii*), coast live oak (*Q. agrifolia*), and the valley oak (*Q. lobata*). For all species, acorns placed in simulated caches germinated and survived well. However, seedlings of all three species were susceptible to summer droughts, especially on south facing slopes (Griffin 1971)—a problem that is likely to continue in the face of climate change. Since Griffin's (1971) work, we know a great deal more about how many of these California oaks are dispersed, especially by jays (see Chapter 10). For many of these species, dispersal may not be the limiting factor, but, instead, the fate of the seedling after germination.

If the intensive activity of humans is not enough, several of the oaks and other plant species in California have been threatened by Sudden Oak Death, a disease caused by the brown algal pathogen, *Phytophthora ramorum*, identified within the last 20 years, but first recognized many years earlier. As described by Grunwald et al. (2012), this pathogen causes lesions on the leaves and bleeding cankers on the stems and bark due to discoloration of the phloem. The poor conductivity in the vascular tissues leads first to crown death followed soon thereafter by death of the tree. Among the oaks, it is primarily the California black oak (*Q. kelloggii*), Santa Cruz Island oak (*Q. parvula*), and the coast live oak that are most affected.

THE RED LIST OF OAKS

I close with a brief look at the IUCN (International Union for Conservation of Nature and Natural Resources) Red List of Oaks (Oldfield and Eastwood 2007). Although this is an evaluation of all oaks worldwide (~508 species), at least 300 species were classified as *Not Evaluated* because it was not possible to acquire the necessary information to assess their status. Of the remaining 208 species, another 33 are considered at some level of threat, but are declared *Data Deficient* due to the need for additional field studies. The remaining species considered globally threatened (excluding those listed as *Data Deficient*) include species listed as *Critically Endangered* (13), *Endangered* (16), and *Vulnerable* (27).

Among the countries with species listed as *Globally Threatened* (from *Near Threatened* to *Critically Endangered*), Mexico and China lead with 27 and 18 species, respectively. Another 10 species are found in both Mexico and/or parts of

Central America. Elsewhere, counts of Globally Threatened Species are found in the United States (8), Turkey and neighboring countries (3), Taiwan (3), Spain and Portugal (4), Malaysia (3), Russia and Georgia (1), and Cyprus (1).

Although the Red List always serves as an important warning for these species of concern, especially when the distribution of one of these species overlaps with other plant and animal species of conservation concern, it is also important to keep a few points in mind. First, as expected, the countries with the greatest number of threatened species of oaks are also those with the highest diversity of oaks and therefore those, by default, with more species of limited range. Where oaks have diversified, we can expect to find more species of limited range and distribution putting them at risk. As Mexico and Southeast Asia have been the center for oak evolution, they too have more species of conservation concern. Secondly, I note that we would also expect Red List species to occur more often on the edge of the oak's range in places such as Cyprus, Central America, Taiwan, and Malaysia where species are more isolated or limited by the conditions favorable for their establishment. Although threatened species are always an important "canary in the coal mine," it is also necessary that we understand the vulnerability of the entire system.

I now close the book with an outline of potential areas for future research. In Chapter 15, I identify new areas of investigation, as well as some of the subjects covered previously that require additional attention.

OAK DISPERSAL AND OAK-ANIMAL INTERACTIONS: LOOKING FORWARD

INTRODUCTION

My goal in this concluding chapter is to first discuss some of the studies on oak forest structure and tree distribution that follows from our many behavioral studies discussed previously. I first review how studies on DNA matching of adult trees and seedlings direct us toward some alternative perspectives on how the seed dispersing community may contribute to the distribution of oaks and the structure of oak forests, why it is essential to rely on a diversity of approaches when evaluating dispersal, and why taking an entire community approach to understanding the dispersal process is often necessary. The genetic studies give us reason to believe that acorns of white oaks are successfully dispersed beyond those of red oak species under certain conditions. We briefly review our current plans to investigate alternative hypotheses about the spatial distribution of red oak species and those of white oak using Geographic Information Systems (GIS). I then review a number of other potential investigations that might logically follow from discussions in this book, including contrasting masting patterns in red and white oak species, the relative importance of keystone dispersal species versus the entire community of dispersal agents on oak dispersal, and the potential for some yet undiscovered mechanisms of diplochory to contribute to oak dispersal. I also emphasize the importance of biogeographic studies for understanding the subtle changes that occur in oak systems across latitude and elevation. Finally, I close with a call for future studies to consider the insightful concept of seed dispersal effectiveness (Schupp 1993; Schupp et al. 2010) as a systematic and comprehensive approach for understanding the process of oak dispersal in different oak ecosystems across the globe.

Our exhaustive studies on oak dispersal by rodents in eastern deciduous forests showed quite definitively that red oak acorns are consistently dispersed over acorns of white oak. Delayed germination in white oaks above all else limits the dispersal of white oak species by rodents. In addition, acorn size seems to influence jay dispersal more than germination schedule, and there was evidence from the literature that jays selectively dispersed small-seeded red oak acorns, such as pin oak (*Q. palustris*) and willow oak (*Q. phellos*), over those of other species, including white oaks (*Q. alba*) (Scarlett and Smith 1991; but see Moore and Swihart 2006). Some time ago, these observations led us to hypothesize that red oaks have a clear dispersal advantage over species of white oaks in deciduous forests of northeastern North America. We predicted that red oaks should establish farther from parent trees and that ultimately this might lead to a markedly different distribution for these two groups of oaks in the forest.

In our initial test of this hypothesis, directed by my friend and colleague, Dr. Peter Smallwood of the University of Richmond, we sought to measure the distribution of seedlings around adult trees of both white oak and red oak species (Smallwood et al. 1998). To do this, we plotted and measured the distribution of both seedlings and adult trees of four species: two white oaks, the white oak and the chestnut oak (*Q. montana*), and two red oaks, the northern red oak (*Q. rubra*) and the black oak (*Q. velutina*). These plots were located in second growth hardwood forest stands of about 100 years of age in the George Washington National Forest of western Virginia. In each plot ($>60,000$ m^2), we mapped and measured the Diameter at Breast Height (as a proxy for age; DBH) of each adult oak, and then, in a smaller 100 m transect in the center of each plot, we counted the number of each species of seedling in 1-m plots along the transects. We then used an analytical computer model called RECRUITS developed by Ribbens et al. (1994) to predict the distribution of seedlings around parent trees of each of the four species. In simple terms, the model predicts the seedling number by the location and size (scaled by DBH) of each potential parent, the distribution of seedlings, and the assumption that each adult tree (scaled by DBH) generates seedling shadows as predicted by the Janzen-Connell model predicting that the density of seedlings will be higher closer to the parent and decline with distance from the parent (see Chapter 1). The model then predicts the potential distribution of the seedlings around potential parents for each of the species studied.

Across the two plots, we mapped 636 adult trees and 616 seedlings. Results from this analysis were striking and strongly supported our hypothesis (Fig. 15.1). The two white oak species were both predicted to have short seedling shadows in which seedlings were clustered near parents with their numbers declining rapidly in short distances from the potential parent. The two red oak species,

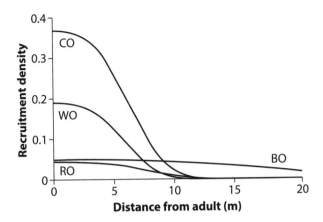

15.1. Predicted recruitment profiles for four oak species. Each line shows the predicted density of seedlings around a 30-cm (dbh) parent, using the recruits model (Ribbens et al. 1994). The model is calibrated for each species using spatial information for both seedlings and adult trees. White oak species are WO = *Q. alba* and CO = *Q. prinus*; red oak species are RO = *Q. rubra* and BO = *Q. velutina*. Note that the distribution of the two white oak species are far more clumped around parent trees, as predicted from the behavior of seed-dispersing mammals. *(From Smallwood et al. 1998)*

in contrast, were calculated to not cluster near potential parents and instead exhibited a flat distribution that extended a considerable distance from parents (Smallwood et al. 1998). These predicted distributions of the seedlings were as expected if the red oak acorns were dispersed and cached farther from parent sources than acorns of white oaks, and if seedlings survived where their acorn was initially cached. The results were consistent with our behavioral studies. Yet, more importantly, these results further suggested that if seedling shadows convert into similar distributions for saplings, and eventually adult trees, we would expect that trees of red oak species would be widely distributed in the forest, whereas those of white oak species would be highly clumped in comparison.

Before testing the potential for this differential dispersal hypothesis to contribute to forest structure, we sought to determine the distribution of seedlings around their "actual" parental sources. The RECRUITS model, of course, was only a mathematical prediction of seedling shadows and was based on the distribution of unrelated adult trees and seedlings. Although the outcome of these results certainly supported our hypothesis, the next logical step was to map related adult trees and their seedlings and determine the distance between actual parental trees and their offspring. To do this, we recruited two geneticists (Dr. William Terzaghi, Wilkes University, and Dr. John Carlson, Pennsylvania State University), ecologist Peter Smallwood, and a small army of students.

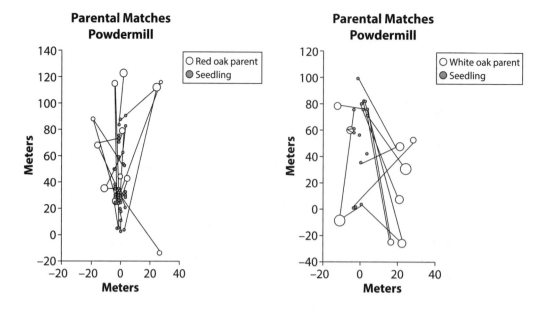

15.2. Spatial relationship between adult oaks (large circles) and positively matched seedlings (small circles). Where possible, lines connect the closest and farthest related seedling to each parent. Seedlings matched parents on at least 7 loci and were excluded on none (see text). Note that dispersal distances for *Q. alba* (a WO species) are similar to those for *Q. rubra* (a RO species). *(From Steele et al. 2007)*

With financial support of the National Science Foundation, we set out to determine the relationship between adult trees and seedlings.

This was done by measuring seedling-parent relationships using microsatellite DNA (Steele et al. 2007) following similar methods outlined in previous studies (e.g., Dow and Ashley 1998, and references therein). We first systematically mapped adult trees and seedlings on large grids in mature oak forests (Fig. 15.2) and, as we did, we collected tissue samples from the seedlings (leaves) and the trees (cambium). These samples were then quick frozen, transported to the lab, stored, and then DNA was extracted. DNA fingerprints were then created for both the seedlings and the adult trees for 10 primer sets (Steele et al. 2007). We then matched seedlings with adult trees when there was a minimum of seven matching microsatellites (see details in Steele et al. 2007). We estimated seedling shadows in two ways. First, we measured the distance between seedlings and adult trees in which we had precise parent-offspring matches. Methods were validated by testing and matching acorns with known maternal trees. Although helpful, this measurement did not distinguish between male (pollen) and maternal (acorn) parents so it limited the conclusions we could

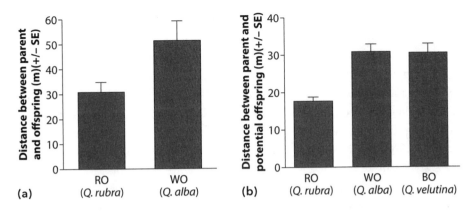

15.3. Shown are **(a)** distances between parent and offspring of red oak (*Q. rubra*) and white oak (*Q. alba*), and **(b)** distance to nearest potential offspring of red oak, white oak, and black oak (*Q. velutina*). (*From Steele et al. 2007*)

draw. In the second method, we measured the minimum distance between offspring and the closest parent that could not be excluded based on DNA data. In many cases, we could not find a potential parent on the site. This allowed us to measure the minimum distances to a potential maternal source, and assumed that parent trees did not die since establishment of the seedling. Given that there was no logging on the site and no recent evidence of dead trees, we considered this an unlikely occurrence.

Although we generated successful DNA fingerprints for 256 seedlings and 105 parents, our conservative measures of parent-seedling matches resulted in only 20 matches for white oak and 83 matches for red oak. Comparisons of the distances between parent and offspring for the two species was a bit of a surprise. As summarized by Steele et al. (2007), matches for red oak ranged from 2.3 m to 112.6 m, and averaged 30.8 m (±4.1 SE) for northern red oaks. The same measures for white oaks ranged from 2.4 m to 109 m, with a mean of 51.3 m (±8.0 SE) (Fig. 15.3a). The distances between seedlings and adult trees were significantly greater for white oaks (t = <minus>2.40, df = 81, P = 0.019). This was clearly opposite of our predictions. When we began our studies, we had assumed from the literature that pollen sources would most likely originate from some distance off our study areas. However, these results gave us pause and we needed to somehow control for this. We therefore increased our sample size and measured the distance between seedlings and the nearest potential parent that could not be excluded as a parent with DNA analyses. Although not a parental match, all other closer trees could be excluded as such. This helped to increase our sample size to 164 red oaks, 76 white oaks, and now included 16 black oaks.

Minimum distances between seedlings and the nearest potential parent source that could not be excluded averaged about 30 m for both black oak and white oak and was less than 20 m for red oak (Fig. 15.3b).

Even more telling was the distribution of seedlings in relation to minimal distances to putative parents. The seedling shadow of red oaks was similar to the dispersal kernels (seed shadow) of rodent dispersed acorns (Fig. 15.4a). It peaked at 10 m and the tail extended to more than 60 m. This was not the case for the other two species, however. The same distributions for white oaks and black oaks were strikingly different (Fig. 15.4bc). That of white oak showed two peaks at about 18 m and 55 m, and the distribution of black oak approximated a normal distribution with a peak at 30 m. Distribution of seedlings of these two species suggested nothing likely to follow from our observations on rodent dispersal.

As we discussed in detail in Steele et al. (2007), these patterns of seedling establishment likely reflect a combination of events that collectively contributed to these unexpected seedling shadows in white and black oak. Differential mortality of cached acorns and young seedlings is certainly a potential contributor to the distribution of seedlings of all three species. However, higher mortality closer to parental sources as predicted by Janzen (1970) and Connell (1971) would not by itself contribute to the observed patterns. Yes, the blue jay (*Cynaocitta cristata*) and a few other species of avian seed predators, including the red-bellied woodpecker (*Melanerpes carolinus*), tufted titmouse (*Baeolophus bicolor*), and white-breasted nuthatches (*Sitta carolinensis*) readily retrieve and consume acorns of these oaks (Richardson et al. 2013). However, evidence that these species disperse and cache acorns within forest patches is indeed lacking. In fact, Moore (2005) failed to observe any evidence of dispersal of acorns by jays when acorns were presented in forest patches. Although there is ample evidence that blue jays regularly avoid the larger seeded acorns of both red oaks and white oaks (Scarlett and Smith 1991; Moore and Swihart 2006; Richardson et al. 2013), Moore and Swihart (2006) found that jays will consume these acorns when those of other preferred species are not available, often only consuming the basal portion of the acorns.

Taken together, these behavioral observations, coupled with the seedling shadows for black and white oaks, suggest an alternative scenario for dispersal of these two oaks within forests. I suggest jays and perhaps a few of these other avian seed predators may regularly retrieve acorns from the canopy of these trees, disperse the acorns short distances, and then discard partially eaten acorns to the ground where some may germinate and establish. As discussed in detail in Chapter 6, acorns are well equipped with chemical (tannin, lipid, and nutrients) gradients that likely promote partial consumption of the fruits by both vertebrate and insect seed predators, as well as the ability to tolerate such damage and still germinate and establish. Although white oaks are generally

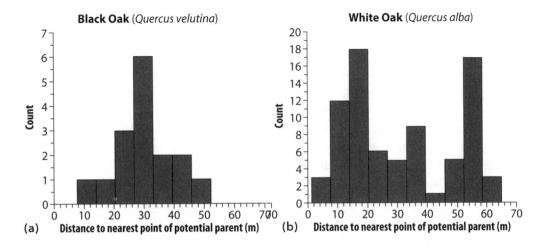

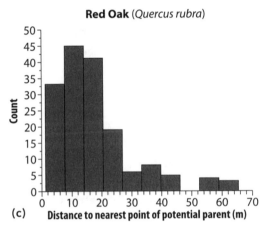

15.4. Distances between potential parent oaks and offspring for three oak species. Data based on the closest parent that DNA fingerprinting did not exclude. *(From Steele et al. 2007)*

less preferred by jays, for example, they are consumed when the fruits of other oaks are not available. Perhaps dispersal of white oaks requires that this species produces mast crops independently of red oaks. In rare years such as this, when only white oaks mast, perhaps then rodents also contribute to dispersal by selectively dispersing individual white oak acorns that are not obviously germinating at the time (Xiao et al. 2010). There is no doubt that it is likely a number of circumstances that collectively contribute to the seedling shadows described here for black and white oak and that much additional research is needed to understand the connection between these seedling distributions and our observed patterns of acorn dispersal.

ON THE DISTRIBUTION OF THE OAKS

Prior to our attempts to genetically match seedlings and adult trees, our behavioral research would have potentially led us to the conclusion that white oak species, at least in the context of red oaks, are at an extreme dispersal disadvantage and are afforded few opportunities for establishment beyond the parent tree. My coauthors and I have frequently made such assertions in the past literature and have further hypothesized that this differential pattern of dispersal should likely contribute to markedly different distributions of red and white oak species in the forest. The data above, however, certainly counters this argument and certainly suggests that white oaks are able to achieve successful dispersal and establishment. Nevertheless, there are clear benefits of working backward from adult tree and sapling distributions just as we have done so with that of seedlings. While an understanding of the structure, spatial distribution, and genetic relatedness of oaks within a forest will not fully reveal how this structure was achieved, it will provide important insights into the dispersal mechanisms and differential mortality that may have contributed to current conditions, and, in turn, lead to additional hypotheses worthy of testing.

MASTING PATTERNS IN WHITE OAKS AND RED OAKS

Just as our behavioral studies with rodents in eastern deciduous forests led us to hypothesize that white oaks have a dispersal disadvantage, especially in the context of red oak species, we also predict that masting patterns should differ between the two genera of oaks. In particular, it would follow that it is likely an advantage for white oak species to mast out of synchrony with red oaks. This certainly appeared to be the case as shown by Koenig and Knops (2002) for California oaks, although these authors expressed uncertainty as to why this would occur. My lab is now exploring several predictions to follow from hypotheses of how masting patterns in white oaks should differ from that of red oak species at numerous sites across eastern North America. Most importantly, we predict that white oaks are more likely to experience mast failures when red oaks mast, overall exhibit more variable crops (e.g., higher CVs of crop production), and produce the most prodigious crops when red oak species experience crop failure. However, because of the various proximate factors that influence acorn masting in both white and red oaks, it is unlikely that differences in masting cycles in the two genera of oaks will necessarily behave exactly as predicted. Most instructive may be comparisons of stands of white and red oak species with those in which only species of one genera are present.

Indeed, much yet remains to be done on this front.

Although the discussion above emphasizes the importance of approaching the study of oak dispersal and regeneration from a community perspective—one that emphasizes all species that influence acorn dispersal and predation—equally compelling is the focused investigations of potential keystone species that can influence acorn fates.

Perhaps the most salient examples of keystone dispersal agents are those that have developed close mutualisms with the pine species (*Pinus* spp.) they both disperse and depend on for food. These include the corvids, such as the Clark's nutcracker (*Nucifraga columbiana*) and a few other species of nutcrackers worldwide and, of course, the jays. As thoroughly and meticulously demonstrated by Vander Wall (1990, and references therein) and Lanner (1996), the Clark's nutcracker of western North America is a "top-of-the-line pinivore" (Lanner 1996, p. 38). It is highly dependent on pine seeds and is well equipped to recover and move huge quantities of seeds considerable distances (several kilometers), scatterhoard these seeds, recover a portion of these caches, and, in the process, contribute significantly to dispersal and establishment of several species. One such mutualist is the white bark pine (*Pinus albicaulis*). Lanner (1996) details accounts by Hutchins (see also Hutchins and Lanner 1982) estimating that in one season a single bird could disperse and cache 98,000 seeds just under 7 km on average. Often dispersing several times more seeds than they need to sustain themselves, many seeds are not recovered and successfully germinate in caches (Vander Wall and Balda 1977).

Whereas the Clark's nutcracker exhibits several adaptations to exploit the pines, such as a heavy bill for retrieving the seeds, a sublingual pouch for transporting large quantities of seeds and the cognitive ability for remembering, relocating, and recovering cached seeds, the pine also possesses its own traits that speak to its relationship with the bird. Armored bracts that remain closed and unavailable to other seed consumers when seeds mature and wingless seeds that can only be dispersed by the birds suggest a strong mutualism that were forged by the heavy arms of coevolution. And, while the Clark's nutcracker's closest mutualist is likely the white barked pine, similar relationships extend to several other pine species as well, allowing the bird to exploit other species when the crop of one species fails (Lanner 1996). Extending well beyond this mutualism, and the nutcracker's contribution to forest structure is an entire community of plants and animals that depend on this ecosystem. Lanner also reviews the similar role that other species of nutcrackers in regions of northern Europe, Siberia, and China have on the forest ecosystems in these regions. However, just as these forests are built on the strength of these bird-conifer mutualisms, as Lanner so insightfully warned in 1996, they are highly susceptible to fire exclusion and the encroachment of shade-tolerant trees, global warming, and the loss of habitat, as well as outbreaks of a parasitic fungus and the bark beetle—all major concerns that are now being realized.

KEYSTONE OAK DISPERSERS

The case of the pines and nutcrackers clearly illustrates how a dispersal agent may impart a keystone effect on an entire forest ecosystem. The oaks, in contrast, are not likely served by a single dispersal agent in most systems. As discussed throughout this book, oaks are almost always dispersed by both rodents and jays, with jays facilitating dispersal of smaller-seeded oaks and the rodents often dispersing a diversity of species of especially red oaks. Long-distance dispersal generally to forest edges and successional habitats or conifer stands where rodent competitors are less abundant appears to be common for jays (see Chapter 10). The rodents' influence on dispersal is largely restricted to within forest stands (see Chapter 9), although the genetic studies reviewed earlier in this chapter suggest another mechanism by which avian dispersers may contribute to oak dispersal within forests.

Based on many of the disperser-oak duos discussed in this book, it would be easy to jump to the conclusion that many of these relationships represent a mutualism achieved by tight coevolutionary interactions. The eastern gray squirrel of the deciduous forests of eastern North America represent such an example. Although the tree and pine squirrels can be classified as keystone seed predators in conifer forests that have likely coevolved with these trees (Steele et al. 2005), it is not as clear that the squirrels necessarily share such a tight mutualism with only the oaks. Squirrels, no doubt, are well adapted for scatterhoarding acorns, maintaining ownership of these caches, and relocating and maximizing returns on these stores. The oaks, in return, benefit tremendously when squirrels fail to recover caches as a result of many factors, perhaps the most significant of which is their death due to the activity of mesic and apex predators. But this relationship between gray squirrels and the oaks extends to other hardwood species (Carya spp., Juglans spp.), some of which are dispersed by only the gray squirrel in parts of their range. In fact, it may very well be that several hardwood species have exerted similar selective pressures on the tree squirrels. We know, for example, that nuts of hardwood species with the highest utility (most preferred) for a gray squirrel are also nuts that are likely to be dispersed to open sites, and that these, more preferred species, exhibit the highest light requirements as seedlings.

Similarly, gray squirrels consistently eat only the basal half of acorns of several oaks species most certainly as a result of the chemical gradients and maybe even the shape of these fruits (see Chapters 5 and 6). However, other rodents, the jays, and even the insect larvae that feed on these acorns are equally affected by these chemical gradients, which likely allow acorns to escape their wrath as well. Whereas a strong evolutionary relationship between the squirrel and the oak exists, the oaks' interactions with other seed dispersers and seed predators cannot be minimized.

The same can be said for the eastern blue jay (*Cyanocitta cristata*) of North America and the Eurasian jay (*Garrulus glandarius*) of much of Europe and Asia. Although both are critical agents of long-distance dispersal for several species of oaks, the selective pressures that other species exert on these acorns is also evident. In nearly all oak systems, from dry forests of Costa Rica, to the diverse forest systems of China and Mexico, to the California oaks, keystone agents of dispersal no doubt exist. However, none are likely to share the tight mutualism exhibited by the nutcrackers and the pines. Oak dispersal instead follows from a suite of diffuse evolutionary interactions that collectively define the process of oak dispersal. My suggestion to future researchers is to avoid the temptation of focusing on just the single, most apparent dispersal agent and instead concentrate on the many more subtle interactions contributing to this process.

THE ROLE OF MESIC AND APEX PREDATORS ON THE DISPERSAL OF OAKS

Perhaps some of the most underestimated interactions contributing to oak dispersal are the factors that lead to failure of cache recovery. Given that the rodents and birds that move acorns possess a superior cognitive ability for recovering their caches (see Chapter 11), it is unlikely, despite its long popularity as an explanation, that seed dispersers simply forget where caches were placed. The truth is that we know little about the factors that contribute to the failure of cache recovery and its contribution to seedling establishment, as it has not been a major focus of investigation.

My laboratory's efforts in this area suggest that even in a high-density population, individual gray squirrels are able to scatterhoard acorns out of reach of conspecifics. When cache owners are experimentally removed from the study area thereby simulating their mortality, their caches are not pilfered by conspecifics (Steele et al. 2011). Only when these cache owners are returned to the study area are most of their scatterhoards recovered. In other studies at the same site, we found that gray squirrels selectively cache acorns of greater value farther from cover (Steele et al. 2014) where squirrels perceived higher risks of predation (Steele et al. 2015).

Although somewhat limited in their focus, these studies collectively suggest that higher predation risks translate into safer cache sites where competitors are less likely to pilfer caches. The question to follow is whether predation of cache owners in turn translates into higher cache survival and seedling establishment. We are now investigating the possibility that mesic and apex predators may contribute to the dispersal process in oaks and other hardwood species. In fact, our preliminary studies in this area suggest that this may be a primary means of oak dispersal and that predators may contribute significantly to establishment of the oaks in suitable habitats. Potentially, it is the squirrels' and jays' predators that offer a critical ecosystem service to the oaks.

DIPLOCHORY: DOES IT OCCUR IN OAKS?

Yet another interaction that is poorly understood in the oak system is that of diplochory—the process by which plant seeds are dispersed in two steps, each by a different dispersal agent. Best articulated by Vander Wall and Longland (2004), the first phase of diplochory usually involves movement of seeds beyond the seed shadow of the parent plant to sites where seed density is lower. The seed is often not the focus of dispersal but often coincidentally moved by the dispersal agent. The second phase often involves further movement and placement of the seed to sites where it is more likely to germinate (e.g., cache site) if not recovered. Vander Wall and Longland (2004) identified five major types of diplochory: wind dispersal followed by rodent scatterhoarding, ballistic dispersal followed by myrmecochory, and endozoochory followed by either dung beetles, myrmecochory, or rodent scatterhoarding. As you may have quickly concluded, most, if not all of these examples, are unlikely by definition to involve oak dispersal.

I suggest, however, that there are a few situations in which oaks may be moved this way. First, there are few species that are likely to consume whole acorns and move them some distance before defecating (i.e., endozoochory). Although numerous species consume acorns, most digest them in the process. Exceptions may be the gray fox (*Urocyon cinereoargenteus*) and the red fox (*Vulpes vulpes*), both omnivores, which are known to regularly consume acorns (Cypher 2003). We have found whole, intact acorns in the fox scat during a mast year, but we have not tested whether acorns that pass through these omnivores' gut can germinate. We have assumed, based on input from Dr. Roger Powell, that black bears (*Ursus americanus*), which are known to regularly consume acorns, are not likely to pass whole acorns in the gut. However, Enders and Vander Wall (2012) have shown that the seeds of other fruits are successfully passed and germinate following bear defecation. Moreover, these seeds are often recovered and then scatterhoarded by deer mice (*Peromyscus maniculatus*). Whether similar means of diplochory occur for the oaks is not known, but if it does, even on rare occasions, it is likely to have a significant effect on dispersal.

A second potential means of oak diplochory that has not been well studied involves jay dispersal followed by secondary dispersal by scatter-hoarding rodents. Although jays selectively disperse acorns to sites with lower rodent activity (see Chapter 10), rodents are in most cases not absent at these sites. Given the distance jays often disperse acorns and the suitability of many of these habitats for acorn germination and oak establishment, pilferage, and re-caching by rodents only needs to occur occasionally for this to significantly drive oak dispersal and establishment. Based on my qualitative observations of jay caching, I suspect that in some habitats, jay caches are particularly susceptible to rodent pilferage. Future studies should consider this as a plausible avenue of successful diplochory.

UNDERSTANDING DIFFERENTIAL MORTALITY AND SEEDLING SURVIVAL IN RELATION TO THE PARENT OAK

Much is known about where oak seedlings do not grow, especially when it involves light, moisture, and soil conditions. However, few studies have investigated oak seed and seedling survival in relation to the parent plant. The Janzen-Connell (J-C) hypotheses, visited several times in this book, is the underlying explanation for why dispersal is so important. Despite its assumptions, specific tests of the J-C model in the oak system are rare. Comita et al. (2014) conducted a comprehensive meta-analyses of tests of the predictions of Janzen (1970) and Connell (1971) across the globe over the four decades since they were first published. They found significant evidence to support both the distance and density predictions to follow from the J-C model. They also found that these effects do not, as predicted by several authors, vary with latitude. And, although these effects do not vary with location, they are stronger in wetter environments. J-C predictions were also stronger for seedlings than for seeds. Among all the studies reviewed by Comita et al. (2014), only two obscure studies centered on oaks: one on the jolcham oak (Q. serrata) and the other on northern red oak and the scarlet (Q. coccinea). The first of these found both a distance and density effect and the second demonstrated only a distance effect. Although these results are important, they come as little surprise given what has been reported on the negative effects of seed density on seed survival, as well as the negative effects of overstory density, nutrient conditions of the soil, and other edaphic factors on seedling survival (Steele and Smallwood 2002, and references therein).

Despite the general support of the J-C model, there is growing evidence that our overall perspective may be biased by focusing primarily on interactions that are occurring above ground. A few recent studies strongly suggest that many of the interactions driving seedling survival in relation to parental sources occurs in the soil, often as a result of interactions with mycorrhizal fungi (Liang et al. 2015; Bennett et al. 2017; Deniau et al. 2018).

In a recent comprehensive study of 550 populations of 55 tree species in temperate forests of North America, Bennett et al. (2017) produced strong evidence that species that depend on ectomycorrhizal fungi, four of which were oaks (northern red oak, black oak, pin oak [*Q. palustris*], and white oak), showed strong positive plant-soil feedback compared with species that are symbiotic with arbuscular mycorrhiza, those in which the symbiont fungus penetrates the cortical cells of the roots. Deniau et al. (2018), in contrast, planted seedlings of pedunculate oak (*Q. robur*) and sessile oak (*Q. petrea*) at different distances from adult trees and applied fungicide to half the seedlings. They found increased mortality of seedlings near conspecifics as a result of a specialist fungal decomposer, which increased seedling quality, thereby increasing intensity of herbivory. They found that ectomycorrhizal fungi were associated with higher rates of herbivory of seedlings, and the effect was stronger closer to conspecific adults. These studies only represent the beginning—there is much to learn about the seedling's relationship with parental and conspecific adults, much of which may occur below the soil surface.

UNDERSTANDING THE COMMUNITY EFFECTS ON DISPERSAL

As evident throughout my overview of the oak dispersal process, it is clear that in any oak ecosystem the patterns of acorn predation and dispersal and the ultimate establishment of oak seedlings does not depend on the activity of a single species, but instead that of a community of insects, mammals, and birds. Nearly all studies to date have emphasized one aspect and usually the impact of a single species or interaction between a few species in the process. A systematic whole system approach has been rarely applied to understanding the process of seed dispersal, especially in the oaks. It is not realistic to assume this can be done in a single investigation; however, there is now enough completed research in some systems where a systematic analyses of past studies can be considered. Perhaps one of the first places to start is to build a systematic framework around what is known in a particular oak forest type, keeping in mind that even within one system there may be considerable variation.

BIOGEOGRAPHIC VARIATION IN OAK-ANIMAL INTERACTIONS

Future research should also be directed at understanding biogeographic variation in patterns of seed dispersal and predation. My lab's comparative studies of acorn dispersal and predation by tree squirrels in North America, north of Mexico, with those in central Mexico exhibit how both similarities and stark contrasts exist in the animals' responses to acorns of red oak and white oak species in these two regions (Steele et al. 2001; see Chapter 5). Although these responses likely follow from differences in climate, soil conditions, and the rodents' need to store acorns in these two regions, the variables are many, making it nearly impossible to identify the specific conditions that alter acorn fates in the two regions. Even the acorn and squirrel species differ in these regions. Controlled comparative studies across latitude are what are most needed to understand how changes in behavior of seed predators and dispersers may vary with climate, soil, and subtle differences in seed characteristics. Current studies in my lab are focusing on variation in masting patterns, seed characteristics (including seed chemistry), and patterns of weevil infestation across a latitudinal gradient from upstate New York to North Carolina. Similar analyses in the Iberian Peninsula (Bogdziewicz et al. 2019) demonstrate how weevil predation, weevil community structure, and acorn size varies across the entire latitudinal gradient of the holm oak (Q. ilex). Larger acorn size in more southernly latitudes allows this species to tolerate seed predation by weevils. More studies like this will help us understand the subtle differences influencing oak-animal interactions that are linked to acorn and oak seedling survival.

One question of particular interest that may be revealed through biogeographic comparisons is how various soil and climatic conditions influence patterns of pilferage of caches and seed fates. My lab has observed limited rates of cache pilferage by gray squirrels when scatterhoarding acorns (Steele et al. 2011) possibly because of the multiple strategies employed by this species when preparing caches (Steele et al. 2008, 2014, 2015). In contrast, in many other systems, pilfering is common and may lead to intense levels of reciprocal pilferage between scatterhoarders (Vander Wall and Jenkins 2003). The ability of rodents to pilfer seed caches often relates to changes in soil moisture (Vander Wall 1998, 2000), and in dry environments, pilferage may be enhanced when it rains, and soil moisture increases the pilferers' advantage (Vander Wall 1998). An important question that remains unanswered is how these conditions may vary across different oak micro-environments and oak ecosystems and, in turn, influence the probability of pilferage and the effectiveness of oak dispersal.

THE CONCEPT OF SEED DISPERSAL EFFECTIVENESS

Finally, I close with a brief reference to the seminal work of Schupp and colleagues on the conceptual framework of seed dispersal effectiveness (SDE). First introduced by Schupp in 1993, SDE is defined as the relative impact a seed disperser has on the reproduction of a plant. In its first configuration, Schupp outlined a hierarchy of measures to assess the relative contributions of frugivores on seed dispersal. In brief, these measures include the quantity of seeds dispersed (which is based on the number of visits to the plant and the number of seeds dispersed per visit) and the quality of the dispersal (which includes the quality of the treatment of seeds when vomited or defecated by the frugivore, as well as the quality of the deposition patterns of the seeds).

More recently Schupp et al. (2010) provided an overview of what is known of SDE based on the more than 4500 papers published on seed dispersal in the 25 years previous to this publication. They then presented a conceptual analysis in the form of a two-dimensional landscape that characterizes the many factors that contribute to the quality and quantity of seed dispersal and overall SDE. This analysis and overview offer a systematic approach for any investigator studying seed dispersal or other forms of mutualism (Schupp et al. 2017). Clearly, the next step in the study of oak dispersal is to develop a detailed hierarchical outline of the many factors that contribute to oak dispersal effectiveness and how it varies within and across various ecosystems dominated by the oaks. It will be this kind of systematic approach that helps us to define the subtle suite of factors that drive oak dispersal and oak regeneration. My colleagues and I are in the process of preparing such an overview, in part, based on the overview and analysis outlined in this book.

LITERATURE CITED

Abbott, H. G., and T. F. Quink. 1970. Ecology of eastern white pine seed caches made by small forest mammals. Ecology **51**:271–278.

Abrams, M. D. 1992. Fire and the development of oak forests. BioScience **42**:346–353.

Abrams, M. D. 2003. Where has all the white oak gone? BioScience **53**:927–939.

Abrams, M. D., and S. E. Johnson. 2012. Long-term impacts of deer exclosures on mixed-oak forest composition at the Valley Forge National Historical Park, Pennsylvania, USA. Journal of the Torrey Botanical Society **139**:167–181.

Abrams, M. D., and F. K. Seischab. 1997. Does the absence of sediment charcoal provide substantial evidence against the fire and oak hypothesis? Journal of Ecology **85**:373–375.

Acacio, V., M. Holmgran, P. Jansen, and O. Schrotter. 2007. Multiple recruitment limitation causes arrested succession in Mediterranean cork oak systems. Ecosystems **10**:1220–1230.

Aizen, M. A., and W. A. Patterson III. 1990. Acorn size and geographical range in the North American oaks (*Quercus* L.). Journal of Biogeography **17**:327–332.

Aizen, M. A., and H. Woodcock. 1992. Latitudinal trends in acorn size in eastern North American species of *Quercus*. Canadian Journal of Botany **70**:1218–1222.

Akresh, M. E., D. I. King, and R. T. Brooks. 2015. Demographic response of a shrubland bird to habitat creation, succession, and disturbance in a dynamic landscape. Forest Ecology and Management **336**:72–80.

Andersen, D., and M. Folk. 1993. *Blarina brevicauda* and *Peromyscus leucopus* reduce overwinter survivorship of acorn weevils in an Indiana hardwood forest. Journal of Mammalogy **74**:656–664.

Anderson, J. T., T. Nuttle, J. S. Saldaña Rojas, T. H. Pendergast, and A. S. Flecker. 2011. Extremely long-distance seed dispersal by an overfished Amazonian frugivore. Proceedings of the Royal Society B: Biological Sciences **278**:3329–3335.

Andersson, M., and J. Krebs. 1978. On the evolution of hoarding behaviour. Animal Behaviour **26**:707–711.

Andresen, E., F. Feer, P. Forget, J. Lambert, P. Hulme, and S. Vander Wall. 2005. The role of dung beetles as secondary seed dispersers and their effect on plant regeneration in tropical rainforests. Pp. 331–350 *in* P. M. Forget, J. E. Lambert, P. E. Hulme, and S. B. Vander Wall, editors. Seed Fate: Predation, Dispersal, and Seedling Establishment. CABI, Wallingford, Oxfordshire, UK.

Arias-LeClaire, H., R. Bonal, D. Garcia-López, and J. M. Espelta. 2018. Role of seed size, phenology, oogenesis and host distribution in the specificity and genetic structure of seed weevils (*Curculio* spp.) in mixed forests. Integrative Zoology **13**:267–279.

Axelrod, R., and W. Hamilton. 1981. The evolution of cooperation. Science **211**:1390–1396.

Balda, R. P., and A. C. Kamil. 1992. Long-term spatial memory in Clark's nutcracker, *Nucifraga columbiana*. Animal Behaviour **44**:761–769.

Baldwin, I. T., and J. C. Schultz. 1983. Rapid changes in tree leaf chemistry induced by damage: Evidence for communication between plants. Science **221**:277–279.

Barkley, C. L., and L. F. Jacobs. 1998. Visual environment and delay affect cache retrieval accuracy in a food-storing rodent. Animal Learning & Behavior **26**:439–447.

Barnea, A., and F. Nottebohm. 1994. Seasonal recruitment of hippocampal neurons in adult free-ranging black-capped chickadees. Proceedings of the National Academy of Sciences of the United States of America **91**:11217–11221.

Barnea, A., and F. Nottebohm. 1996. Recruitment and replacement of hippocampal neurons in young and adult chickadees: An addition to the theory of hippocampal learning. Proceedings of the National Academy of Sciences of the United States of America **93**:714–718.

Barnea, A., and V. Pravosudov. 2011. Birds as model to study neurogenesis: Bridging evolutionary, comparative and neuroethological approaches. European Journal of Neuroscience **34**:884–907.

Barnett, R. J. 1977. The effect of burial by squirrels on germination and survival of oak and hickory nuts. American Midland Naturalist **98**:319–330.

Bartlow, A. W., S. J. Agosta, R. Curtis, X. Yi, and M. A. Steele. 2018a. Acorn size and tolerance to seed predators: The multiple roles of acorns as food for seed predators, fruit for dispersal and fuel for growth. Integrative Zoology **13**:251–266.

Bartlow, A. W., M. Kachmar, N. Lichti, R. K. Swihart, J. A. Stratford, and M. A. Steele. 2011. Does multiple seed loading in blue jays result in selective dispersal of smaller acorns? Integrative Zoology **6**:235–243.

Bartlow, A. W., N. I. Lichti, R. Curtis, R. K. Swihart, and M. A. Steele. 2018b. Re-
    caching of acorns by rodents: Cache management in eastern deciduous forests
    of North America. Acta Oecologica 92:117–122.

Basil, J. A., A. C. Kamil, R. Balda, and K. V. Fite. 1996. Differences in hippocampal
    volume among food storing corvids. Brain, Behavior and Evolution 47:156–164.

Bate-Smith, E. 1973. Haemanalysis of tannins: The concept of relative astringency.
    Phytochemistry 12:907–912.

Bednekoff, P. A., and R. P. Balda. 1996a. Observational spatial memory in Clark's
    nutcrackers and Mexican jays. Animal Behaviour 52:833–839.

Bednekoff, P. A., and R. P. Balda. 1996b. Social caching and observational spatial
    memory in pinyon jays. Behaviour 133:807–826.

Bednekoff, P. A., R. P. Balda, A. C. Kamil, and A. G. Hile. 1997. Long-term spatial
    memory in four seed-caching corvid species. Animal Behaviour 53:335–341.

Benkman, C. W. 1995a. The impact of tree squirrels (*Tamiasciurus*) on limber pine
    seed dispersal adaptations. Evolution 49:585–592.

Benkman, C. W. 1995b. Wind dispersal capacity of pine seeds and the evolution of
    different seed dispersal modes in pines. Oikos 73:221–224.

Benkman, C. W. 1999. The selection mosaic and diversifying coevolution between
    crossbills and lodgepole pine. American Naturalist 153:S75-S91.

Benkman, C. W. 2003. Divergent selection drives the adaptive radiation of
    crossbills. Evolution 57:1176–1181.

Benkman, C. W., W. C. Holimon, and J. W. Smith. 2001. The influence of a
    competitor on the geographic mosaic of coevolution between crossbills and
    lodgepole pine. Evolution 55:282–294.

Bennett, A. T. D. 1993. Spatial memory in a food storing corvid: 1. Near tall
    landmarks are primarily used. Journal of Comparative Physiology A:
    Neurothology, Sensory, Neural and Behavioral Physiology 173:193–207.

Bennett, J. A., H. Maherali, K. O. Reinhart, Y. Lekberg, M. M. Hart, and J.
    Klironomos. 2017. Plant-soil feedbacks and mycorrhizal type influence
    temperate forest population dynamics. Science 355:181–184.

Bernays, E. 1981. Plant tannins and insect herbivores: An appraisal. Journal of
    Ecological Entomology 6:353–360.

Beucke, K., and P. Choate. 2009. Notes on the feeding behavior of *Mycotrupes
    lethroides* (Westwood) (Coleoptera: Geotrupidae), a flightless North American
    beetle. Coleopterists Bulletin 63:228–229.

Bieberich, J., M. Lauerer, and G. Aas. 2016. Acorns of introduced *Quercus rubra*
    are neglected by European jay but spread by mice. Annals of Forest Research
    59:249–258.

Biegler, R., A. McGregor, J. R. Krebs, and S. D. Healy. 2001. A larger hippocampus
    is associated with longer-lasting spatial memory. Proceedings of the National
    Academy of Sciences of the United States of America 98:6941.

Bjorkman Anne, D., and M. Vellend. 2010. Defining historical baselines for conservation: Ecological changes since European settlement on Vancouver Island, Canada. Conservation Biology **24**:1559–1568.

Bobiec, A, A. Reif, and K. Öllerer. 2018. Seeking the oakscape beyond the forest: A landscape approach to the oak regeneration in Europe. Landscape Ecology **33**:513–528.

Bogdziewicz, M., R. Bonal, J. M. Espelta, E. Kalemba, M. A. Steele, and R. Zwolak. 2018d. Invasive oaks escape pre-dispersal insect seed predation and trap enemies in their seeds. Integrative Zoology **13**:228–237.

Bogdziewicz, M., E. E. Crone, M. A. Steele, and R. Zwolak. 2017a. Effects of nitrogen deposition on reproduction in a masting tree: Benefits of higher seed production are trumped by negative biotic interactions. Journal of Ecology **105**:310–320.

Bogdziewicz, M., J. M. Espelta, and R. Bonal. 2018a. Tolerance to seed predation mediated by seed size increases at lower latitudes in a Mediterranean oak. Annals of Botany **20**:1–8.

Bogdziewicz, M., J. M. Espelta, A. Muñoz, J. M. Aparicio, and R. Bonal. 2018b. Effectiveness of predator satiation in masting oaks is negatively affected by conspecific density. Oecologia **186**:983–993.

Bogdziewicz, M., M. Fernández-Martínez, R. Bonal, J. Belmonte, and J. M. Espelta. 2017b. The Moren effect and environmental vetoes: Phenological synchrony and drought drive seed production in a Mediterranean oak. Proceedings of the Royal Society B: Biological Sciences **284**:2017.1784.

Bogdziewicz, M., S. Marino, R. Bonal, R. Zwolak, and M. A. Steele. 2018c. Rapid aggregative and reproductive responses of weevils to masting of North American oaks counteract predator satiation. Ecology **99**:2575–2582.

Bogdziewicz, M., J. Szymkowiak, I. Kasprzyk, Ł. Grewling, Z. Borowski, K. Borycka, W. Kantorowicz, D. Myszkowska, K. Piotrowicz, and M. Ziemianin. 2017c. Masting in wind-pollinated trees: System-specific roles of weather and pollination dynamics in driving seed production. Ecology **98**:2615–2625.

Bogdziewicz, M., M. Zywiec, J. Espelta, M. Fernandez-Martinez, R. Calama, M. Ledwon, E. McIntire, and E. Crone. 2019. Environmental veto synchronizes mast seeding in four contrasting tree species. American Naturalist **194**:246–259.

Bonal, R., J. M. Espelta, and A. P. Vogler. 2011. Complex selection on life-history traits and the maintenance of variation in exaggerated rostrum length in acorn weevils. Oecologia **167**:1053–1061.

Bonal, R., M. Hernández, J. M. Espelta, A. Muñoz, and J. M. Aparicio. 2015. Unexpected consequences of a drier world: Evidence that delay in late summer rains biases the population sex ratio of an insect. Royal Society Open Science **2**:150198.

Bonal, R., M. Hernández, J. Ortego, A. Muñoz, and J. M. Espelta. 2012. Positive cascade effects of forest fragmentation on acorn weevils mediated by seed size enlargement. Insect Conservation and Diversity 5:381–388.

Bonal, R., and A. Muñoz. 2007. Multi-trophic effects of ungulate predation on acorn weevils. Oecologia 152:533–540.

Bonal, R., and A. Muñoz. 2008. Seed growth suppression constrains the growth of seed parasites: Premature acorn abscission reduces *Curculio elephas* larval size. Ecological Entomology 33:31–36.

Bonal, R., A. Muñoz, and M. Díaz. 2007. Satiation of predispersal seed predators: The importance of considering both plant and seed levels. Evolutionary Ecology 21:367–380.

Bonal, R., A. Muñoz, and J. M. Espelta. 2010. Mismatch between the timing of oviposition and the seasonal optimum: The stochastic phenology of Mediterranean acorn weevils. Ecological Entomology 35:270–278.

Bonner, F., and J. A. Vozzo. 1987. Seed Biology and Technology of *Quercus*. General Technical Report SO-66. US Department of Agriculture, Forest Service, Southern Forest Experiment Station, New Orleans, LA.

Bossema, I. 1979. Jays and oaks: An eco-ethological study of a symbiosis. Behaviour 70:1–117.

Bossema, I., and W. Pot. 1974. Het terugvinden van verstopt voedsel door de Vlaamse gaai (*Garrulus g. glandarius* L.). De Levende Natuur 77:265–279.

Boucher, D. H. 1981. Seed predation by mammals and forest dominance by *Quercus oleoides*, a tropical lowland oak. Oecologia 49:409–414.

Boucher, D. H. 1983. *Quercus oleoides* (roble encino, oak). Pp. 319–322 *in* D. H. Janzen, editor. Costa Rican Natural History. University of Chicago Press, Chicago, IL.

Boucher, D. H., and V. L. Sork. 1979. Early drop of nuts in response to insect infestation. Oikos 33:440–443.

Boutin, S., K. W. Larsen, and D. Berteaux. 2000. Anticipatory parental cares: Acquiring resources for offspring prior to conception. Proceedings of the Royal Society B: Biological Sciences 267:2081–2085.

Boutin, S., L. A. Wauters, A. G. McAdam, M. M. Humphries, G. Tosi, and A. A. Dhondt. 2006. Anticipatory reproduction and population growth in seed predators. Science 314:1928–1930.

Bradford, D. F., and C. C. Smith. 1977. Seed predation and seed number in *Scheelea* palm fruits. Ecology 58:667–673.

Bried, J. T., W. A. Patterson III, and N. A. Gifford. 2014. Why pine barrens restoration should favor barrens over pine. Restoration Ecology 22:442–446.

Brodbeck, D. R. 1994. Memory for spatial and local cues: A comparison of a storing and a non-storing species. Animal Learning & Behavior 22:119–133.

Brodin, A. 2010. The history of scatter hoarding studies. Philosophical Transactions of the Royal Society of London B: Biological Sciences **365**:869–881.

Bronstein, J. L. 1994. Conditional outcomes in mutualistic interactions. Trends in Ecology & Evolution **9**:214–217.

Bugnyar, T., and B. Heinrich. 2006. Pilfering ravens, *Corvus corax*, adjust their behaviour to social context and identity of competitors. Animal Cognition **9**:369–376.

Bugnyar, T., and K. Kotrschal. 2002. Observational learning and the raiding of food caches in ravens, *Corvus corax*: Is it "tactical" deception? Animal Behaviour **64**:185–195.

Burger, W. 1983. *Quercus costaricensis*. Pp. 318–319 *in* D. H. Janzen, editor. Costa Rican Natural History. University of Chicago Press, Chicago, IL.

Cahalane, V. H. C. 1942. Caching and recovery of food by the western fox squirrel. Journal of Wildlife Management **6**:338–352.

Callahan, H. S., K. Del Fierro, A. E. Patterson, and H. Zafar. 2008. Impacts of elevated nitrogen inputs on oak reproductive and seed ecology. Global Change Biology **14**:285–293.

Cancio, I., A. González-Robles, J. M. Bastida, J. Isla, A. J. Manzaneda, T. Salido, and P. J. Rey. 2017. Landscape degradation affects red fox (*Vulpes vulpes*) diet and its ecosystem services in the threatened *Ziziphus lotus* scrubland habitats of semiarid Spain. Journal of Arid Environments **145**:24–34.

Canelo, T., Á. Gaytán, G. González-Bornay, and R. Bonal. 2018. Seed loss before seed predation: Experimental evidence of the negative effects of leaf feeding insects on acorn production. Integrative Zoology **13**:238–250.

Cao, L., Z. Dong, W-J. Liu, J-J. Lei, Z. Shen, and Y-Q. Yang. 2013. Relationship between insect infestation and seed rain dynamics of oriental cork oak (*Quercus variabilis*). Journal of Henan Agricultural Sciences **42**:77–81.

Cao, L., Z. Xiao, Z. Wang, C. Guo, J. Chen, and Z. Zhang. 2011. High regeneration capacity helps tropical seeds to counter rodent predation. Oecologia **166**:997–1007.

Cao, L., C. Yan, and B. Wang. 2018. Differential seed mass selection on hoarding decisions among three sympatric rodents. Behavioral Ecology and Sociobiology **72**:1–9.

Carlo, T. A. 2005. Interspecific neighbors change seed dispersal pattern of an avian-dispersed plant. Ecology **86**:2440–2449.

Cavender-Bares, J., A. González-Rodríguez, A. Pahlich, K. Koeler, and N. Deacon. 2011. Phylogeography and climatic niche evolution in live oaks (*Quercus* series *Virentes*) from the tropics to the temperate zone. Journal of Biogeography **38**:962–981.

Cavender-Bares, J., S. Kothari, J. E. Meireles, M. A. Kaproth, P. S. Manos, and A. L. Hipp. 2018. The role of diversification in community assembly of the oaks

(*Quercus* L.) across the continental U.S. American Journal of Botany **105**:565–586.

Chancellor, L. V., T. C. Roth, L. D. LaDage, and V. V. Pravosudov. 2011. The effect of environmental harshness on neurogenesis: A large-scale comparison. Developmental Neurobiology **71**:246–252.

Charnov, E. L. 1976. Optimal foraging, the marginal value theorem. Theoretical Population Biology **9**:129–136.

Cheng, J., and H. Zhang. 2011. Seed-hoarding of Edward's long-tailed rats (*Leopoldamys edwardsi*) in response to weevil infestation in cork oak (*Quercus variabilis*). Current Zoology **57**:50–55.

Cheng, K., S. J. Shettleworth, J. Huttenlocher, and J. J. Rieser. 2007. Bayesian integration of spatial information. Psychological Bulletin **133**:625.

Chengjiu, H., Z. Yongtian, and B. Bartholomew. 1999. Fagaceae. Pp. 314–400 *in* W. Zhengyi, P. H. Raven, and D. Hong, editors. Flora of China, vol. 4, Cycadaceae through Fagaceae. Science Press, Beijing, China, and Missouri Botanical Garden, St. Louis, MO.

Chidumayo, E. N., and D. J. Gumbo. 2013. The environmental impacts of charcoal production in tropical ecosystems of the world: A synthesis. Energy for Sustainable Development **17**:86–94.

Choi, K. S., J. K. Kim, W. I. Bae. 1993. Life history of the oak nut weevil *Mechoris ursulus* (Roelofs) in Korea. [In Korean, abstract in English.] Research Reports of the Forestry Research Institute (Seoul) **47**:153–157.

Choo, J. P. 2005. The avifauna and wild fruits of two equatorial rainforest sites: An inter-tropical comparison. PhD dissertation, Rutgers University, New Brunswick, NJ.

Chung-MacCoubrey, A. L., A. E. Hagerman, and R. L. Kirkpatrick. 1997. Effects of tannins on digestion and detoxification activity in gray squirrels (*Sciurus carolinensis*). Physiological Zoology **70**:270–277.

Clark, J. S. 1998. Why trees migrate so fast: Confronting theory with dispersal biology and the paleorecord. American Naturalist **152**:204–224.

Clark, J. S., C. Fastie, G. Hurtt, S. T. Jackson, C. Johnson, G. A. King, M. Lewis, J. Lynch, S. Pacala, and C. Prentice. 1998a. Reid's paradox of rapid plant migration: Dispersal theory and interpretation of paleoecological records. BioScience **48**:13–24.

Clark, J. S., E. Macklin, and L. Wood. 1998b. Stages and spatial scales of recruitment limitation in southern Appalachian forests. Ecological Monographs **68**:213–235.

Clark, J. S., and P. D. Royall. 1996. Local and regional sediment charcoal evidence for fire regimes in presettlement north-eastern North America. Journal of Ecology **84**:365–382.

Clark, J. S., M. Silman, R. Kern, E. Macklin, and J. HilleRisLambers. 1999. Seed dispersal near and far: Patterns across temperate and tropical forests. Ecology 80:1475–1494.

Clary, D., and D. M. Kelly. 2011. Sociality and cognition: Fact or fiction? Insight from the cache protection behaviours of Clark's nutcrackers. Canadian Journal of Experimental Psychology/Revue canadienne de psychologie expérimentale 65:306–306.

Clary, D., and D. M. Kelly. 2016. Clark's nutcracker (*Nucifraga columiana*) flexibly adapt caching behavior to a cooperative context. Frontiers in Pschychology 7:1–9.

Clayton, N. S. 1995. The neuroethological development of food-storing memory: A case of use it, or lose it! Behavioural Brain Research 70:95–102.

Clayton, N. S. 1996. Development of food-storing and the hippocampus in juvenile marsh tits (*Parus palustris*). Behavioural Brain Research 74:153–159.

Clayton, N. S. 1998. Memory and the hippocampus in food-storing birds: A comparative approach. Neuropharmacology 37:441–452.

Clayton, N. S., and A. Dickinson. 1998. Episodic-like memory during cache recovery by scrub jays. Nature 395:272.

Clayton, N. S., D. P. Griffiths, N. J. Emery, and A. Dickinson. 2001. Elements of episodic-like memory in animals. Philosophical Transactions of the Royal Society of London B: Biological Sciences 356:1483–1491.

Clayton, N. S., and J. R. Krebs. 1994. Memory for spatial and object-specific cues in food-storing and non-storing birds. Journal of Comparative Physiology A: Sensory, Neural and Behavioral Physiology 174:371–379.

Clayton, N. S., and J. R. Krebs. 1995. Memory in food-storing birds: From behaviour to brain. Current Opinion in Neurobiology 5:149–154.

Clayton, N. S., J. C. Reboreda, and A. Kacelnik. 1997. Seasonal changes of hippocampus volume in parasitic cowbirds. Behavioural Processes 41:237–243.

Clements, D. R. 2013. Translocation of rare plant species to restore Garry oak ecosystems in western Canada: Challenges and opportunities. Botany 91:283–291.

Comita, L. S., S. A. Queenborough, S. J. Murphy, J. L. Eck, K. Xu, M. Krishnadas, N. Beckman, and Y. Zhu. 2014. Testing predictions of the Janzen–Connell hypothesis: A meta-analysis of experimental evidence for distance- and density-dependent seed and seedling survival. Journal of Ecology 102:845–856.

Connell, J. H. 1971. On the role of natural enemies in preventing competitive exclusion in some marine animals and in rain forest trees. Dynamics of Populations 298:312.

Corlett, R. T. 2013. The shifted baseline: Prehistoric defaunation in the tropics and its consequences for biodiversity conservation. Biological Conservation 163:13–21.

Correia, R., M. Bugalho, A. Franco, and J. Palmeirim. 2018. Contribution of spatially explicit models to climate change adaptation and mitigation plans for a priority forest habitat. Mitigation and Adaptation Strategies for Global Change **23**:371–386.

Cousens, R. D., J. Hill, K. French, and I. D. Bishop. 2010. Towards better prediction of seed dispersal by animals. Functional Ecology **24**:1163–1170.

Cowie, R. J., J. R. Krebs, and D. F. Sherry. 1981. Food storing by marsh tits. Animal Behaviour **29**:1252–1259.

Craft, K. J., and M. V. Ashley. 2007. Landscape genetic structure of bur oak (*Quercus macrocarpa*) savannas in Illinois. Forest Ecology and Management **239**:13–20.

Craft, K. J., and M. V. Ashley. 2010. Pollen-mediated gene flow in isolated and continuous stands of bur oak, *Quercus macrocarpa* (Fagaceae). American Journal of Botany **97**:1999–2006.

Crawley, M. J. 1985. Reduction of oak fecundity by low-density herbivore populations. Nature **314**:163.

Crawley, M. J., and C. R. Long. 1995. Alternate bearing, predator satiation and seedling recruitment in *Quercus robur* L. Journal of Ecology **83**:683–696.

Crone, E. E., and J. M. Rapp. 2014. Resource depletion, pollen coupling, and the ecology of mast seeding. Annals of the New York Academy of Sciences **1322**:21–34.

Cuevas-Reyes, P., G. Pérez-López, Y. Maldonado-López, and A. González-Rodríguez. 2017. Effects of herbivory and mistletoe infection by *Psittacanthus calyculatus* on nutritional quality and chemical defense of *Quercus deserticola* along Mexican forest fragments. Plant Ecology **218**:687–697.

Curt, T., W. Adra, and L. Borgniet. 2009. Fire-driven oak regeneration in French Mediterranean ecosystems. Forest Ecology and Management **258**:2127–2135.

Cypher, B. L. 2003. Foxes. Pp. 511–546 *in* J. A. Chapman and G. A. Feldhammer, editors. Wild Mammals of North America: Biology, Management, and Conservation. Johns Hopkins University Press, Baltimore, MD.

Dalgleish, H. J., and R. K. Swihart. 2012. American chestnut past and future: Implications of restoration for resource pulses and consumer populations of eastern US forests. Restoration Ecology **20**:490–497.

Dally, J. M., N. S. Clayton, and N. J. Emery. 2006. The behaviour and evolution of cache protection and pilferage. Animal Behaviour **72**:13–23.

Dally, J. M., N. J. Emery, and N. S. Clayton. 2005. Cache protection strategies by western scrub-jays, *Aphelocoma californica*: Implications for social cognition. Animal Behaviour **70**:1251–1263.

Damschen, E. I., N. M. Haddad, J. L. Orrock, J. J. Tewksbury, and D. J. Levey. 2006. Corridors increase plant species richness at large scales. Science **313**:1284–1286.

Darley-Hill, S., and W. C. Johnson. 1981. Acorn dispersal by the blue jay (*Cyanocitta cristata*). Oecologia **50**:231–232.

Davis, F. W., M. Borchert, L. E. Harvey, and J. C. Michaelsen. 1991. Factors affecting seedling survivorship of blue oak (*Quercus douglasii* H. & A.) in central California. Pp. 81–86 *in* R. B. Sandiford, technical coordinator. Proceedings of the Symposium on Oak Woodlands and Hardwood Rangeland Management, October 31–November 2, 1990, Davis, California. General Technical Report PSW-GTR-126. US Department of Agriculture, Forest Service, Pacific Southwest Research Station, Berkeley, CA.

Deacon, N. J., and J. Cavender-Bares. 2015. Limited pollen dispersal contributes to population genetic structure but not local adaptation in *Quercus oleoides* forests of Costa Rica. PloS One **10**:e0138783.

Deen, R. T., and J. D. Hodges. 1991. Oak regeneration in abandoned fields: Presumed role of the blue jay. Pp. 84–93 *in* S. S. Coleman and D. G. Neary, compilers and editors. Proceedings of the Sixth Biennial Southern Silvicultural Research Conference, Memphis, Tennessee, October 30–November 1, 1990, vol. 1. Southeastern Forest Experiment Station, Asheville, NC.

DeGange, A. R., J. W. Fitzpatrick, J. N. Layne, and G. E. Woolfenden. 1989. Acorn harvesting by Florida scrub jays. Ecology **70**:348–356.

Delcourt, P. A., and H. R. Delcourt. 1987a. Late-Quaternary dynamics of temperate forests: Applications of paleoecology to issues of global environmental change. Quaternary Science Reviews **6**:129–146.

Delcourt, P. A., and H. R. Delcourt. 1987b. Long-term forest dynamics of the temperate zone. Pp. 374–398 *in* Long-Term Forest Dynamics of the Temperate Zone: A Case Study of Late-Quaternary Forests in Eastern North America. Ecological Studies 63. Springer-Verlag, New York, NY.

Delgado, M. M., and L. F. Jacobs. 2017. Caching for where and what: Evidence for a mnemonic strategy in a scatter-hoarder. Royal Society Open Science **4**:170958.

Den Ouden, J., P. A. Jansen, and R. Smit. 2005. Jays, mice and oaks: Predation and dispersal of *Quercus robur* and *Q. petraea* in north-western Europe. Pp. 223–240 *in* J. E. Lambert, P. E. Hulme, and S. E. Vander Wall, editors. Seed Fate: Predation, Dispersal and Seedling Establishment. CABI, Wallingford, Oxfordshire, UK.

Deng, M., X.-L. Jiang, A. L. Hipp, P. S. Manos, and M. Hahn. 2018. Phylogeny and biogeography of East Asian evergreen oaks (*Quercus* section *Cyclobalanopsis*; Fagaceae): Insights into the Cenozoic history of evergreen broad-leaved forests in subtropical Asia. Molecular Phylogenetics and Evolution **119**:170–181.

Deniau, M., V. Jung, C. Le Lann, H. Kellner, B. Béchade, T. Morra, and A. Prinzing. 2018. Janzen–Connell patterns can be induced by fungal-driven decomposition and offset by ectomycorrhizal fungi accumulated under a closely related canopy. Functional Ecology **32**:785–798.

Denk, T., G. Grimm, P. Manos, M. Deng, and A. Hipp. 2017. An updated infrageneric classification of the oaks: Review of previous taxonomic schemes and synthesis of evolutionary patterns. Pp. 13–38 *in* E. Gil-Pelegrín, J. J. Peguero-Pina, and D. Sancho-Knapik, editors. Oaks Physiological Ecology: Exploring the Functional Diversity of Genus *Quercus* L. Springer, Cham, Switzerland.

Dennis, A. J., R. J. Green, and E. W. Schupp. 2007. Seed Dispersal: Theory and Its Application in a Changing World. CABI, Wallingford, Oxfordshire, UK.

Dey, D. 2002. Fire history and postsettlement disturbance. Pp, 46–59 *in* W. J. McShea and W. M. Healy, editors. Oak Forest Ecosystems: Ecology and Management for Wildlife. Johns Hopkins University Press, Baltimore, MD.

Dittel, J. W., R. Perea, and S. B. Vander Wall. 2017. Reciprocal pilfering in a seed-caching rodent community: Implications for species coexistence. Behavioral Ecology and Sociobiology **71**:1–8.

Dittel, J. W., and S. B. Vander Wall. 2018. Effects of rodent abundance and richness on cache pilfering. Integrative Zoology **13**:331–338.

Dixon, M. D., W. C. Johnson, and C. S. Adkisson. 1997a. Effects of caching on acorn tannin levels and blue jay dietary performance. Condor **99**:756–764.

Dixon, M. D., W. C. Johnson, and C. S. Adkisson. 1997b. Effects of weevil larvae on acorn use by blue jays. Oecologia **111**:201–208.

Donald, J. L., and S. Boutin. 2011. Intraspecific cache pilferage by larder-hoarding squirrels (*Tamiasciurus hudsonicus*). Journal of Mammalogy **92**:1013–1020.

Dow, B., and M. Ashley. 1998. High levels of gene flow in bur oak revealed by paternity analysis using microsatellites. Journal of Heredity **89**:62–70.

Dunning, C. E., T. D. Paine, and R. A. Redak. 2002. Insect-oak interactions with coast live oak (*Quercus agrifolia*) and Engelmann oak (*Q. engelmannii*) at the acorn and seedling stage. Pp. 205–281 *in* R. B. Standiford, D. Creary, and K. J. Purcell, technical coordinators. Proceedings of the Fifth Symposium on Oak Woodlands: Oaks in California's Challenging Landscape. General Technical Report PSW-GTR-184, US Department of Agriculture, Forest Service, Pacific Southwest Research Station, Albany, CA.

Dunning, C. E., R. A. Redak, and T. D. Paine. 2003. Preference and performance of a generalist insect herbivore on *Quercus agrifolia* and *Quercus engelmannii* seedlings from a southern California oak woodland. Forest Ecology and Management **174**:593–603.

Ehrlich, P. R., and P. H. Raven. 1964. Butterflies and plants: A study in coevolution. Evolution **18**:586–608.

Ellis, A. M., T. Václavík, and R. K. Meentemeyer. 2010. When is connectivity important? A case study of spatial pattern of sudden oak death. Oikos **119**:485–493.

Emlen, M. 1966. The role of time and energy in food preference. American Naturalist **100**:611–617.

Enders, M. S., and S. B. Vander Wall. 2012. Black bears *Ursus americanus* are effective seed dispersers, with a little help from their friends. Oikos **121**:589–596.

Erickson, W. 1996. Classification and interpretation of Garry oak (*Quercus garryana*) plant communities and ecosystems in southwestern British Columbia. Master's thesis, University of Victoria, Victoria, BC.

Espelta, J. M., H. Arias-LeClaire, M. Fernández-Martinez, E. Doblas-Miranda, A. Muñoz, and R. Bonal. 2017. Beyond predator satiation: Masting but also the effects of rainfall stochasticity on weevils drive acorn predation. Ecosphere **8**:e01836.

Espelta, J. M., R. Bonal, and B. Sánchez-Humanes. 2009a. Pre-dispersal acorn predation in mixed oak forests: Interspecific differences are driven by the interplay among seed phenology, seed size and predator size. Journal of Ecology **97**:1416–1423.

Espelta, J. M., P. Cortés, R. Molowny-Horas, and J. Retana. 2009b. Acorn crop size and pre-dispersal predation determine inter-specific differences in the recruitment of co-occurring oaks. Oecologia **161**:559–568.

Estrada, A., and T. H. Fleming. 1986. Frugivores and Seed Dispersal. Springer Netherlands, Dordrecht, Netherlands.

Farris, E., L. Canopoli, E. Cucca, S. Landi, A. Maccioni, and R. Filigheddu. 2017. Foxes provide a direct dispersal service to Phoenician junipers in Mediterranean coastal environments: Ecological and evolutionary implications. Plant Ecology and Evolution **150**:117–128.

Feeny, P. 1970. Seasonal changes in oak leaf tannins and nutrients as a cause of spring feeding by winter moth caterpillars. Ecology **51**:565–581.

Feldhamer, G. 2002. Acorns and white-tailed deer: Interrelationships in forest ecosystems. Pp. 215–223 *in* W. J. McShea and W. M. Healy, editors. Oak Forest Ecosystems: Ecology and Management for Wildlife. Johns Hopkins University Press, Baltimore, MD.

Fenner, M. 2000. Seeds: The Ecology of Regeneration in Plant Communities. CABI, Wallingford, Oxfordshire, UK.

Fleck, D. C., and G. E. Woolfenden. 1997. Can acorn tannin predict scrub-jay caching behavior? Journal of Chemical Ecology **23**:793–806.

Fleming, T. H. 1993a. *Heteromys desmarestianus* (raton semiespinosa, spiny pocket mouse). Pp. 474–475 *in* D. H. Janzen, editor. Costa Rican Natural History. University of Chicago Press, Chicago, IL.

Fleming, T. H. 1993b. *Liomys salvani* (raton semiespinosa, guardafiesta, Salvin's spiny pocket mouse). Pp. 475–476 *in* D. H. Janzen, editor. Costa Rican Natural History. University of Chicago Press, Chicago, IL.

Fleming, T. H., and A. Estrada. 1993. Frugivory and Seed Dispersal: Ecological and Evolutionary Aspects. Springer Netherlands, Dordrecht, Netherlands.

Forget, P. M., J. E. Lambert, P. E. Hulme, and S. B. Vander Wall, editors. 2005. Seed

Fate: Predation, Dispersal, and Seedling Establishment. CABI, Wallingford, Oxfordshire, UK.

Foster, D. R., S. Clayden, D. A. Orwig, B. Hall, and S. Barry. 2002. Oak, chestnut and fire: Climatic and cultural controls of long-term forest dynamics in New England, USA. Journal of Biogeography 29:1359–1379.

Fox, J. 1982. Adaptation of gray squirrel behavior to autumn germination by white oak acorns. Evolution 36:800–809.

Frank, J., S. Anglin, E. Carrington, D. Taylor, B. Viratos, and D. Southworth. 2009. Rodent dispersal of fungal spores promotes seedling establishment away from mycorrhizal networks on *Quercus garryana*. Botany 87:821–829.

Freinkel, S. 2007. American Chestnut: The Life, Death, and Rebirth of a Perfect Tree. University of California Press, Berkeley, CA.

Fuchs, M. 1998. Seedling ecology of Garry oaks in British Columbia and dispersal of Garry oak acorns by Stellar's jays. Master's thesis, University of British Columbia, Vancouver, BC.

Fuchs, M., P. Krannitz, and A. Harestad. 2000. Factors affecting emergence and first-year survival of seedlings of Garry oaks (*Quercus garryana*) in British Columbia, Canada. Forest Ecology and Management 137:209–219.

Fukumoto, H., and H. Kajimura. 2011. Effects of asynchronous acorn production by co-occurring *Quercus* trees on resource utilization by acorn-feeding insects. Journal of Forest Research 16:62–67.

Futuyma, D., and M. Kirkpatrick. 2017. Evolution, 4th edition. Sinauer, Sunderland. MA.

Gallardo, P., L. Moyano, and A. M. Cárdenas. 2011. Incidence of acorn-boring insects in the area of environmental improvement associated to the Breña II dam (central Sierra Morena, Cordoba). [In Spanish, abstract in English.] Boletín de Sanidad Vegetal, Plagas 37:69–78.

Galloway, J. N., A. R. Townsend, J. W. Erisman, M. Bekunda, Z. Cai, J. R. Freney, L. A. Martinelli, S. P. Seitzinger, and M. A. Sutton. 2008. Transformation of the nitrogen cycle: Recent trends, questions, and potential solutions. Science 320:889–892.

Gálvez, D., B. Kranstauber, R. W. Kays, and P. A. Jansen. 2009. Scatter hoarding by the Central American agouti: A test of optimal cache spacing theory. Animal Behaviour 78:1327–1333.

Gao, X., and S. Sun. 2005. Effects of the small forest carnivores on the recruitment and survival of Liaodong oak (*Quercus wutaishanica*) seedlings. Forest Ecology and Management 206:283–292.

Garrison, W., and C. Augspurger. 1983. Double- and single-seeded acorns of bur oak (*Quercus macrocarpa*): Frequency and some ecological consequences. Bulletin of the Torrey Botanical Club 110:154–160.

Garry Oak Recovery Team. 2003. Species at Risk in Garry Oak and Associated

Ecosystems in British Columbia. Garry Oak Ecosystems Recovery Team, Victoria, BC.

Geist, V. 1987. Bergmann's rule is invalid. Canadian Journal of Zoology **65**:1035–1038.

Gerhardt, F. 2005. Food pilfering in larder-hoarding red squirrels (*Tamiasciurus hudsonicus*). Journal of Mammalogy **86**:108–114.

Gibbs, S. E. B., S. E. G. Lea, and L. F. Jacobs. 2007. Flexible use of spatial cues in the southern flying squirrel (*Glaucomys volans*). Animal Cognition **10**:203–209.

Gibson, B. M., and A. C. Kamil. 2001a. Tests for cognitive mapping in Clark's nutcrackers (*Nucifraga columbiana*). Journal of Comparative Psychology **115**:403–417.

Gibson, B. M., and A. C. Kamil. 2001b. Search for a hidden goal by Clark's nutcrackers (*Nucifraga columbiana*) is more accurate inside than outside a landmark array. Animal Learning & Behavior **29**:234–249.

Gibson, L. P. 1964. Biology and life history of acorn-infesting weevils of the genus *Conotrachelus* (Coleoptera: Curculionidae). Annals of the Entomological Society of America **57**:521–526.

Gibson, L. P. 1972. Insects That Damage White Oak Acorns. Research Paper NE 220. US Department of Agriculture, Forest Service, Northeastern Forest Experiment Station, Upper Darby, PA.

Gibson, L. P. 1982. Insects That Damage Northern Red Oak Acorns. Research Paper NE 492. US Department of Agriculture, Forest Service, Northeastern Forest Experiment Station, Broomall, PA.

Gibson, L. P. 1985. Description and key to larvae of *Curculio* spp. of eastern United States and Canada (Coleoptera: Curculionidae). Proceedings of the Entomological Society of Washington **87**:554–563.

Gil, L., and M. C. Varela. 2008. Technical guidelines for genetic conservation and use: Cork oak (*Quercus suber*). EUFORGEN Networks. Bioversity International, Rome, Italy. http://www.euforgen.org/fileadmin/bioversity/publications/pdfs/1323_Cork_oak_Quercus_suber_.pdf.

Glendenning, R. 1944. The Garry oak in British Columbia: An interesting example of discontinuous distribution. Canadian Field Naturalist **58**:61–65.

Godoy, J. A., and P. Jordano. 2001. Seed dispersal by animals: Exact identification of source trees with endocarp DNA microsatellites. Molecular Ecology **10**:2275–2283.

Gómez, J., R. Márquez, and C. Puerta-Piñero. 2003. Interacciones entre la encina *Quercus ilex* y el ratón de campo *Apodemus sylvaticus*: Mutualismo o antagonismo? Pp. 14–24 *in* Proceedings of VII AEET Conference. Asociación Española de Ecología Terrestre, Barcelona, Spain.

Gómez, J. M. 2003. Spatial patterns in long-distance dispersal of *Quercus ilex* acorns by jays in a heterogeneous landscape. Ecography **26**:573–584.

Gómez, J. M. 2004. Bigger is not always better: Conflicting selective pressures on seed size in *Quercus ilex*. Evolution 58:71–80.

González-Rodríguez, V., and R. Villar. 2012. Post-dispersal seed removal in four Mediterranean oaks: Species and microhabitat selection differ depending on large herbivore activity. Ecological Research 27:587–594.

Goodrum, P. D. 1959. Acorns in the diet of wildlife. Pp. 54–57 *in* James W. Webb, editor. Proceedings of the Thirteenth Annual Conference, Southeastern Association of Game and Fish Commissioners, October 25–27, 1959, Baltimore, Maryland. Southeastern Association of Game and Fish Commissioners, New Orleans, LA.

Goodyear, A. J., and A. C. Kamil. 2004. Clark's nutcrackers (*Nucifraga columbiana*) and the effects of goal-landmark distance on overshadowing. Journal of Comparative Psychology 118:258–264.

Gould, K. L., K. E. Gilbertson, A. J. Hrvol, J. C. Nelson, A. L. Seyfer, R. M. Brantner, and A. C. Kamil. 2013. Differences in relative hippocampus volume and number of hippocampus neurons among five corvid species. Brain, Behavior and Evolution 81:56–70.

Gould, K. L., D. M. Kelly, and A. C. Kamil. 2010. What scatter-hoarding animals have taught us about small-scale navigation. Philosophical Transactions of the Royal Society of London B: Biological Sciences 365:901.

Gould, S. J. 1993. A special fondness for beetles. Natural History 102:4–8.

Gould-Beierle, K. L., and A. C. Kamil. 1996. The use of local and global cues by Clark's nutcrackers, *Nucifraga columbiana*. Animal Behaviour 52:519–528.

Gould-Beierle, K. L., and A. C. Kamil. 1999. The effect of proximity on landmark use in Clark's nutcrackers. Animal Behaviour 58:477–488.

Govindan, B. N. 2013. The role of resource predictability in the metapopulation dynamics of insects. PhD dissertation, Purdue University, West Lafayette, IN.

Govindan, B. N., and R. K. Swihart. 2014. Community structure of acorn weevils (*Curculio*): Inferences from multispecies occupancy models. Canadian Journal of Zoology 93:31–39.

Grand, J., and S. A. Cushman. 2003. A multi-scale analysis of species-environment relationships: Breeding birds in a pitch pine–scrub oak (*Pinus rigidi–Quercus ilicifolia*) community. Biological Conservation 112:307–317.

Greenler, S. M. 2018. Acorn dispersal and oak regeneration in a managed landscape. Master's thesis, Purdue University, West Lafayette, IN.

Griffin, J. 1971. Oak regeneration in the upper Carmel Valley, California. Ecology 52:862–868.

Grivet, D., P. E. Smouse, and V. L. Sork. 2005. A novel approach to an old problem: Tracking dispersed seeds. Molecular Ecology 14:3585–3595.

Grodzinski, U., and N. S. Clayton. 2010. Problems faced by food-caching corvids and the evolution of cognitive solutions. Philosophical Transactions of the

Royal Society of London B: Biological Sciences **365**:977–987.

Grodzinski, U., A. Watanabe, and N. S. Clayton. 2012. Peep to pilfer: What scrub-jays like to watch when observing others. Animal Behaviour **83**:1253–1260.

Grunwald, N., M. Garbelotto, E. Goss, K. Heungens, and S. Prospero. 2012. Emergence of the sudden oak death pathogen *Phytophthora ramorum*. Trends in Microbiology **20**:131–138.

Guzmán, B., C. M. Rodríguez-López, A. Forrest, E. Cano, and P. Vargas. 2015. Protected areas of Spain preserve the neutral genetic diversity of *Quercus ilex* L. irrespective of glacial refugia. Tree Genetics and Genomes **11**:1–18.

Haarsma, A.-J., and R. Kaal. 2016. Predation of wood mice (*Apodemus sylvaticus*) on hibernating bats. Population Ecology **58**:567–576.

Hadj-Chikh, L. Z., M. A. Steele, and P. D. Smallwood. 1996. Caching decisions by grey squirrels: A test of the handling time and perishability hypotheses. Animal Behaviour **52**:941–948.

Hagerman, A. E. 1987. Radial diffusion method for determining tannin in plant extracts. Journal of Chemical Ecology **13**:437–449.

Hall, J. G. 1981. A field study of the Kaibab squirrel in Grand Canyon National Park. Wildlife Monographs **75**:3–54.

Hallwachs, W. 1994. The clumsy dance between agoutis and plants: Scatterhoarding by Costa Rican dry forest agoutis (*Dasyprocta punctata*: Dasyproctidae: Rodentia). PhD dissertation, Cornell University, Ithaca, NY.

Hampe, A., M.-H. Pemonge, and R. J. Petit. 2013. Efficient mitigation of founder effects during the establishment of a leading-edge oak population. Proceedings of the Royal Society B: Biological Sciences **280**:2013.1070.

Harms, K. E., S. J. Wright, O. Calderón, A. Hernández, and E. A. Herre. 2000. Pervasive density-dependent recruitment enhances seedling diversity in a tropical forest. Nature **404**:493–495.

Harries, H. C., and C. R. Clement. 2013. Long-distance dispersal of the coconut palm by migration within the coral atoll ecosystem. Annals of Botany **113**:565–570.

Harris, D. B. 2009. Review of negative effects of introduced rodents on small mammals on islands. Biological Invasions **11**:1611–1630.

Harris, L. 1984. The Fragmented Forest: Island Biogeography Theory and the Preservation of Biotic Diversity. University of Chicago Press, Chicago, IL.

Havera, S. P., and K. E. Smith. 1979. A nutritional comparison of selected fox squirrel foods. Journal of Wildlife Management **43**:691–704.

Healy, S. D., N. S. Clayton, and J. R. Krebs. 1994. Development of hippocampal specialization in 2 species of tit (*Parus* spp.). Behavioural Brain Research **61**:23–28.

Healy, S. D., S. R. de Kort, and N. S. Clayton. 2005. The hippocampus, spatial memory and food hoarding: A puzzle revisited. Trends in Ecology & Evolution **20**:17–22.

Healy, S. D., and J. R. Krebs. 1992. Food storing and the hippocampus in corvids: Amount and volume are correlated. Proceedings of the Royal Society B: Biological Sciences **248**:241.

Healy, S. D., and K. R. Krebs. 1993. Development of hippocampal specialization in a food-storing bird. Behavioural Brain Research **53**:127–131.

Heaney, L. R. 1983. *Sciurus granatensis* (ardilla rojay, ardilla chisa, red-tailed squirrel). Pp. 489–490 *in* D. H. Janzen, editor. Costa Rican Natural History. University of Chicago Press, Chicago, IL.

Henriques, J., M. J. Barrento, L. Bonifácio, A. A. Gomes, A. Lima, and E. Sousa. 2014. Factors affecting the dispersion of *Biscogniauxia mediterranea* in Portuguese cork oak stands. Silva Lusitana **22**:83–97.

Herrera, C. M. 2002. Seed dispersal by vertebrates. Pp. 185–208 *in* C. M. Herrera and O. Pellmyr, editors. Plant-Animal Interactions: An Evolutionary Approach. Blackwell Science, Oxford, UK, and Malden, MA.

Herrera, C. M., P. Jordano, L. López-Soria, and J. A. Amat. 1994. Recruitment of a mast-fruiting, bird-dispersed tree: Bridging frugivore activity and seedling establishment. Ecological Monographs **64**:315–344.

Herrera, J. 1995. Acorn predation and seedling production in a low-density population of cork oak (*Quercus suber* L.). Forest Ecology and Management **76**:197–201.

Herrera-Arroyo, M. L., V. L. Sork, A. González-Rodríguez, V. Rocha-Ramirez, E. Vega, and K. Oyama. 2013. Seed-mediated connectivity among fragmented populations of *Quercus castanea* (Fagaceae) in a Mexican landscape. American Journal of Botany **100**:1663–1671.

Higaki, M. 2016. Prolonged diapause and seed predation by the acorn weevil, *Curculio robustus*, in relation to masting of the deciduous oak *Quercus acutissima*. Entomologia Experimentalis et Applicata **159**:338–346.

Hipp, A. L., P. S. Manos, A. González-Rodríguez, M. Hahn, M. Kaproth, J. D. McVay, S. V. Avalos, and J. Cavender-Bares. 2018. Sympatric parallel diversification of major oak clades in the Americas and the origins of Mexican species diversity. New Phytologist **217**:439–452.

Hirsch, B. T., R. Kays, and P. A. Jansen. 2013. Evidence for cache surveillance by a scatter-hoarding rodent. Animal Behaviour **85**:1511–1516.

Hirsch, B. T., R. Kays, V. E. Pereira, and P. A. Jansen. 2012. Directed seed dispersal towards areas with low conspecific tree density by a scatter-hoarding rodent. Ecology Letters **15**:1423–1429.

Hopewell, L. J., L. A. Leaver, S. E. G. Lea, and A. J. Wills. 2010. Grey squirrels (*Sciurus carolinensis*) show a feature-negative effect specific to social learning. Animal Cognition **13**:219–227.

Hoshizaki, K., and P. E. Hulme. 2002. Mast seeding and predator-mediated indirect interactions in a forest community: Evidence from post-dispersal fate of rodent-generated caches. Pp. 227–240 *in* D. J. Levey, W. R. Silva, and M. Galetti, editors. Seed Dispersal and Frugivory: Ecology, Evolution, and Conservation. CABI, Wallingford, Oxfordshire, UK.

Hoshizaki, K., and H. Miguchi. 2005. Influence of forest composition on tree seed predation and rodent responses: A comparison of monodominant and mixed temperate forests in Japan. Pp. 253–267 *in* P. M. Forget, J. E. Lambert, P. E. Hulme, and S. B. Vander Wall, editors. Seed Fate: Predation, Dispersal and Seedling Establishment. CABI, Wallingford, Oxfordshire, UK.

Hou, X., X. Yi, Y. Yang, and W. Liu. 2010. Acorn germination and seedling survival of *Q. variabilis*: Effects of cotyledon excision. Annals of Forest Science **67**:711p1–711p7.

Howard, L. F., T. D. Lee, and R. T. Eckert. 2011. Forest community composition and dynamics of the Ossipee Pine Barrens, New Hampshire. Journal of the Torrey Botanical Society **138**:434–452.

Howe, H. F. 1993. Aspects of variation in a neotropical seed dispersal system. Pp. 149–162 *in* T. H. Fleming and A. Estrada, editors. Frugivory and Seed Dispersal: Ecological and Evolutionary aspects. Springer Netherlands, Dordrecht, Netherlands.

Howe, H. F., and G. F. Estabrook. 1977. On intraspecific competition for avian dispersers in tropical trees. American Naturalist **111**:817–832.

Howe, H. F., and J. Smallwood. 1982. Ecology of seed dispersal. Annual Review of Ecology and Systematics **13**:201–228.

Howe, H. F., and L. C. Westley. 1988. Ecological Relationships of Plants and Animals. Oxford University Press, New York, NY.

Huang, Z. Y., Y. Wang, H. M. Zhang, F. Q. Wu, and Z. B. Zhang. 2011. Behavioural responses of sympatric rodents to complete pilferage. Animal Behaviour **81**:831–836.

Hubbard, J. A., and G. R. McPherson. 1997. Acorn selection by Mexican jays: A test of a tri-trophic symbiotic relationship hypothesis. Oecologia **110**:143–146.

Hubbell, S. P. 1980. Seed predation and the coexistence of tree species in tropical forests. Oikos **35**:214–229.

Hutchins, H. E., and R. M. Lanner. 1982. The central role of Clark's nutcracker in the dispersal and establishment of whitebark pine. Oecologia **55**:192–201.

Hutchinson, G. E. 1959. Homage to Santa Rosalia or why are there so many kinds of animals? American Naturalist **93**:145–159.

Iida, S. 2006. Dispersal patterns of *Quercus serrata* acorns by wood mice in and around canopy gaps in a temperate forest. Forest Ecology and Management **227**:71–78.

Isagi, Y., K. Sugimura, A. Sumida, and H. Ito. 1997. How does masting happen and synchronize? Journal of Theoretical Biology **187**:231–239.

Ito, F., and S. Higashi. 1991. An indirect mutualism between oaks and wood ants via aphids. Journal of Animal Ecology **60**:463–470.

Jacobs, L. F. 1987. Food-storing decisions in the gray squirrel (*Sciurus carolinensis*). PhD dissertation, Princeton University, Princeton, NJ.

Jacobs, L. F. 1992. Memory for cache locations in Merriam kangaroo rats. Animal Behaviour **43**:585–593.

Jacobs, L. F., S. J. C. Gaulin, D. F. Sherry, and G. E. Hoffman. 1990. Evolution of spatial cognition: Sex-specific patterns of spatial-behavior predict hippocampal size. Proceedings of the National Academy of Sciences of the United States of America **87**:6349–6352.

Jacobs, L. F., and E. R. Liman. 1991. Grey squirrels remember the locations of buried nuts. Animal Behaviour **41**:103–110.

Jacobs, L. F., and W. D. Spencer. 1994. Natural space-use patterns and hippocampal size. Brain, Behavior and Evolution **44**:125–132.

Jansen, P. A., F. Bongers, and L. Hemerik. 2004. Seed mass and mast seeding enhance dispersal by a neotropical scatter-hoarding rodent. Ecological Monographs **74**:569–589.

Jansen, P. A., B. T. Hirsch, W.-J. Emsens, V. Zamora-Gutiérrez, M. Wikelski, and R. Kays. 2012. Thieving rodents as substitute dispersers of megafaunal seeds. Proceedings of the National Academy of Sciences of the United States of America **109**:12610–12615.

Janzen, D. H. 1970. Herbivores and the number of tree species in tropical forests. American Naturalist **104**:501–528.

Janzen, D. H. 1971. Seed predation by animals. Annual Review of Ecology and Systematics **2**:465–492.

Janzen, D. H. 1980. When is it coevolution? Evolution **34**:611–612.

Janzen, D. H. 1982a. Fruit traits, and seed consumption by rodents, of *Crescentia alata* (Bignoniaceae) in Santa Rosa National Park, Costa Rica. American Journal of Botany **69**:1258–1268.

Janzen, D. H. 1982b. Seeds in tapir dung in Santa Rosa National Park, Costa Rica. Brenesia **19–20**:129–135.

Janzen, D. [H.] 1983. *Odocoileus virginianus* (venado, venada coal blanca, white-tailed deer), Pp. 481–483 *in* D. H. Janzen, editor. Costa Rican Natural History. University of Chicago Press, Chicago, IL.

Janzen, D. H. 1984. Dispersal of small seeds by big herbivores: Foliage is the fruit. American Naturalist **123**:338–353.

Janzen, D. H., and P. S. Martin. 1982. Neotropical anachronisms: The fruits the gomphotheres ate. Science **215**:19–27.

Janzen, D. [H.], and D. Wilson. 1983. Mammals. Pp. 426–442 *in* D. H. Janzen, editor. Costa Rican Natural History. University of Chicago Press, Chicago, IL.

Jara-Guerrero, A., G. Escribano-Avila, C. Ivan Espinosa, M. De la Cruz, and M. Méndez. 2017. White-tailed deer as the last megafauna dispersing seeds in neotropical dry forests: the role of fruit and seed traits. Biotropica **50**:169–177.

Jensen, R. 2011. The Origins of Oaks. The Oaks of Chevithorne Barton, Devon, UK. http://www.oaksofchevithornebarton.com.

Johnson, J. S., R. S. Cantrell, C. Cosner, F. Hartig, A. Hastings, H. S. Rogers, E. W. Schupp, K. Shea, B. J. Teller, X. Yu, D. Zurell, and G. Pufal. 2019. Rapid changes in seed dispersal traits may modify plant responses to global change. AoB Plants **11**:plz020.

Johnson, P. S., S. R. Shifley, R. Rogers, D. C. Dey, and J. M. Kabrick. 2009. The Ecology and Silviculture of Oaks. CABI, Wallingford, Oxfordshire, UK.

Johnson, W. C., and C. S. Adkisson. 1985. Dispersal of beech nuts by blue jays in fragmented landscapes. American Midland Naturalist **138**:319–324.

Johnson, W. C., C. S. Adkisson, T. R. Crow, and M. D. Dixon. 1997. Nut caching by blue jays (*Cyanocitta cristata* L.): Implications for tree demography. American Midland Naturalist **138**:357–370.

Johnson, W. C., L. Thomas, and C. S. Adkisson. 1993. Dietary circumvention of acorn tannins by blue jays: Implications for oak demography. Oecologia **94**:159–164.

Johnson, W. C., and T. Webb III. 1989. The role of blue jays (*Cyanocitta cristata* L.) in the postglacial dispersal of fagaceous trees in eastern North America. Journal of Biogeography **16**:561–571.

Jones, C. G., R. S. Ostfeld, M. P. Richard, E. M. Schauber, and J. O. Wolff. 1998. Chain reactions linking acorns to gypsy moth outbreaks and Lyme disease risk. Science **279**:1023–1026.

Jones, J. E., E. Antoniadis, S. J. Shettleworth, and A. C. Kamil. 2002. A comparative study of geometric rule learning by nutcrackers (*Nucifraga columbiana*), pigeons (*Columba livia*) and jackdaws (*Corvus monedula*). Journal of Comparative Psychology **116**:350–356.

Jones, J. E., and A. C. Kamil. 2001. The use of relative and absolute bearings by Clark's nutcrackers, *Nucifraga columbiana*. Animal Learning & Behavior **29**:120–132.

Jones, T., N. Coops, and T. Sharma. 2011. Exploring the utility of hyperspectral imagery and LiDAR data for predicting *Quercus garryana* ecosystem distribution and aiding in habitat resortation. Restoration Ecology **19**:245–256.

Jordano, P. 2000. Fruits and frugivory. Pp. 125–166 *in* M. Fenner, editor. Seeds: the Ecology of Regeneration in Plant Communities. CABI, Wallingford, Oxfordshire, UK.

Jordano, P. 2017. What is long-distance dispersal? And a taxonomy of dispersal events. Journal of Ecology **105**:75–84.

Källander, H. 2007. Food hoarding and use of stored food by rooks, *Covus frugilegus*. Bird Study **54**:192–198.

Kamil, A. C., and R. P. Balda. 1985. Cache recovery and spatial memory in Clark's nutcrackers (*Nucifraga columbiana*). Journal of Experimental Psychology: Animal Behavior Processes **11**:95.

Kamil, A. C., R. P. Balda, and S. Good. 1999. Patterns of movement and orientation during caching and recovery by Clark's nutcrackers, *Nucifraga columbiana*. Animal Behaviour **57**:1327–1335.

Kamil, A. C., and K. Cheng. 2001. Way-finding and landmarks: The multiple-bearings hypothesis. Journal of Experimental Biology **204**:103–113.

Kamil, A. C., A. J. Goodyear, and K. Cheng. 2001. The use of landmarks by Clark's nutcrackers: First tests of a new model. Journal of Navigation **54**:429–435.

Kamil, A. C. , and J. E. Jones. 1997. The seed-storing corvid Clark's nutcracker learns geometric relationships among landmarks. Nature **390**:276–279.

Kato, J. 1985. Food and hoarding behavior in Japanese squirrels. Japanese Journal of Forestry **35**:13–20.

Keator, G. 1998. The Life of an Oak. Heyday Books, Berkeley, CA, and California Oak Foundation, Oakland, CA.

Keith, J. O. 1965. The Abert's squirrel and its dependence on ponderosa pine. Ecology **46**:150–163.

Kelley, L. A., and N. S. Clayton. 2017. California scrub-jays reduce visual cues available to potential pilferers by matching food colour to caching substrate. Biology Letters **13**:20170242.

Kellner, K. F., N. I. Lichti, and R. K. Swihart. 2016. Midstory removal reduces effectiveness of oak (*Quercus*) acorn dispersal by small mammals in the Central Hardwood Forest region. Forest Ecology and Management **375**:182–190.

Kellner, K. F., J. K. Riegel, and R. K. Swihart. 2014. Effects of silvicultural disturbance on acorn infestation and removal. New Forests **45**:265–281.

Kellner, K. F., and R. K. Swihart. 2017. Simulation of oak early life history and interactions with disturbance via an individual-based model, SOEL. PLoS One **12**:e0179643.

Kelly, D. 1994. The evolutionary ecology of mast seeding. Trends in Ecology & Evolution **9**:465–470.

Kelly, D., W. D. Koenig, and A. M. Liebhold. 2008a. An intercontinental comparison of the dynamic behavior of mast seeding communities. Population Ecology **50**:329–342.

Kelly, D., and V. L. Sork. 2002. Mast seeding in perennial plants: Why, how, where? Annual Review of Ecology and Systematics **33**:427–447.

Kelly, D. M., S. Kippenbrock, J. Templeton, and A. C. Kamil. 2008b. Use of a geometric rule or absolute vectors: Landmark use by Clark's nutcrackers (*Nucifraga columbiana*). Brain Research Bulletin **76**:293–299.

Kelly, L. A., and N. S. Clayton. 2017. California scrub jays reduce visual cues available to potential pilferers by matching food colour to caching substrate. Biology Letters **13**:2017.0242.

Kissling, D. W., D. E. Pattemore, and M. Hagen. 2014. Challenges and prospects in the telemetry of insects. Biological Reviews **89**:511–530.

Klinger, R., and M. Rejmánek. 2009. The numerical and functional responses of a granivorous rodent and the fate of neotropical tree seeds. Ecology **90**:1549–1563.

Koenig, W. D., and M. V. Ashley. 2003. Is pollen limited? The answer is blowing in the wind. Trends in Ecology & Evolution **18**:157–159.

Koenig, W. D., and S. H. Faeth. 1998. Effects of storage on tannin and protein content of cached acorns. Southwestern Naturalist **43**:170–175.

Koenig, W. D., K. A. Funk, T. S. Kraft, W. J. Carmen, B. C. Barringer, and J. M. H. Knops. 2012. limitation in a wind-pollinated tree. Journal of Ecology **100**:758–763.

Koenig, W. D., and M. K. Heck. 1988. Ability of two species of oak woodland birds to subsist on acorns. Condor **90**:705–708.

Koenig, W. D. and J. M. H. Knops. 2002. The behavioral ecology of masting in oaks. Pp. 129–148 *in* W. J. McShea and W. M. Healy, editors. Oak Forest Ecosystems. Johns Hopkins University Press, Baltimore, MD.

Koenig, W. D., and J. M. H. Knops. 2005. The mystery of masting in trees: Some trees reproduce synchronously over large areas, with widespread ecological effects, but how and why? American Scientist **93**:340–347.

Koenig, W. D., and J. M. H. Knops. 2013. Large-scale spatial synchrony and cross-synchrony in acorn production by two California oaks. Ecology **94**:83–93.

Koenig, W. D., J. M. H. Knops, W. J. Carmen, and I. S. Pearse. 2015. What drives masting? The phenological synchrony hypothesis. Ecology **96**:184–192.

Koenig, W. D., J. M. H. Knops, J. L. Dickinson, and B. Zuckerberg. 2009a. Latitudinal decrease in acorn size in bur oak (*Quercus macrocarpa*) is due to environmental constraints, not avian dispersal. Botany **87**:349–356.

Koenig, W. D., A. H. Krakauer, W. B. Monahan, J. Haydock, J. M. H. Knops, and W. J. Carmen. 2009b. Mast-producing trees and the geographical ecology of western scrub-jays. Ecography **32**:561–570.

Koenig, W. D., R. L. Mumme, W. J. Carmen, and M. T. Stanback. 1994b. Acorn production by oaks in central coastal California: Variation within and among years. Ecology **75**:99–109.

Kollmann, J., and H.-P. Schill. 1996. Spatial patterns of dispersal, seed predation and germination during colonization of abandoned grassland by *Quercus petraea* and *Corylus avellana*. Vegetatio **125**:193–205.

Koprowski, J. L. 1991. Response of fox squirrels and gray squirrels to a late spring–early summer food shortage. Journal of Mammalogy **72**:367–372.

Koprowski, J. L. 1994. *Sciurus carolinensis*. Mammalian Species **480**:1–9.

Koprowski, J. L., E. Goldstein, K. Bennett, and C. P. Mendes. 2016. Family Sciuridae (tree, flying and ground squirrels, chipmunks, marmots and prairie dogs).

Pp. 648–837 *in* D. E. Wilson and R. A. Mittermeier, chief editors. Handbook of the Mammals of the World, vol. 6, Lagomorphs and Rodents I. Lynx Edicions, Barcelona, Spain.

Korstian, C. F. 1927. Factors controlling germination and early survival in oaks. Yale University, School of Forestry Bulletin No. 19. Yale University, New Haven, CT.

Krebs, J. R., N. S. Clayton, S. D. Healy, D. A. Cristol, S. N. Patel, and A. R. Jolliffe. 1996. The ecology of the avian brain: Food-storing memory and the hippocampus. Ibis **138**:34–46.

Krebs, J. R., D. F. Sherry, S. D. Healy, V. H. Perry, and A. L. Vaccarino. 1989. Hippocampal specialization of food-storing birds. Proceedings of the National Academy of Sciences of the United States of America **86**:1388–1392.

Kremer, A., A. G. Abbott, J. E. Carlson, P. S. Manos, C. Plomion, P. Sisco, M. E. Staton, S. Ueno, and G. G. Vendramin. 2012. Genomics of Fagaceae. Tree Genetics & Genomes **8**:583–610.

Kulahci, I. G., and R. Bowman. 2011. Recaching decisions of Florida scrub-jays are sensitive to ecological conditions. Ethology **117**:700–707.

Kurek, P., and D. Dobrowolska. 2016. Acorns dispersal by jays (*Garrulus glandarius*) onto clear-cuts and under the forest canopy. [In Polish, abstract in English.] Sylwan **160**:512–518.

Landis, R., J. Gurevitch, G. Fox, W. Fang, and D. Taub. 2005. Variation in recruitment and early demography in *Pinus rigida* following crown fire in the pine barrens of Long Island, New York. Journal of Ecology **93**:607–617.

Lanner, R. M. 1996. Made for Each Other: A Symbiosis of Birds and Pines. Oxford University Press, New York, NY.

Larsen, K. W., and S. Boutin. 1994. Movements, survival, and settlement of red squirrels (*Tamiasciurus hudsonicus*) offspring. Ecology **75**:214–223.

Lavenex, P., M. Shiflett, R. Lee, and L. Jacobs. 1998. Spatial versus nonspatial relational learning in free-ranging fox squirrels (*Sciurus niger*). Journal of Comparative Psychology **112**:127.

Lavenex, P., M. A. Steele, and L. F. Jacobs. 2000a. Sex differences, but no seasonal variations in the hippocampus of food-caching squirrels: A stereological study. Journal of Comparative Neurology **425**:152–166.

Lavenex, P., M. A. Steele, and L. F. Jacobs. 2000b. The seasonal pattern of cell proliferation and neuron number in the dentate gyrus of wild adult eastern grey squirrels. European Journal of Neuroscience **12**:643–648.

Lavenex, P. B., P. Lavenex, and N. S. Clayton. 2001. Comparative studies of postnatal neurogenesis and learning: A critical review. Avian and Poultry Biology Reviews **12**:103–125.

Lawton, M. F. 1983. *Cyanocorax morio*. Pp. 573–574 *in* D. H. Janzen, editor. Costa Rican Natural History. University of Chicago Press, Chicago, IL.

Leaver, L. A., L. Hopewell, C. Caldwell, and L. Mallarky. 2007. Audience effects on food caching in grey squirrels (*Sciurus carolinensis*): Evidence for pilferage avoidance strategies. Animal Cognition **10**:23–27.

Leaver, L. A., K. Jayne, and S. E. Lea. 2017. Behavioral flexibility versus rules of thumb: How do grey squirrels deal with conflicting risks? Behavioral Ecology **28**:186–192.

Lee, D. W., G. T. Smith, A. D. Tramontin, K. K. Soma, E. A. Brenowitz, and N. S. Clayton. 2001. Hippocampal volume does not change seasonally in a non food-storing songbird. Neuroreport **12**:1925–1928.

Legg, E. W., and N. S. Clayton. 2014. Eurasian jays (*Garrulus glandarius*) conceal caches from onlookers. Animal Cognition **17**:1223–1226.

Legg, E. W., L. Ostojić, and N. S. Clayton. 2016. Caching at a distance: A cache protection strategy in Eurasian jays. Animal Cognition **19**:753–758.

Lenda, M., J. H. Knops, P. Skórka, D. Moroń, and M. Woyciechowski. 2018. Cascading effects of changes in land use on the invasion of the walnut *Juglans regia* in forest ecosystems. Journal of Ecology **106**:671–686.

Lengyel, S., A. D. Gove, A. M. Latimer, J. D. Majer, and R. R. Dunn. 2010. Convergent evolution of seed dispersal by ants, and phylogeny and biogeography in flowering plants: A global survey. Perspectives in Plant Ecology, Evolution and Systematics **12**:43–55.

Leverkus, A. B., J. M. Rey Benayas, and J. Castro. 2016. Shifting demographic conflicts across recruitment cohorts in a dynamic post-disturbance landscape. Ecology **97**:2628–2639.

Levey, D. J., and C. W. Benkman. 1999. Fruit-seed disperser interactions: Timely insights from a long-term perspective. Trends in Ecology & Evolution **14**:41–43.

Levey, D. J., B. M. Bolker, J. J. Tewksbury, S. Sargent, and N. M. Haddad. 2005. Effects of landscape corridors on seed dispersal by birds. Science **309**:146.

Levey, D. J., W. R. Silva, and M. Galetti. 2002. Seed Dispersal and Frugivory: Ecology, Evolution, and Conservation. CABI, Wallingford, Oxfordshire, UK.

Levey, D. J., J. J. Tewksbury, and B. M. Bolker. 2008. Modelling long-distance seed dispersal in heterogeneous landscapes. Journal of Ecology **96**:599–608.

Lewis, A. R. 1980. Patch use by gray squirrels and optimal foraging. Ecology **61**:1371–1379.

Lewis, I. M. 1911. The seedling of "*Quercus virginiana*." Plant World **14**:119–123.

Lewis, J. L., and A. C. Kamil. 2006. Interference effects in the memory for serially presented locations in Clark's nutcrackers, *Nucifraga columbiana*. Journal of Experimental Psychology: Animal Behavior Processes **32**:407.

Liang, M., X. Liu, R. S. Etienne, F. Huang, Y. Wang, and S. Yu. 2015. Arbuscular mycorrhizal fungi counteract the Janzen–Connell effect of soil pathogens. Ecology **96**:562–574.

Lichti, N. I. 2012. Implications of context-dependent scatterhoarding for seed survival and dispersal in North American oaks (*Quercus*). PhD dissertation, Purdue University, West Lafayette, IN.

Lichti, N. I., M. A. Steele, and R. K. Swihart. 2017. Seed fate and decision-making processes in scatter-hoarding rodents. Biological Reviews **92**:474–504.

Lichti, N. I., M. A. Steele, H. Zhang, and R. K. Swihart. 2014. Mast species composition alters seed fate in North American rodent-dispersed hardwoods. Ecology **95**:1746–1758.

Lindstedt, S. L., and M. S. Boyce. 1985. Seasonality, fasting endurance, and body size in mammals. American Naturalist **125**:873–878.

Little, E. L., and L. A. Viereck. 1971. Atlas of United States Trees, vol. 1, Conifers and Important Hardwoods. US Department of Agriculture, Forest Service, Miscellaneous Publications. US Government Printing Office, Washington, DC.

Liu, C., G. Liu, Z. Shen, and X. Yi. 2013. Effects of disperser abundance, seed type, and interspecific seed availability on dispersal distance. Acta Theriologica **58**:267–278.

Lockie, J. 1956. The food and feeding behaviour of the jackdaw, rook and carrion crow. Journal of Animal Ecology **25**:421–428.

LoGiudice, K., R. S. Ostfeld, K. A. Schmidt, and F. Keesing. 2003. The ecology of infectious disease: Effects of host diversity and community composition on Lyme disease risk. Proceedings of the National Academy of Sciences of the United States of America **100**:567–571.

López-de-Heredia, U., P. Jiménez, P. Díaz-Fernández, and L. Gil. 2005. The Balearic Islands: A reservoir of cpDNA genetic variation for evergreen oaks. Journal of Biogeography **32**:939–949.

Lucas, J. R., A. Brodin, S. R. de Kort, and N. S. Clayton. 2004. Does hippocampal size correlate with the degree of caching specialization? Proceedings of the Royal Society B: Biological Sciences **271**:2423.

Luo, Y., Z. Yang, M. A. Steele, Z. B. Zhang, J. A. Stratford, and H. M. Zhang. 2014. Hoarding without reward: Rodent responses to repeated episodes of complete cache loss. Behavioural Processes **106**:36–43.

MacArthur, R. H., and E. R. Pianka. 1966. On optimal use of a patchy environment. American Naturalist **100**:603–609.

MacArthur, R. H., and E. O. Wilson. 1963. An equilibrium theory of insular zoogeography. Evolution **17**:373–387.

MacArthur, R. H., and E. O. Wilson. 1967. Island Biogeography, 1st edition. Princeton University Press, Princeton, NJ.

Macdonald David, W. 1976. Food caching by red foxes and some other carnivores. Zeitschrift für Tierpsychologie **42**:170–185.

MacDougall, A. 2001. Invasive perennial grasses in *Quercus garryana* meadows of southwestern British Columbia: Prospects for restoration. Pp. 159–168 *in* R. B. Standiford, D. McCreary, and K. L. Purcell, technical coordinators. Proceedings of the Fifth Symposium on Oak Woodlands: Oaks in California's Changing Landscape, October 22–25, San Diego, California. General Technical Report PSW 184. US Department of Agriculture, Forest Service, Pacific Southwest Research Station, San Diego, CA.

Maeto, K., and K. Ozaki. 2003. Prolonged diapause of specialist seed-feeders makes predator satiation unstable in masting of *Quercus crispula*. Oecologia **137**:392–398.

Maghnia, F. Z., H. Sanguin, Y. Abbas, M. Verdinelli, B. Kerdouh, N. El Ghachtouli, E. Lancellotti, S. E. Bakkali Yakhlef, and R. Duponnois. 2017. Impact of cork oak management on the ectomycorrhizal fungal diversity associated with *Quercus suber* in the Mâamora forest (Morocco). [In French, abstract in English.] Comptes rendus biologies **340**:298–305.

Major, K., P. Nosko, C. Kuehne, D. Campbell, and J. Bauhus. 2013. Regeneration dynamics of non-native northern red oak (*Quercus rubra* L.) populations as influenced by environmental factors: A case study in managed hardwood forests of southwestern Germany. Forest Ecology and Management **291**:144–153.

Male, L. H., and T. V. Smulders. 2007. Hyper-dispersed cache distributions reduce pilferage: A field study. Animal Behaviour **73**:717–726.

Male, L. H., and T. V. Smulders. 2008. Hyper-dispersed cache distributions reduce pilferage: A laboratory study. Journal of Avian Biology **39**:170–177.

Mancilla-Leytón, J. M., J. Cambrollé, and A. M. Vicente. 2012. The impact of the common rabbit on cork oak regeneration in SW Spain. Plant Ecology: An International Journal **213**:1503–1510.

Mangini, A. C., and R. W. Perry. 2004. The insect guild of white oak acorns: Its effect on mast quality in the Ozark and Ouachita National Forests. Pp.79–82 *in* J. M. Guldin, technical compiler. Ouachita and Ozark Mountains Symposium: Ecosystem Management Research. General Technical Report SRS-74. US Department of Agriculture, Forest Service, Southern Research Station, Asheville, NC.

Manos, P. S. 1997. *Quercus* section *Protobalanus*. Pp. 470–471 *in* N. R. Morin, editor. Flora of North America, North of Mexico, vol. 3, Magnoliophyta: Magnoliidae and Hamamelidae. Oxford University Press, New York, NY.

Manos, P. S., C. H. Cannon, and S.-H. Oh. 2008. Phylogenetic relationships and taxonomic status of the paleoendemic Fagaceae of western North America: Recognition of a new genus, *Notholithocarpus*. Madroño **55**:181–190.

Marsico, T. D., J. J. Hellmann, and J. Romero-Severson. 2009. Patterns of seed dispersal and pollen flow in *Quercus garryana* (Fagaceae) following post-glacial climatic changes. Journal of Biogeography **36**:929–941.

Martin, J. S., and M. M. Martin. 1982. Tannin assays in ecological studies: Lack of correlation between phenolics, proanthocyanidins and protein-precipitating constituents in mature foliage of six oak species. Oecologia **54**:205–211.

Martin, M. A., and D. G. Wallace. 2007. Selective hippocampal cholinergic deafferentation impairs self-movement cue use during a food hoarding task. Behavioural Brain Research **183**:78–86.

Martin, R. E., R. H. Pine, and A. F. DeBlase. 2011 A Manual of Mammalogy: With Keys to Families of the World, 3rd edition. Waveland Press, Long Grove, IL.

Martin, S. J., and R. E. Clark. 2007. The rodent hippocampus and spatial memory: From synapses to systems. Cellular and Molecular Life Sciences **64**:401–431.

McCune, J. L., M. G. Pellatt, and M. Vellend. 2013. Multidisciplinary synthesis of long-term human-ecosystem interactions: A perspective from the Garry oak ecosystem of British Columbia. Biological Conservation **166**:293–300.

McCune, J. L., M. Vellend, and M. G. Pellatt. 2015. Combining phytolith analysis with historical ecology to reveal the long-term, local-scale dynamics within a savannah-forest landscape mosaic. Biodiversity and Conservation **24**:609–626.

McEuen, A. B., and M. A. Steele. 2005. Atypical acorns appear to allow seed escape after apical notching by squirrels. American Midland Naturalist **154**:450–458.

McQuade, D. B., E. H. Williams, and H. B. Eichenbaum. 1986. Cues used for localizing food by the gray squirrel (*Sciurus carolinensis*). Ethology **72**:22–30.

McShea, W. J., and G. Schwede. 1993. Variable acorn crops: Responses of white-tailed deer and other mast consumers. Journal of Mammalogy **74**:999–1006.

Menu, F. 1993. Strategies of emergence in the chestnut weevil *Curculio elephas* (Coleoptera: Curculionidae). Oecologia **96**:383–390.

Menu, F., and D. Debouzie. 1993. Coin-flipping plasticity and prolonged diapause in insects: Example of the chestnut weevil *Curculio elephas* (Coleoptera: Curculionidae). Oecologia **93**:367–373.

Menu, F., and E. Desouhant. 2002. Bet-hedging for variability in life cycle duration: Bigger and later-emerging chestnut weevils have increased probability of a prolonged diapause. Oecologia **132**:167–174.

Menu, F., J.-P. Roebuck, and M. Viala. 2000. Bet-hedging diapause strategies in stochastic environments. American Naturalist **155**:724–734.

Merritt, J. F., M. Lima, and F. Bozinovic. 2001. Seasonal regulation in fluctuating small mammal populations: Feedback structure and climate. Oikos **94**:505–514.

Mezquida, E., and C. Benkman. 2005. The geographic selection mosaic for squirrels, crossbills and Aleppo pine. Journal of Evolutionary Biology **18**:348–357.

Mikailou, S., K. Keenleyside, K. Adare, B. Reader, M. Plante, and P. Deering. 2009. Protecting native biodiversity from high-impact invasive species through the protected areas of Parks Canada. Biodiversity **10**:51–55.

Miller, J. C., P. E. Hanson, and D. N. Kimberling. 1991. Development of the gypsy moth (Lepidoptera: Lymantriidae) on Garry oak and red alder in western North America. Environmental Entomology **20**:1097–1101.

Moffett, M. W. 1989. Life in a nutshell. National Geographic **175**:782–796.

Mogensen, H. L. 1975. Ovule abortion in *Quercus* (Fagaceae). American Journal of Botany **62**:160–165.

Moore, J. E. 2005. Ecology of animal-mediated seed dispersal in the fragmented central hardwoods region. PhD dissertation, Purdue University, West Lafayette, IN.

Moore, J. E., A. B. McEuen, R. K. Swihart, T. A. Contreras, and M. A. Steele. 2007. Determinants of seed removal distance by scatter-hoarding rodents in deciduous forests. Ecology **88**:2529–2540.

Moore, J. E., and R. K. Swihart. 2006. Nut selection by captive blue jays: Importance of availability and implications for seed dispersal. Condor **108**:377–388.

Moore, J. E., and R. K. Swihart. 2008. Factors affecting the relationship between seed removal and seed mortality. Canadian Journal of Zoology **86**:378–385.

Moore, P. D. 1984. Plant science: Evolution of the oaks. Nature **307**:598.

Moran, E. V., and J. S. Clark. 2012a. Between-site differences in the scale of dispersal and gene flow in red oak. PLoS One **7**:e36492.

Moran, E. V., and J. S. Clark. 2012b. Causes and consequences of unequal seedling production in forest trees: A case study in red oaks. Ecology **93**:1082–1094.

Morán-López, T., M. Fernández, C. L. Alonso, D. Flores-Renteriá, F. Valladres, and M. Díaz. 2015. Effects of forest fragmentation on the oak-rodent mutualism. Oikos **124**:1482–1491.

Morán-López, T., J. J. Robledo-Arnuncio, M. Díaz, J. M. Morales, A. Lazáro-Nogal, Z. Lorenzo, and F. Valladres. 2016b. Determinants of functional connectivity of holm oak woodlands: Fragment size and mouse foraging behavior. Forest Ecology and Management **368**:111–122.

Morán-López, T., T. Wiegand, J. M. Morales, F. Valladares, and M. Díaz. 2016a. Predicting forest management effects on oak-rodent mutualisms. Oikos **125**:1445–1457.

Moreira, X., I. M. Pérez-Ramos, L. Abdala-Roberts, and K. A. Mooney. 2017. Functional responses of contrasting seed predator guilds to masting in two Mediterranean oak species. Oikos **126**:1042–1050.

Morin, N. R., editor. 1997. Flora of North America, North of Mexico, vol. 3, Magnoliophyta: Magnoliidae and Hamamelidae. Oxford University Press, New York, NY.

Muhamed, H., E. Lingua, J. Maalouf, and R. Michalet. 2015. Shrub-oak seedling spatial associations change in response to the functional composition of neighboring shrubs in coastal dune forest communities. Annals of Forest Science **72**:231–241.

Mulder, P. G., E. Stafne, W. Reid, and R. Grantham. 2007. Monitoring adult weevil populations in pecan and fruit trees in Oklahoma. Oklahoma Cooperative Extension Service Fact Sheet EPP-7190. Division of Agricultural Sciences and Natural Resources, Oklahoma State University, Stillwater, OK.

Muller-Landau, H. C., S. J. Wright, O. Calderón, R. Condit, and S. P. Hubbell. 2008. Interspecific variation in primary seed dispersal in a tropical forest. Journal of Ecology **96**:653–667.

Munger, J. C. 1984. Long-term yield from harvester ant colonies: Implications for horned lizard foraging strategy. Ecology **65**:1077–1086.

Muñoz, A., and R. Bonal. 2007. Rodents change acorn dispersal behavior in response to ungulate presence. Oikos **116**:1631–1638.

Muñoz, A., and R. Bonal. 2008a. Are you strong enough to carry that seed? Seed size/body size ratios influence seed choices by rodents. Animal Behaviour **76**:709–715.

Muñoz, A., and R. Bonal. 2008b. Seed choice by rodents: Learning or inheritance? Behavioral Ecology and Sociobiology **62**:913–922.

Muñoz, A., and R. Bonal. 2011. Linking seed dispersal to cache protection strategies. Journal of Ecology **99**:1016–1025.

Muñoz, A., R. Bonal, and J. M. Espelta. 2014. Acorn–weevil interactions in a mixed-oak forest: Outcomes for larval growth and plant recruitment. Forest Ecology and Management **322**:98–105.

Myczko, L., L. Dylewski, P. Zduniak, T. H. Sparks, and P. Tryjanowski. 2014. Predation and dispersal of acorns by European jay (*Garrulus glandarius*) differs between a native (pedunculate oak *Quercus robur*) and an introduced oak species (northern red oak *Quercus rubra*) in Europe. Forest Ecology and Management **331**:35–39.

Myers, J. A., M. Vellend, S. Gardescu, and P. L. Marks. 2004. Seed dispersal by white-tailed deer: Implications for long-distance dispersal, invasion, and migration of plants in eastern North America. Oecologia **139**:35–44.

Nakashima, Y., E. Inoue, M. Inoue-Murayama, and J. R. A. Sukor. 2010. Functional uniqueness of a small carnivore as seed dispersal agents: A case study of the common palm civets in the Tabin Wildlife Reserve, Sabah, Malaysia. Oecologia **164**:721–730.

Naoe, S., I. Tayasu, Y. Sakai, T. Masaki, K. Kobayashi, A. Nakajima, Y. Sato, K. Yamazaki, H. Kiyokawa, and S. Koike. 2019. Downhill seed dispersal by temperate mammals: A potential threat to plant escape from global warming. Scientific Reports **9**:1–11.

Nathan, R. 2006. Long-distance dispersal of plants. Science **313**:786–788.

Nathan, R., G. G. Katul, G. Bohrer, A. Kuparinen, M. B. Soons, S. E. Thompson, A. Trakhtenbrot, and H. S. Horn. 2011. Mechanistic models of seed dispersal by wind. Theoretical Ecology **4**:113–132.

Nathan, R., G. G. Katul, H. S. Horn, S. M. Thomas, R. Oren, R. Avissar, S. W. Pacala, and S. A. Levin. 2002. Mechanisms of long-distance dispersal of seeds by wind. Nature **418**:409–412.

Nixon, K. C. 1985. A biosystematic study of *Quercus* series *Virentes* (the live oaks) with phylogenetic analysis of *Fagales, Fagaceae,* and *Quercus*. PhD dissertation, University of Texas at Austin, Austin, TX.

Nixon, K. C. 1993a. Infrageneric classification of *Quercus* (Fagaceae) and typification of sectional names. Annales des sciences forestières **50**:25–34.

Nixon, K. C. 1993b. The genus *Quercus* in Mexico. Pp. 447–458 *in* T. P. Ramamoorthy, R. Bye, A. Lot, and J. Fa, editors. Biological Diversity of Mexico: Origins and Distribution. Oxford University Press, Oxford, UK.

Nixon, K. C. 1997a. Fagaceae. Pp. 436–437 *in* N. R. Morin, editor. Flora of North America, North of Mexico, vol. 3, Magnoliophyta: Magnoliidae and Hamamelidae. Oxford University Press, New York, NY.

Nixon, K. C. 1997b. *Quercus*. Pp. 445–447 *in* N. R. Morin, editor. Flora of North America, North of Mexico, vol. 3, Magnoliophyta: Magnoliidae and Hamamelidae. Oxford University Press, New York, NY.

Nixon, K. C. 2006. Global and neotropical distribution and diversity of oak (genus *Quercus*) and oak forests. Pp. 3–13 *in* M. M. Caldwell, G. Heldmaier, R. B. Jackson, O. K. Lange, H. A. Mooney, E. D. Schulze, and U. Sommer, editors. Ecology and Conservation of Neotropical Montane Oak Forests. Springer-Verlag, Berlin, Germany.

Nixon, K. C., and C. H. Muller. 1997. *Quercus linnaeus* section *Quercus*. Pp. 471–506 *in* N. R. Morin, editor. Flora of North America, North of Mexico, vol. 3, Magnoliophyta: Magnoliidae and Hamamelidae. Oxford University Press, New York, NY.

Nowacki, G. J., and M. D. Abrams. 2008. The demise of fire and "mesophication" of forests in the eastern United States. BioScience **58**:123–138.

Oldfield, S., and A. Eastwood. 2007. The Red List of Oaks. Flora & Fauna International, Cambridge, UK.

Ortego J., R. Bonal, and A. Muñoz. 2010. Genetic consequences of habitat fragmentation in long-lived tree species: The case of the Mediterranean holm oak. (*Q. ilex* L.). Journal of Heredity **101**:717–726.

Ortego, J., R. Bonal, A. Muñoz, and J. M. Aparico. 2014. Extensive pollen immigration and no evidence of disrupted mating patterns or reproduction in a highly fragmented holm oak stand. Journal of Plant Ecology **7**:384–395.

Ostfeld, R. S. 1997. The ecology of Lyme-disease risk: Complex interactions between seemingly unconnected phenomena determine risk of exposure to this expanding disease. American Scientist **85**:338–346.

Ostfeld, R. S., C. G. Jones, and J. O. Wolff. 1996. Of mice and mast. BioScience **46**:323–330.

Ostfeld, R. S., and F. Keesing. 2000. Pulsed resources and community dynamics of consumers in terrestrial ecosystems. Trends in Ecology & Evolution **15**:232–237.

Ostojić, L., E. W. Legg, K. F. Brecht, F. Lange, C. Deininger, M. Mendl, and N. S. Clayton. 2017. Correspondence: Current desires of conspecific observers affect cache-protection strategies in California scrub jays and Eurasian jays. Current Biology **27**:51–53.

Packer, A., and K. Clay. 2003. Soil pathogens and *Prunus serotina* seedling and sapling growth near conspecific trees. Ecology **84**:108–119.

Pan, Y., M. Li, X. Yi, Q. Zhao, C. Lieberwirth, Z. Wang, and Z. Zhang. 2013. Scatter hoarding and hippocampal cell proliferation in Siberian chipmunks. Neuroscience **255**:76–85.

Parchman, T. L., and C. W. Benkman. 2002. Diversifying coevolution between crossbills and black spruce on Newfoundland. Evolution **56**:1663–1672.

Parchman, T. L., C. W. Benkman, and E. T. Mezquida. 2007. Coevolution between Hispaniolan crossbills and pine: Does more time allow for greater phenotypic escalation at lower latitude? Evolution **61**:2142–2153.

Patel, S. N., N. S. Clayton, and J. R. Krebs. 1997. Spatial learning induces neurogenesis in the avian brain. Behavioural Brain Research **89**:115–128.

Paulsen, T. R., L. Colville, I. Kranner, M. I. Daws, G. Högstedt, V. Vandvik, and K. Thompson. 2013. Physical dormancy in seeds: A game of hide and seek? New Phytologist **198**:496–503.

Paulsen, T. R., G. Högstedt, K. Thompson, V. Vandvik, and S. Eliassen. 2014. Conditions favouring hard seededness as a dispersal and predator escape strategy. Journal of Ecology **102**:1475–1484.

Pausas, J., E. Ribeiro, S. Dias, J. Pons, and C. Beseler. 2006. Regeneration of a marginal *Quercus suber* forest in the eastern Iberian Peninsula. Journal of Vegetation Science **17**:729–738.

Pavlik, B. M., P. C. Muick, S. G. Johnson, and M. Popper. 2002. Oaks of California. Cachuma Press, Los Olivos, CA, and the California Oak Foundation, Oakland, CA.

Pearse, I. S., W. D. Koenig, K. A. Funk, and M. B. Pesendorfer. 2015. Pollen limitation and flower abortion in a wind-pollinated, masting tree. Ecology **96**:587–593.

Pearse, I. S., W. D. Koenig, and D. Kelly. 2016. Mechanisms of mast seeding: Resources, weather, cues, and selection. New Phytologist **212**:546–562.

Peguero, G., R. Bonal, D. Sol, A. Muñoz, V. L. Sork, and J. M. Espelta. 2017. Tropical insect diversity: Evidence of greater host specialization in seed-feeding weevils. Ecology **98**:2180–2190.

Pélisson, P-F., M-C. Bel-Venner, B. Rey, L. Burgevin, F. Martineau, F. Fourel, C. Lecuyer, F. Menu, and S. Venner. 2012. Contrasted breeding strategies in four sympatric sibling insect species: When a proovigenic and capital breeder copes with a stochastic environment. Functional Ecology **26**:198–206.

Pellatt, M. G., and Z. Gedalof. 2014. Environmental change in Garry oak (*Quercus garryana*) ecosystems: The evolution of an eco-cultural landscape. Biodiversity and Conservation **23**:2053–2067.

Perea, R., D. López, A. San Miguel, and L. Gil. 2012. Incorporating insect infestation into rodent seed dispersal: Better if the larva is still inside. Oecologia **170**:723–733.

Perea, R., A. San Miguel, and L. Gil. 2011a. Acorn dispersal by rodents: The importance of re-dispersal and distance to shelter. Basic and Applied Ecology **12**:432–439.

Perea, R., A. San Miguel, and L. Gil. 2011b. Flying vs. climbing: Factors controlling arboreal seed removal in oak-beech forests. Forest Ecology and Management **262**:1251–1257.

Perea, R., A. San Miguel, and L. Gil. 2011c. Leftovers in seed dispersal: Ecological implications of partial seed consumption for oak regeneration. Journal of Ecology **99**:194–201.

Perea, R., A. San Miguel, and L. Gil. 2014. Interacciones planta-animal en la regeneración de *Quercus pyrenaica*: Ecología y gestión. Ecosistemas: Revista de la Asociación Española de Ecología Terrestre **23**:18–26.

Pérez-Izquierdo, L., and F. Pulido. 2013. Spatiotemporal variation in acorn production and damage in a Spanish holm oak (*Quercus ilex*) dehesa. Forest Systems **22**:106–113.

Pérez-Ramos, I. M., and T. Marañón. 2008. Factors affecting post-dispersal seed predation in two coexisting oak species: microhabitat, burial and exclusion of large herbivores. Forest Ecology and Management **255**:3506–3514.

Pérez-Ramos, I. M., T. Marañón, J. M. Lobo, and J. R. Verdú. 2007. Acorn removal and dispersal by the dung beetle *Thorectes lusitanicus*: Ecological implications. Ecological Entomology **32**:349–356.

Pérez-Ramos, I. M., I. R. Urbieta, T. Marañón, M. A. Zavala, and R. K. Kobe. 2008. Seed removal in two coexisting oak species: ecological consequences of seed size, plant cover and seed-drop timing. Oikos **117**:1386–1396.

Pesendorfer, M. B., and W. D. Koenig. 2016. The effect of within-year variation in acorn crop size on seed harvesting by avian hoarders. Oecologia **181**:97–106.

Pesendorfer, M. B., and W. D. Koenig. 2017. Competing for seed dispersal: Evidence for the role of avian seed hoarders in mediating apparent predation among oaks. Functional Ecology **31**:622–631.

Pesendorfer, M. B., W. D. Koenig, I. S. Pearse, J. M. H. Knops, and K. A. Funk. 2016a. Individual resource limitation combined with population-wide pollen availability drives masting in the valley oak (*Quercus lobata*). Journal of Ecology **104**:637–645.

Pesendorfer, M. B., T. S. Sillett, W. D. Koenig, and S. A. Morrison. 2016b. Scatter-hoarding corvids as seed dispersers for oaks and pines: A review of a widely distributed mutualism and its utility to habitat restoration. Condor **118**:215–237.

Pesendorfer, M. B., T. S. Sillett, and S. A. Morrison. 2017. Spatially biased dispersal of acorns by a scatter-hoarding corvid may accelerate passive restoration of oak habitat on California's largest island. Current Zoology **63**:363–367.

Pesendorfer, M. B., T. S. Sillett, S. A. Morrison, and A. C. Kamil. 2016c. Context-dependent seed dispersal by a scatter-hoarding corvid. Journal of Animal Ecology **85**:798–805.

Plomion, C., J.-M. Aury, J. Amselem, T. Leroy, F. Murat, S. Duplessis, S. Faye, N. Francillonne, K. Labadie, G. Le Provost, et al. 2018. Oak genome reveals facets of long lifespan. Nature Plants **4**:440.

Pons, J., and J. G. Pausas. 2006. Oak regeneration in heterogeneous landscapes: The case of fragmented *Quercus suber* forests in the eastern Iberian Peninsula. Forest Ecology and Management **231**:196–204.

Pons, J., and J. G. Pausas. 2007a. Acorn dispersal estimated by radio-tracking. Oecologia **153**:903–911.

Pons, J., and J. G. Pausas. 2007b. Not only size matters: Acorn selection by the European jay (*Garrulus glandarius*). Acta Oecologia **31**:353–360.

Pons, J., and J. G. Pausas. 2007c. Rodent acorn selection in a Mediterranean oak landscape. Ecological Research **22**:535–541.

Powell, J. A., and N. E. Zimmermann. 2004. Multiscale analysis of active seed dispersal contributes to resolving Reid's Paradox. Ecology **85**:490–506.

Pravosudov, V. V. 2007. The relationship between environment, corticosterone, food caching, spatial memory, and the hippocampus in chickadees. Pp. 25–41 *in* K. A. Otter, editor. Ecology and Behavior of Chickadees and Titmice: An Integrated Approach. Oxford University Press, Oxford, UK, and New York, NY.

Pravosudov, V. V. 2008. Mountain chickadees discriminate between potential cache pilferers and non-pilferers. Proceedings of the Royal Society B: Biological Sciences **275**:55–61.

Pravosudov, V. V., and S. R. de Kort. 2006. Is the western scrub-jay (*Aphelocoma californica*) really an underdog among food-hoarding corvids when it comes to hippocampal volume and caching propensity? Brain, Behavior and Evolution **67**:1–9.

Pravosudov, V. V., P. Lavenex, and N. S. Clayton. 2002. Changes in spatial memory mediated by experimental variation in food supply do not affect hippocampal anatomy in mountain chickadees (*Poecile gambeli*). Journal of Neurobiology **51**:142–148.

Pravosudov, V., T. Roth II, M. Forister, L. D. LaDage, R. Kramer, F. Schilkey, and A. van der Linden. 2013. Differential hippocampal gene expression is associated with climate-related natural variation in memory and the hippocampus in food-caching chickadees. Molecular Ecology **22**:397–408.

Preston, S. D., and L. F. Jacobs. 2009. Mechanisms of cache decision making in fox squirrels (*Sciurus niger*). Journal of Mammalogy **90**:787–795.

Price, K., and S. Boutin. 1993. Territorial bequeathal by red squirrel mothers. Behavioral Ecology **4**:144–150.

Purves, D. W., M. A. Zavala, K. Ogle, F. Prieto, and J. M. R. Benayas. 2007. Environmental heterogeneity, bird-mediated directed dispersal, and oak woodland dynamics in Mediterranean Spain. Ecological Monographs **77**:77–97.

Pyke, G. H. 1978. Optimal foraging in bumblebees and coevolution with their plants. Oecologia **36**:281–293.

Ramos, S., F. Vázquez, and T. Ruiz. 2013. Ecological implications of acorn size at the individual tree level in *Quercus suber* L. ISRN Botany **2013**:1–6.

Ramos-Palacios, C. R., and E. I. Badano. 2014. The relevance of burial to evade acorn predation in an oak forest affected by habitat loss and land use changes. Botanical Sciences **92**:299–308.

Raven, P., G. Johnson, K. Mason, J. Losos, and S. Singer. 2017. Biology, 11th edition. McGraw Hill, New York, NY.

Reichman, O. 1988. Caching behaviour by eastern woodrats, *Neotoma floridana*, in relation to food perishability. Animal Behaviour **36**:1525–1532.

Reichman, O., A. Fattaey, and K. Fattaey. 1986. Management of sterile and mouldy seeds by a desert rodent. Animal Behaviour **34**:221–225.

Reid, C. 1899. Origin of the British Flora. Dulau, London, UK.

Ribbens, E., J. A. Silander Jr., and S. W. Pacala. 1994. Seedling recruitment in forests: Calibrating models to predict patterns of tree seedling dispersion. Ecology **75**:1794–1806.

Ribeiro, J., and E. Vieira. 2016. Microhabitat selection for caching and use of potential landmarks for seed recovery by a neotropical rodent. Journal of Zoology **300**:274–280.

Richardson, K. B., N. I. Lichti, and R. K. Swihart. 2013. Acorn-foraging preferences of four species of free-ranging avian seed predators in eastern deciduous forests. Condor **115**:863–873.

Robbins, C., T. Hanley, A. Hagerman, O. Hjeljord, D. Baker, C. Schwartz, and W. Mautz. 1987. Role of tannins in defending plants against ruminants: Reduction in protein availability. Ecology **68**:98–107.

Rolando, A. 1998. Factors affecting movements and home ranges in the jay (*Garrulus glandarius*). Journal of Zoology **246**:249–257.

Ronce, O. 2007. How does it feel to be like a rolling stone? Ten questions about dispersal evolution. Annual Review of Ecology, Evolution, and Systematics **38**:231–253.

Roth, T. C., A. Brodin, T. V. Smulders, L. D. LaDage, and V. V. Pravosudov. 2010. Is bigger always better? A critical appraisal of the use of volumetric analysis in the study of the hippocampus. Philosophical Transactions of the Royal Society of London B: Biological Sciences **365**:915.

Roth, T. C., L. D. LaDage, C. A. Freas, and V. V. Pravosudov. 2012. Variation in memory and the hippocampus across populations from different climates: A common garden approach. Proceedings of the Royal Academy B: Biological Sciences **279**:402–410.

Roth, T. C., and V. V. Pravosudov. 2009. Hippocampal volumes and neuron numbers increase along a gradient of environmental harshness: A large-scale comparison. Proceedings of the Royal Society B: Biological Sciences **276**:401.

Rougon, C., and D. Rougon. 2001. Impact of Curculionidae on the regeneration of oaks in the Loire Valley. Symbiosis **4**:39–46.

Ruiz-Carbayo, H., R. Bonal, J. Pino, and J. M. Espelta. 2018. Zero-sum landscape effects on acorn predation associated with shifts in granivore insect community in new holm oak (*Q. ilex*) forests. Diversity and Distributions **24**:521–534.

Russell, E. W. B. 1987. Pre-blight distribution of *Castanea dentata* (Marsh.) Borkh. Bulletin of the Torrey Botanical Club **114**:183–190.

Sadaghian, B., A. A. Dordaei, and M. Nikdel. 2007. Damage assessment of acorn pests in Arasbaran forests. Iranian Journal of Forest and Range Protection Research **4**:113–118.

Sargent, S. 1990. Neighborhood effects on fruit removal by birds: A field experiment with *Viburnum dentatum* (Caprifoliaceae). Ecology **71**:1289–1298.

Satake, A., and Y. Iwasa. 2000. Pollen coupling of forest trees: Forming synchronized and periodic reproduction out of chaos. Journal of Theoretical Biology **203**:63–84.

Sawaya, G. M., A. S. Goldberg, M. A. Steele, and H. J. Dalgleish. 2018. Environmental variation shifts the relationship between trees and scatterhoarders along the continuum from mutualism to antagonism. Integrative Zoology **13**:319–330.

Scarlett, T. L., and K. G. Smith. 1991. Acorn preference of urban blue jays (*Cyanocitta cristata*) during fall and spring in northwestern Arkansas. Condor **93**:438–442.

Scheller, R. M., S. Van Tuyl, K. Clark, N. G. Hayden, J. Hom, and D. J. Mladenoff. 2008. Simulation of forest change in the New Jersey Pine Barrens under current and colonial conditions. Forest Ecology and Management **225**:1489–1500.

Schmidt, K. A., and R. S. Ostfeld. 2003. Songbird populations in fluctuating environments: Predator responses to pulsed resources. Ecology **84**:406–415.

Schmidt, K. A., and R. S. Ostfeld. 2008a. Natural history note: Eavesdropping squirrels reduce their future value of food under the perceived presence of cache robbers. American Naturalist **171**:386–393.

Schmidt, K. A., and R. S. Ostfeld. 2008b. Numerical and behavioral effects within a pulse-driven system: Consequences for shared prey. Ecology **89**:635–646.

Schnurr, J. L., R. S. Ostfeld, and C. D. Canham. 2002. Direct and indirect effects of masting on rodent populations and tree seed survival. Oikos **96**:402–410.

Schultz, J. C. 1988. Many factors influence the evolution of herbivore diets, but plant chemistry is central. Journal of Ecology **69**:896–897.

Schultz, J. C. 1989. Tannin-insect interactions. Pp. 417–433 *in* R. W. Hemingway and J. J. Karchesy, editors. Chemistry and Significance of Condensed Tannins.

Springer, New York, NY, and London, UK.

Schultz, J. C., and I. T. Baldwin. 1982. Oak leaf quality declines in response to defoliation by gypsy-moth larvae. Science **217**:149–150.

Schultz, J. C., I. T. Baldwin, and P. J. Nothnagle. 1981. Hemoglobin as a binding substrate in the quantitative analysis of plant tannins. Journal of Agricultural and Food Chemistry **29**:823–826.

Schupp, E. W. 1993. Quantity, quality and the effectiveness of seed dispersal by animals. Vegetatio **107**:15–29.

Schupp, E. W. 2007. The suitability of a site for seed dispersal is context-dependent. Pp. 445–462 *in* A. J. Dennis, R. J. Green, E. W. Schupp, R. A. Green, and D. A. Westcott, editors. Seed Dispersal: Theory and Its Application in a Changing World. CABI, Wallingford, Oxfordshire, UK.

Schupp, E. W., P. Jordano, and J. M. Gómez. 2010. Seed dispersal effectiveness revisited: A conceptual review. New Phytologist **188**:333–353.

Schupp, E. W., P. Jordano, and J. M. Gómez. 2017. A general framework for effectiveness concepts in mutualisms. Ecology Letters **20**:577–590.

Scofield, D. G., V. L. Sork, and P. E. Smouse. 2010. Influence of acorn woodpecker social behaviour on transport of coast live oak (*Quercus agrifolia*) acorns in a southern California oak savanna. Journal of Ecology **98**:561–572.

Selås, V. 2017. Autumn irruptions of Eurasian jay (*Garrulus glandarius*) in Norway in relation to acorn production and weather. Ornis Fennica **94**:92–100.

Shaw, R. C., and N. S. Clayton. 2012. Eurasian jays, *Garrulus glandarius*, flexibly switch caching and pilfering tactics in response to social context. Animal Behaviour **84**:1191–1200.

Shaw, R. C., and N. S. Clayton. 2013. Careful cachers and prying pilferers: Eurasian jays (*Garrulus glandarius*) limit auditory information available to competitors. Proceedings of the Royal Society B: Biological Sciences **280**:2012–2238.

Shaw, R. C., and N. S. Clayton. 2014. Pilfering Eurasian jays use visual and acoustic information to locate caches. Animal Cognition **17**:1281–1288.

Sherry, D. F. 1984. Food storage by black-capped chickadees. Animal Behaviour **32**:451–464.

Sherry, D. F., and J. S. Hoshooley. 2007. Neurobiology of spatial behavior. Pp. 9–24 *in* K. Otter, editor. Ecology and Behavior of Chickadees and Titmice: An Integrated Approach. Oxford University Press, Oxford, UK, and New York, NY.

Sherry, D. F., and J. S. Hoshooley. 2010. Seasonal hippocampal plasticity in food-storing birds. Philosophical Transactions of the Royal Society of London B: Biological Sciences **365**:933.

Sherry, D. F., L. F. Jacobs, and S. J. Gaulin. 1992. Spatial memory and adaptive specialization of the hippocampus. Trends in Neurosciences **15**:298–303.

Sherry, D. F., J. R. Krebs, and R. J. Cowie. 1981. Memory for the location of stored food in marsh tits. Animal Behaviour **29**:1260–1266.

Sherry, D. F., and A. L. Vaccarino. 1989. Hippocampus and memory for food caches in black-capped chickadees. Behavioral Neuroscience **103**:308.

Sherry, D. F., A. L. Vaccarino, K. Buckenham, and R. S. Herz. 1989. The hippocampal complex of food-storing birds. Brain, Behavior and Evolution **34**:308–317.

Shettleworth, S. J., and J. R. Krebs. 1982. How marsh tits find their hoards: The roles of site preference and spatial memory. Journal of Experimental Psychology: Animal Behavior Processes **8**:354.

Short, H. L. 1976. Composition and squirrel use of acorns of black and white oak groups. Journal of Wildlife Management **40**:479–483.

Shumway, D. L., M. D. Abrams, and C. M. Ruffner. 2001. A 400-year history of fire and oak recruitment in an old-growth oak forest in western Maryland, USA. Canadian Journal of Forest Research **31**:1437–1443.

Shuttleworth, C. M., P. W. W. Lurz, and J. Gurnell, editors. 2016. The Grey Squirrel: Ecology & Management of an Invasive Species in Europe. European Squirrel Initiative, Woodbridge, UK.

Siepielski, A. M., and C. W. Benkman. 2007. Selection by a predispersal seed predator constrains the evolution of avian seed dispersal in pines. Functional Ecology **21**:611–618.

Siepielski, A. M., and C. W. Benkman. 2008. A seed predator drives the evolution of a seed dispersal mutualism. Proceedings of the Royal Society B: Biological Sciences **275**:1917–1925.

Signell, S. A., M. D. Abrams, J. C. Hovis, and S. W. Henry. 2005. Impact of multiple fires on stand structure and tree regeneration in central Appalachian oak forests. Forest Ecology and Management **218**:146–158.

Sipes, A., Jr., N. I. Lichti, and R. Swihart. 2013. Acorn germination is not enhanced near cache sites relative to random locations. Canadian Journal of Zoology **91**:529–532.

Smallwood, P. D., and W. D. Peters. 1986. Grey squirrel food preferences: the effects of tannin and fat concentration. Ecology **67**:168–174.

Smallwood, P. D., M. A. Steele, and S. H. Faeth. 2001. The ultimate basis of the caching preferences of rodents, and the oak-dispersal syndrome: tannins, insects, and seed germination. American Zoologist **41**:840–851.

Smallwood, P. D., M. A. Steele, E. Ribbens, and W. J. McShea. 1998. Detecting the effect of seed hoarders on the distribution of seedlings of tree species: Gray squirrels (*Sciurus carolinensis*) and oaks (*Quercus*) as a model system. Pp. 211–222 *in* M. A. Steele, J. F. Merritt, and D. A. Zegers, editors. Ecology and Evolutionary Biology of Tree Squirrels: Proceedings of the International Colloquium on the Ecology of Tree Squirrels, Powdermill Biological Station, Carnegie Museum of Natural History, 22–28 April 1994. Virginia Museum of Natural History Special Publication 6. Virginia Museum of Natural History, Martinsville, VA.

Smit, C., M. Díaz, and P. Jansen. 2009. Establishment limitation on holm oak (*Quercus ilex* subsp. *ballota* (Desf.) Samp) in a Mediterranean savanna-forest ecosystem. Annals of Forest Science **66**:1–7.

Smith, C. C. 1965. Interspecific competition in the genus of tree squirrels *Tamiasciurus*. PhD dissertation, University of Washington, Seattle, WA.

Smith, C. C. 1968. The adaptive nature of social organization in the genus of three squirrels *Tamiasciurus*. Ecological Monographs **38**:31–64.

Smith, C. C. 1970. The coevolution of pine squirrels (*Tamiasciurus*) and conifers. Ecological Monographs **40**:349–371.

Smith, C. C. 1975. The coevolution of seeds and seed predators. Pp. 53–77 *in* L. E. Gilbert and P. H. Raven, editors. Coevolution of Animals and Plants. University of Texas Press, Austin, TX.

Smith, C. C. 1981. The indivisible niche of *Tamiasciurus*: An example of nonpartitioning of resources. Ecological Monographs **51**:343–363.

Smith, C. C. 1998. The evolution of reproduction in trees: Its effect on squirrel ecology and behaviour. Pp. 203–209 *in* M. A. Steele, J. F. Merritt, and D. A. Zegers, editors. Ecology and Evolutionary Biology of Tree Squirrels: Proceedings of the International Colloquium on the Ecology of Tree Squirrels, Powdermill Biological Station, Carnegie Museum of Natural History, 22–28 April 1994. Virginia Museum of Natural History Special Publication 6. Virginia Museum of Natural History, Martinsville, VA.

Smith, C. C., and R. P. Balda. 1979. Competition among insects, birds, and mammals for conifer seeds. American Zoologist **19**:1065–1083.

Smith, C. C., and D. Follmer. 1972. Food preferences of squirrels. Ecology **53**:82–91.

Smith, C. C., and O. J. Reichman. 1984. The evolution of food caching by birds and mammals. Annual Review of Ecology and Systematics **15**:329–351.

Smith, T. M., and R. L. Smith. 2012. Elements of Ecology, 8th edition. Pearson Benjamin Cummings, San Francisco.

Smulders, T. V. 2006. A multi-disciplinary approach to understanding hippocampal function in foodhoarding birds. Reviews in the Neurosciences **17**:53–70.

Smulders, T. V., A. D. Sasson, and T. J. Devoogd. 1995. Seasonal variation in hippocampal volume in a food-storing bird, the black-capped chickadee. Journal of Neurobiology **27**:15–25.

Snell, R. S., N. G. Beckman, E. Fricke, B. A. Loiselle, C. S. Carvalho, L. R. Jones, N. I. Lichti, N. Lustenhouwer, S. J. Schreiber, C. Strickland, et al. 2019. Consequences of intraspecific variation in seed dispersal for plant demography, communities, evolution and global change. AoB Plants **11**:plz016.

Snow, D. W. 1971. Evolutionary aspects of fruit-eating by birds. Ibis **113**:194–202.

Snyder, M. A. 1992. Selective herbivory by Abert's squirrel mediated by chemical variability in ponderosa pine. Ecology **73**:1730–1741.

Snyder, M. A. 1993. Interactions between Abert's squirrels and ponderosa pine:

The relationship between selective herbivory and host plant fitness. American Naturalist **141**:866–879.

Snyder, M. A. 1998. Abert's squirrel (*Sciurus aberti*) in ponderosa pine (*Pinus ponderosa*): Directional selection, diversifying selection. Pp. 195–202 *in* M. A. Steele, J. F. Merritt, and D. A. Zegers, editors. Ecology and Evolutionary Biology of Tree Squirrels: Proceedings of the International Colloquium on the Ecology of Tree Squirrels, Powdermill Biological Station, Carnegie Museum of Natural History, 22–28 April 1994. Virginia Museum of Natural History Special Publication 6. Virginia Museum of Natural History, Martinsville, VA.

Sone, K., S. Hiroi, D. Nagahama, C. Ohkubo, E. Nakano, S.-I. Murao, and K. Hata. 2002. Hoarding of acorns by granivorous mice and its role in the population processes of *Pasania edulis* (Makino) Makino. Ecological Research **17**:553–564.

Sorensen, A. E. 1986. Seed dispersal by adhesion. Annual Review of Ecology and Systematics **17**:443–463.

Soria, F. J., E. Cano, and M. E. Ocete. 1999. Evaluation of damage by *Curculio elephas* Gyllenhal (Coleoptera, Curculionidae) and *Cydia* spp. (Lepidoptera, Tortricidae) on acorns of cork oak (*Quercus suber* Linné). [In Spanish, abstract in English.] Boletin de Sanidad Vegetal, Plagas 25:69–74.

Sork, V. L. 1984. Examination of seed dispersal and survival in red oak, *Quercus rubra* (Fagaceae), using metal-tagged acorns. Ecology **65**:1020–1022.

Sork, V. L., and J. Bramble. 1993. Prediction of acorn crops in three species of North American oaks: *Quercus alba*, *Q. rubra* and *Q. velutina*. Annals of Forest Science **50**:128–136.

Sork, V. L., J. Bramble, and O. Sexton. 1993a. Ecology of mast-fruiting in three species of North American deciduous oaks. Ecology **74**:528–541.

Sork, V. L., K. A. Stowe, and C. Hochwender. 1993b. Evidence for local adaptation in closely adjacent subpopulations of northern red oak (*Quercus rubra* L.) expressed as resistance to leaf herbivores. American Naturalist **142**:928–936.

Soto, A., Z. Lorenzo, and L. Gil. 2007. Differences in fine-scale genetic structure and dispersal in *Quercus ilex* L. and *Quercus suber* L.: Consequences for regeneration for Mediterranean open woods. Heredity **99**:601–607.

Stapanian, M. A., and C. C. Smith. 1978. A model for seed scatterhoarding: Coevolution of fox squirrels and black walnuts. Ecology **59**:884–896.

Stapanian, M. A., and C. C. Smith. 1984. Density-dependent survival of scatterhoarded nuts: An experimental approach. Ecology **65**:1387–1396.

Stapanian, M. A., and C. C. Smith. 1986. How fox squirrels influence the invasion of prairies by nut-bearing trees. Journal of Mammalogy **67**:326–332.

States, J., and P. Wettstein. 1998. Food habits and evolutionary relationships of the tassel-eared squirrel (*Sciurus aberti*). Pp. 185–194 *in* M. A. Steele, J. F. Merritt, and D. A. Zegers, editors. Ecology and Evolutionary Biology of Tree Squirrels: Proceedings of the International Colloquium on the Ecology of Tree Squirrels,

Powdermill Biological Station, Carnegie Museum of Natural History, 22–28 April 1994. Virginia Museum of Natural History Special Publication 6. Virginia Museum of Natural History, Martinsville, VA.

Steele, M. A. 1988. Patch use and foraging ecology of the fox squirrel: Tests of theoretical predictions. PhD dissertation, Wake Forest University, Winston-Salem, NC.

Steele, M. A. 1998. *Tamiasciurus hudsonicus*. Mammalian Species **586**:1–9.

Steele, M. A. 1999 *Tamiasciurus douglasii*. Mammalian Species **630**:1–8.

Steele, M. A. 2002. Acorn dispersal by birds and mammals. Pp. 182–195 *in* W. J. McShea and W. M. Healy, editors. Oak Forest Ecosystems: Ecology and Management for Wildlife. Johns Hopkins University Press, Baltimore, MD.

Steele, M. A. 2008. Evolutionary interactions between tree squirrels and trees: A review and synthesis. Current Science **95**:271–276.

Steele, M. A., M. Bugdal, A. Yuan, A. Bartlow, J. Buzalewski, N. Lichti, and R. Swihart. 2011. Cache placement, pilfering, and a recovery advantage in a seed-dispersing rodent: Could predation of scatter hoarders contribute to seedling establishment? Acta Oecologica **37**:554–560.

Steele, M. A., J. E. Carlson, P. D. Smallwood, A. B. McEuen, T. A. Contreras, and W. B. Terzaghi. 2007. Linking seed and seedling shadows: A case study in the oaks (*Quercus*). Pp. 322–339 *in* A. J. Dennis, R. J. Green, and E. W. Schupp, editors. Seed Dispersal: Theory and Its Application in a Changing World. CABI, Wallingford, Oxfordshire, UK.

Steele, M. A., T. A. Contreras, L. Z. Hadj-Chikh, S. J. Agosta, P. D. Smallwood, and C. N. Tomlinson. 2013. Do scatter hoarders trade off increased predation risks for lower rates of cache pilferage? Behavioral Ecology **25**:206–215.

Steele, M. A., K. Gavel, and W. Bachman. 1998. Dispersal of half-eaten acorns by gray squirrels: Effects of physical and chemical seed characteristics. Pp. 223–231 *in* M. A. Steele, J. F. Merritt, and D. A. Zegers, editors. Ecology and Evolutionary Biology of Tree Squirrels: Proceedings of the International Colloquium on the Ecology of Tree Squirrels, Powdermill Biological Station, Carnegie Museum of Natural History, 22–28 April 1994. Virginia Museum of Natural History Special Publication 6. Virginia Museum of Natural History, Martinsville, VA.

Steele, M. A., L. Z. Hadj-Chikh, and J. Hazeltine. 1996. Caching and feeding decisions by *Sciurus carolinensis*: Responses to weevil-infested acorns. Journal of Mammalogy **77**:305–314.

Steele, M. A., S. L. Halkin, P. D. Smallwood, T. J. McKenna, K. Mitsopoulos, and M. Beam. 2008. Cache protection strategies of a scatter-hoarding rodent: Do tree squirrels engage in behavioural deception? Animal Behaviour **75**:705–714.

Steele, M. A., T. Knowles, K. Bridle, and E. L. Simms. 1993. Tannins and partial consumption of acorns: Implications for dispersal of oaks by seed predators. American Midland Naturalist **130**:229–238.

Steele, M. A., and J. L. Koprowski. 2001. North American Tree Squirrels. Smithsonian Institution Press, Washington, DC.

Steele, M. A., N. Lichti, and R. K. Swihart. 2010. Avian-mediated seed dispersal: An overview and synthesis with an emphasis on temperate forests of central and eastern U.S. Pp. 28–43 *in* T. L. Master, M. C. Brittingham, R. M. Ross, R. S. Mulvihill, and J. E. Huffman, editors. Avian Ecology and Conservation: A Pennsylvania Focus with National Implications. Pennsylvania Academy of Science, Easton, PA.

Steele, M. A., S. Manierre, T. Genna, T. A. Contreras, P. D. Smallwood, and M. E. Pereira. 2006. The innate basis of food-hoarding decisions in grey squirrels: Evidence for behavioural adaptations to the oaks. Animal Behaviour **71**:155–160.

Steele, M. A., G. Rompré, J. A. Stratford, H. Zhang, M. Suchocki, and S. Marino. 2015. Scatterhoarding rodents favor higher predation risks for cache sites: The potential for predators to influence the seed dispersal process. Integrative Zoology **10**:257–266.

Steele, M. [A.], and P. [D.] Smallwood. 1994. What are squirrels hiding? Natural History **103**:40.

Steele, M. A., P. D. Smallwood, A. Spunar, and E. Nelsen. 2001a. The proximate basis of the oak dispersal syndrome: Detection of seed dormancy by rodents. American Zoologist **41**:852–864.

Steele, M. A., G. Turner, P. D. Smallwood, J. O. Wolff, and J. Radillo. 2001b. Cache management by small mammals: Experimental evidence for the significance of acorn-embryo excision. Journal of Mammalogy **82**:35–42.

Steele, M. A., and L. A. Wauters. 2016. Diet and food hoarding in eastern grey squirrels (*Sciurus carolinensis*): Implications for an invasive advantage. Pp. 97–114 *in* C. M. Shuttleworth, P. W. W. Lurz, and J. Gurnell, editors. The Grey Squirrel: Ecology & Management of an Invasive Species in Europe. European Squirrel Initiative, Woodbridge, UK.

Steele, M. A., L. Wauters, and K. Larsen. 2005. Selection, predation and dispersal of seeds by tree squirrels in temperate and boreal forests: Are tree squirrels keystone granivores? Pp. 205–219 *in* P. M. Forget, J. E. Lambert, P. E. Hulme, and S. B. Vander Wall, editors. Seed Fate: Predation, Dispersal, and Seedling Establishment. CABI, Wallingford, Oxfordshire, UK.

Steele, M. A., and P. D. Weigl. 1992. Energetics and patch use in the fox squirrel: Responses to prey density and patch profitability. American Midland Naturalist **128**:156–167.

Steffen, D. E., N. W. Lafon, and G. W. Norman. 2002. Turkeys, acorns, and oaks. Pp. 241–255 *in* W. J. McShea and W. M. Healy, editors. Oak Forest Ecosystems: Ecology and Management for Wildlife. Johns Hopkins University Press, Baltimore, MD.

Stein, J., D. Binion, and R. Acciavatti. 2003. Field Guide to Native Oak Species in Eastern North America. FHTET 03-01. US Department of Agriculture, Forest Service, Forest Health Technology Enterprise Team, Morgantown, WV.

Stephens, D. W., and J. R. Krebs. 1986. Foraging Theory. Princeton University Press, Princeton, NJ.

Steppan, S. J., B. L. Storz, and R. S. Hoffman. 2004. Nuclear DNA phylogeny of the squirrels (Mammalia: Rodentia) and the evolution of arboreality from c-myc and RAG1. Molecular Phylogenetics and Evolution. 30:703–719.

Steyaert, S., S. C. Frank, S. Puliti, R. Badia, M. P. Arnberg, J. Beardsley, A. Okelsrud, and R. Blaalid. 2018. Special delivery: Scavengers direct seed dispersal towards ungulate carcasses. Biology Letters 14:20180388. http://dx.doi.org/10.1098/rsbl.2018.0388.

Stiles, E. W. 1980. Patterns of fruit presentation and seed dispersal in bird-disseminated woody plants in the eastern deciduous forest. American Naturalist 116:670–688.

Stiles, E. W. 2000. Animals as seed dispersers. Pp. 111–123 in M. Fenner, editor. Seeds: The Ecology of Regeneration in Communities. CABI, Wallingford, Oxfordshire, UK.

Stiling, P., D. Moon, G. Hymus, and B. Drake. 2004. Differential effects of elevated CO_2 on acorn density, weight, germination, and predation among three oak species in a scrub-oak forest. Global Change Biology 10:228–232.

Sun, S., X. Gao, and L. Chen. 2004. High acorn predation prevents the regeneration of Quercus liaotungensis in the Dongling Mountain region of North China. Restoration Ecology 12:335–342.

Sundaram, M. 2016. The role of seed attributes in eastern gray squirrel foraging. PhD dissertation, Purdue University, West Lafayette, IN.

Sundaram, M, N. I. Lichti, N. J. O. Widmar, and R. K. Swihart. 2018. Eastern gray squirrels are consistent shoppers of seed traits: Insights from discrete choice experiments. Integrative Zoology 13:280–296.

Sundaram, M., J. R. Willoughby, N. I. Lichti, M. A. Steele, and R. K. Swihart. 2015. Segregating the effects of seed traits and common ancestry of hardwood trees on eastern gray squirrel foraging decisions. PLoS One 10:e0130942.

Sunyer, P., J. M. Espelta, R. Bonal, and A. Muñoz. 2014. Seeding phenology influences wood mouse seed choices: The overlooked role of timing in the foraging decisions by seed-dispersing rodents. Behavioral Ecology and Sociobiology 68:1205–1213.

Sunyer, P., A. Muñoz, R. Bonal, and J. M. Espelta. 2013. The ecology of seed dispersal by small rodents: A role for predator and conspecific scents. Functional Ecology 27:1313–1321.

Suselbeek, L., V. M. A. P. Adamczyk, F. Bongers, B. A. Nolet, H. H. T. Prins, S. E. van Wieren, and P. A. Jansen. 2014a. Scatter hoarding and cache pilferage by superior competitors: An experiment with wild boar, *Sus scrofa*. Animal Behaviour **96**:107–115.

Suselbeek, L., W.-J. Emsens, B. T. Hirsch, R. Kays, J. M. Rowcliffe, V. Zamora-Gutierrez, and P. A. Jansen. 2014b. Food acquisition and predator avoidance in a neotropical rodent. Animal Behaviour **88**:41–48.

Suselbeek, L., P. A. Jansen, H. H. T. Prins, and M. A. Steele. 2013. Tracking rodent-dispersed large seeds with Passive Integrated Transponder (PIT) tags. Methods in Ecology and Evolution **4**:513–519.

Swaine, M., and T. Beer. 1977. Explosive seed dispersal in *Hura crepitans* L. (Euphorbiaceae). New Phytologist **78**:695–708.

Swihart, R. K., and N. A. Slade. 1985. Testing for independence of observations in animal movements. Ecology **66**:1176–1184.

Tamura, N., T. Katsuki, and F. Hayashi. 2005. Walnut seed dispersal: Mixed effects of tree squirrels and field mice with different hoarding ability. Pp. 241–252 *in* P. M. Forget, J. E. Lambert, P. E. Hulme, and S. B. Vander Wall, editors. Seed Fate: Predation, Dispersal, and Seedling Establishment. CABI, Wallingford, Oxfordshire, UK.

Terborgh, J., N. Pitman, M. Silman, H. Schichter, and P. Núñez. 2002. Maintenance of tree diversity in tropical forests. Pp. 1–17 *in* D. J. Levey, W. R. Silva, and M. Galetti, editors. Seed Dispersal and Frugivory: Ecology, Evolution and Conservation. CABI, Wallingford, Oxfordshire, UK.

Tewksbury, J. J., D. J. Levey, N. M. Haddad, S. Sargent, J. L. Orrock, A. Weldon, B. J. Danielson, J. Brinkerhoff, E. I. Damschen, and P. Townsend. 2002. Corridors affect plants, animals, and their interactions in fragmented landscapes. Proceedings of the National Academy of Sciences of the United States of America **99**:12923.

Theimer, T. C. 2005. Rodent scatterhoarders as conditional mutualists. Pp. 283–296 *in* P. M. Forget, J. E. Lambert, P. E. Hulme, and S. B. Vander Wall, editors. Seed Fate: Predation, Dispersal, and Seedling Establishment. CABI, Wallingford, Oxfordshire, UK.

Theimer, T. C., and C. A. Gehring. 2007. Mycorrhizal plants and vertebrate seed and spore dispersal: Incorporating mycorrhizas into the seed dispersal paradigm. Pp. 463–478 *in* A. J. Dennis, R. J. Green, and E. W. Schupp, editors. Seed Dispersal: Theory and Its Application in a Changing World. CABI, Wallingford, Oxfordshire, UK.

Thompson, D. C. 1978. Regulation of a northern grey squirrel (*Sciurus carolinensis*) population. Ecology **59**:708–715.

Thorington, R. W., Jr., J. L. Koprowski, M. A. Steele, and J. F. Whatton. 2012. Squirrels of the World. Johns Hopkins University Press, Baltimore, MD.

Tinbergen, N. 1974. The Animal in Its World: Explorations of an Ethologist, 1932–1972, vol. 1, Field Studies. Harvard University Press, Cambridge, MA.

Tomback, D. F. 1977. The behavioral ecology of the Clark's nutcracker (*Nucifraga columbiana*) in the eastern Sierra Nevada. PhD dissertation, University of California, Santa Barbara, Santa Barbara, CA.

Tomback, D. F. 1980. How nutcrackers find their seed stores. Condor **82**:10–19.

Tomback, D. F. 1982. Dispersal of whitebark pine seeds by Clark's nutcracker: A mutualism hypothesis. Journal of Animal Ecology **51**:451–467.

Tong, X., Y.-X. Zhang, R. Wang, M. Inbar, and X.-Y. Chen. 2017. Habitat fragmentation alters predator satiation of acorns. Journal of Plant Ecology **10**:67–73.

Toomey, M. B., R. Bowman, and G. E. Woolfenden. 2007. The effects of social context on the food-caching behavior of Florida scrub-jays (*Aphelocoma coerulescens*). Ethology **113**:521–527.

Torchin, M. E., K. D. Lafferty, A. P. Dobson, V. J. McKenzie, and A. M. Kuris. 2003. Introduced species and their missing parasites. Nature **421**:628.

Tornick, J. K., S. N. Rushia, and B. M. Gibson. 2016. Clark's nutcrackers (*Nucifraga columbiana*) are sensitive to distance, but not lighting when caching in the presence of a conspecific. Behavioural Processes **123**:125–133.

Tulving, E. 1984. Précis of elements of episodic memory. Behavioral and Brain Sciences **7**:223–238.

Ussery, J. G., and P. G. Krannitz. 1998. Control of Scot's broom (*Cytisus scoparius* (L.) Link.: The relative conservation merits of pulling versus cutting. Northwest Science **72**:268–273.

Vandermeer, J., B. Hoffman, S. L. Krantz-Ryan, U. Wijayratne, J. Buff, and V. Franciscus. 2001. Effect of habitat fragmentation on gypsy moth (*Lymantria dispar* L.) dispersal: The quality of the matrix. American Midland Naturalist **145**:188–193.

Van der Pijl, L. 1972. Principles of Dispersal in Higher Plants. Berlin: Springer-Verlag, Berlin.

Van Dersal, W. R. 1940. Utilization of oaks by birds and mammals. Journal of Wildlife Management **4**:404–428.

Vander Wall, S. B. 1982. An experimental analysis of cache recovery in Clark's nutcracker. Animal Behaviour **30**:84–94.

Vander Wall, S. B. 1990. Food Hoarding in Animals. University of Chicago Press, Chicago, IL.

Vander Wall, S. B. 1992. The role of animals in dispersing a "wind-dispersed" pine. Ecology **73**:614–621.

Vander Wall, S. B. 1995a. Influence of substrate water on the ability of rodents to find buried seeds. Journal of Mammology **76**:851–856.

Vander Wall, S. B. 1995b. The effects of seed value on the caching behavior of yellow pine chipmunks. Oikos **74**:533–537.

Vander Wall, S. B. 1998. Foraging success of granivorous rodents: Effects of variation in seed and soil water on olfaction. Ecology **79**:233–241.

Vander Wall, S. B. 2000. The influence of environmental conditions on cache recovery and cache pilferage by yellow pine chipmunks (*Tamias amoenus*) and deer mice (*Peromyscus maniculatus*). Behavioral Ecology **11**:544–549.

Vander Wall, S. B. 2001. The evolutionary ecology of nut dispersal. Botanical Review **67**:74–117.

Vander Wall, S. B. 2002. Masting in animal-dispersed pines facilitates seed dispersal. Ecology **83**:3508–3516.

Vander Wall, S. B. 2010. How plants manipulate the scatter-hoarding behaviour of seed-dispersing animals. Philosophical Transactions of the Royal Society of London B: Biological Sciences **365**:989–997.

Vander Wall, S. B., and R. P. Balda. 1977. Coadaptations of the Clark's nutcracker and the pinon pine for efficient seed harvest and dispersal. Ecological Monographs **47**:89–111.

Vander Wall, S. B., and R. P. Balda. 1981. Ecology and evolution of food-storage behavior in conifer-seed-caching corvids. Zeitschrift für Tierpsychologie **56**:217–242.

Vander Wall, S. B., and S. H. Jenkins. 2003. Reciprocal pilferage and the evolution of food-hoarding behavior. Behavioral Ecology **14**:656–667.

Vander Wall, S. B., and J. W. Joyner. 1998. Recaching of Jeffrey pine (*Pinus jeffreyi*) seeds by yellow pine chipmunks (*Tamias amoenus*): Potential effects on plant reproductive success. Canadian Journal of Zoology **76**:154–162.

Vander Wall, S. B., and W. S. Longland. 2004. Diplochory: are two seed dispersers better than one? Trends in Ecology & Evolution **19**:155–161.

Vander Wall, S. B., and C. M. Moore. 2016. Interaction diversity of North American seed-dispersal mutualisms. Global Ecology and Biogeography **25**:1377–1386.

Vanoni, M., H. Bugmann, M. Nötzli, and C. Bigler. 2016. Drought and frost contribute to abrupt growth decreases before tree mortality in nine temperate tree species. Forest Ecology and Management **382**:51–63.

Vaughan, M. 2002. Oak trees, acorns, and bears. Pp. 224–240 *in* W. J. McShea and W. M. Healy, editors. Oak Forest Ecosystems: Ecology and Management for Wildlife. Johns Hopkins University Press, Baltimore, MD.

Venner, S., P.-F. Pélisson, M.-C. Bel-Venner, F. Débias, E. Rajon, and F. Menu. 2011. Coexistence of insect species competing for a pulsed resource: Toward a unified theory of biodiversity in fluctuating environments. PLoS One **6**:e18039.

Verdú, J. R., J. L. Casas, V. Cortez, B. Gallego, and J. M. Lobo. 2013a. Acorn consumption improves the immune response of the dung beetle *Thorectes lusitanicus*. PLoS One **8**:e69277.

Verdú, J. R., J. L. Casas, J. M. Lobo, and C. Numa. 2010. Dung beetles eat acorns to increase their ovarian development and thermal tolerance. PLoS One **5**:e10114.

Verdú, J. R., J. M. Lobo, C. Numa, I. M. Pérez-Ramos, E. Galante, and T. Marañón. 2007. Acorn preference by the dung beetle, *Thorectes lusitanicus*, under laboratory and field conditions. Animal Behaviour **74**:1697–1704.

Verdú, J. R., C. Numa, J. M. Lobo, and I. M. Pérez-Ramos. 2011. Acorn preference under field and laboratory conditions by two flightless Iberian dung beetle species (*Thorectes baraudi* and *Jekellus nitidus*): implications for recruitment and management of oak forests in central Spain. Ecological Entomology **36**:104–110.

Vogt, A. R. 1974. Physiological importance of changes in endogenous hormones during red oak acorn stratification. Forest Science **20**:187–191.

Vranckx, G., H. Jacquemyn, J. Mergeay, K. Cox, V. Kint, B. Muys, and O. Honnay. 2014. Transmission of genetic variation from the adult generation to naturally established seedling cohorts in small forest stands of pedunculated oak (*Quercus robur* L.). Forest Ecology and Management **312**:19–27.

Wada, N., M. Murakami, and K. Yoshida. 2000. Effects of herbivore-bearing adult trees of the oak *Quercus crispula* on the survival of their seedlings. Ecological Research **15**:219–227.

Waisman, A. S., and L. F. Jacobs. 2008. Flexibility of cue use in the fox squirrel (*Sciurus niger*). Animal Cognition **11**:625–636.

Waite, R. 1985. Food caching and recovery by farmland corvids. Bird Study **32**:45–49.

Waitman, B. A., S. B. Vander Wall, and T. C. Esque. 2012. Seed dispersal and seed fate in Joshua tree (*Yucca brevifolia*). Journal of Arid Environments **81**:1–8.

Wang, G., J. O. Wolff, S. H. Vessey, N. A. Slade, J. W. Witham, J. F. Merritt, M. L. Hunter, and S. P. Elias. 2009. Comparative population dynamics of *Peromyscus leucopus* in North America: Influences of climate, food, and density dependence. Population Ecology **51**:133–142.

Wang, X., Z. Xiao, Z. Zhang, and H. Pan. 2008. Insect seed predation and its relationships with seed crop and seed size of *Quercus mongolica*. [In Chinese, abstract in English.] Acta Entomologica Sinica **51**:161.

Warren, R. J., J. P. Love, and M. A. Bradford. 2017. Nest-mediated seed dispersal. Plant Ecology **218**:1213–1220.

Wauters, L. A., and P. Casale. 1996. Long-term scatterhoarding by Eurasian red squirrels (*Sciurus vulgaris*). Journal of Zoology (London) **238**:195–207.

Webb, S. L. J. 1986. Potential role of passenger pigeons and other vertebrates in the rapid Holocene migrations of nut trees. Quaternary Research **26**:367–375.

Weckerly, F. W., K. E. Nicholson, and R. D. Semlitsch. 1989a. Experimental test of discrimination by squirrels for insect-infested and noninfested acorns. American Midland Naturalist **122**:412–415.

Weckerly, F. W., D. W. Sugg, and R. D. Semlitsch. 1989b. Germination success of acorns (*Quercus*): Insect predation and tannins. Canadian Journal of Forest Research **19**:811–815.

Weeks, H. P., Jr., and C. M. Kirkpatrick. 1976. Adaptations of white-tailed deer to naturally occurring sodium deficiencies. Journal of Wildlife Management **40**:610–625.

Weeks, H. P., Jr., and C. M. Kirkpatrick. 1978. Salt preferences and sodium drive phenology in fox squirrels and woodchucks. Journal of Mammalogy 59:531–542.

Weigl, P. D., L. Sherman, M. A. Steele, and D. S. Weaver. 1998. Geographic variation in fox squirrels (*Sciurus niger*): A consideration of size clines, feeding economics, and body size. Pp. 171–184 *in* M. A. Steele, J. F. Merritt, and D. A. Zegers, editors. 1998. Ecology and Evolutionary Biology of Tree Squirrels: Proceedings of the International Colloquium on the Ecology of Tree Squirrels, Powdermill Biological Station, Carnegie Museum of Natural History, 22–28 April 1994. Virginia Museum of Natural History Special Publication 6. Virginia Museum of Natural History, Martinsville, VA.

Weigl, P. D., M. A. Steele, L. J. Sherman, J. Ha, and T. Sharpe. 1989. The Ecology of the Fox Squirrel in North Carolina: Implications for Survival in the Southeast. Bulletin No. 24. Tall Timbers Research Station, Tallahassee, FL.

Wenny, D. G., and D. J. Levey. 1998. Directed seed dispersal by bellbirds in a tropical cloud forest. Proceedings of the National Academy of Sciences of the United States of America 95:6204.

White, D. W., and E. W. Stiles. 1992. Bird dispersal of fruits of species introduced into eastern North America. Canadian Journal of Botany 70:1689–1696.

Wiens, J. A. 1992. What is landscape ecology, really? Landscape Ecology 7:149–150.

Wikelski, M., D. Moskowitz, J. S. Adelman, J. Cochran, D. S. Wilcove, and M. L. May. 2006. Simple rules guide dragonfly migration. Biology Letters 2:325–329.

Wikelski, M., J. Moxley, A. Eaton-Mordas, M. M. López-Uribe, R. Holland, D. Moskowitz, D. W. Roubik, and R. Kays. 2010. Large-range movements of neotropical orchid bees observed via radio telemetry. PLoS One 5:e10738.

Wilson, M., and A. Traveset. 2000. The ecology of seed dispersal. Pp. 85–110 *in* M. Fenner, editor. Seeds: The Ecology of Regeneration in Plant Communities. CABI, Wallingford, Oxfordshire, UK.

Wolfe, L. 2002. Why alien invaders succeed: Support for the escape-from-enemy hypothesis. American Naturalist 160:705–711.

Wolff, J. O. 1996. Population fluctuations of mast-eating rodents are correlated with production of acorns. Journal of Mammalogy 77:850–856.

Worrell, R., C. Rosique, and R. Ennos. 2014. Long-distance colonization of oak dispersed by jays in Highland Scotland. Scottish Forestry 68:24–30.

Wróbel, A., and R. Zwolak. 2013. The choice of seed tracking method influenced fate of beech seeds dispersed by rodents. Plant Ecology 214:471–475.

Xia, K., W. L. Harrower, R. Turkington, H.-Y. Tan, and Z.-K. Zhou. 2016. Pre-dispersal strategies by *Quercus schottkyana* to mitigate the effects of weevil infestation of acorns. Scientific Reports 6:37520.

Xiao, Z., G. Chang, and Z. Zhang. 2008. Testing the high-tannin hypothesis with scatter-hoarding rodents: Experimental and field evidence. Animal Behaviour 75:1235–1241.

Xiao, Z., X. Gao, M. Jang, and Z. Zhang. 2009. Behavioral adaptation of Pallas's squirrel to germination schedule and tannins in acorns. Behavioral Ecology 20:1050–1055.

Xiao, Z., X. Gao, M. A. Steele, and Z. Zhang. 2010. Frequency-dependent selection by tree squirrels: Adaptive escape of nondormant white oaks. Behavioral Ecology 21:169–175.

Xiao, Z., X. Gao, and Z. Zhang. 2013a. The combined effects of seed perishability and seed size on hoarding decisions by Pére David's rock squirrels. Behavioral Ecology and Sociobiology 67:1067–1075.

Xiao, Z., M. K. Harris, and Z. Zhang. 2007. Acorn defenses to herbivory from insects: Implications for the joint evolution of resistance, tolerance and escape. Forest Ecology and Management 238:302–308.

Xiao, Z., P. A. Jansen, and Z. Zhang. 2006. Using seed-tagging methods for assessing post-dispersal seed fate in rodent-dispersed trees. Forest Ecology and Management 223:18–23.

Xiao, Z., X. Mi, M. Holyoak, W. Xie, K. Cao, X. Yang, X. Huang, and C. J. Krebs. 2017. Seed-predator satiation and Janzen–Connell effects vary with spatial scales for seed-feeding insects. Annals of Botany 119:109–116.

Xiao, Z., and Z. Zhang. 2006. Nut predation and dispersal of harland tanoak (*Lithocarpus harlandii*) by scatter-hoarding rodents. Acta Oecologica 29:205–213.

Xiao, Z., and Z. Zhang. 2012. Behavioral responses to acorn germination by tree squirrels in an old forest where white oak have long been extirpated. Animal Behaviour 83:945–951.

Xiao, Z., Z. Zhang, and C. J. Krebs. 2013b. Long-term seed survival and dispersal dynamics in a rodent-dispersed tree: Testing the predator satiation hypothesis and the predator dispersal hypothesis. Journal of Ecology 101:1256–1264.

Xiao, Z., Z. Zhang, and Y. Wang. 2004. Dispersal and germination of big and small nuts of *Quercus serrata* in a subtropical broad-leaved evergreen forest. Forest Ecology and Management 195:141–150.

Xiao, Z., Z. Zhang, and Y. Wang. 2005a. Effects of seed size on dispersal distance in five rodent-dispersed fagaceous species. Acta Oecologica 28:221–229.

Xiao, Z., Z. Zhang, and Y. Wang. 2005b. The effects of seed abundance on seed predation and dispersal by rodents in *Castanopsis fargesii* (Fagaceae). Plant Ecology 177:249–257.

Yang, Y., and X. Yi. 2012. Partial acorn consumption by small rodents: Implication for regeneration of white oak (*Quercus mongolica*). Plant Ecology 213:197–205.

Yang, Y. Q., Z. Y. Wang, C. Yan, Y. H. Zhang, D. Y. Zhang, and X. F. Yi. 2018. Selective predation on acorn weevils by seed-caching Siberian chipmunk (*Tamias sibiricus*) in a tripartite interaction. Oecologia 188:149–158.

Yi, X., A. W. Bartlow, R. Curtis, S. J. Agosta, and M. A. Steele. 2019. Responses of seedling growth and survival to post-germination cotyledon removal: An investigation among seven oak species. Journal of Ecology **107**:1–11.

Yi, X., R. Curtis, A. W. Bartlow, S. J. Agosta, and M. A. Steele. 2013a. Ability of chestnut oak to tolerate acorn pruning by rodents. Naturwissenschaften **100**:81–90.

Yi, X., G. Liu, M. A. Steele, Z. Shen, and C. Liu. 2013b. Directed seed dispersal by a scatter-hoarding rodent: The effects of soil water content. Animal Behaviour **86**:851–857.

Yi, X., M. A. Steele, J. A. Stratford, Z. Wang, and Y. Yang. 2016. The use of spatial memory for cache management by a scatter-hoarding rodent. Behavioral Ecology and Sociobiology **70**:1527–1534.

Yi, X., M. A. Steele, and Z. Zhang. 2012a. Acorn pericarp removal as a cache management strategy of the Siberian chipmunk (*Tamias sibiricus*). Ethology **118**:87–94.

Yi, X., and Z. Wang. 2015. Context-dependent seed dispersal determines acorn survival of sympatric oak species. Plant Ecology **216**:123–132.

Yi, X., Z. Wang, C. Liu, G. Liu, and M. Zhang. 2015. Acorn cotyledons are larger than their seedlings need: Evidence from artificial cutting experiments. Scientific Reports **5**:8112.

Yi, X., Z. Xiao, and Z. Zhang. 2008. Seed dispersal of Korean pine (*Pinus koraiensis*) labeled by two different tags in a northern temperate forest, northeast China. Ecological Research **23**:379–384.

Yi, X., and Y. Yang. 2010a. Apical thickening of epicarp is responsible for embryo protection in acorns of *Quercus variabilis*. Israel Journal of Ecology and Evolution **56**:153–164.

Yi, X., and Y. Yang. 2010b. Large acorns benefit seedling recruitment by satiating weevil larvae in *Quercus aliena*. Plant Ecology **209**:291–300.

Yi, X., Y. Yang, R. Curtis, A. W. Bartlow, S. J. Agosta, and M. A. Steele. 2012b. Alternative strategies of seed predator escape by early-germinating oaks in Asia and North America. Ecology and Evolution **2**:487–492.

Yu, X., H. Zhou, and T. Luo. 2003. Spatial and temporal variations in insect-infested acorn fall in a *Quercus liaotungensis* forest in North China. Ecological Research **18**:155–164.

Zhang, B., X. Chen, M. A. Steele, J. Li, and G. Chang. 2018. Effects of insect-infestation on rodent-mediated dispersal of *Quercus aliena*: Results from field and enclosure experiments. Integrative Zoology **14**:104–113.

Zhang, B., Z. J. Shi, A. M. Lien, G. Chang, X. N. Chen, and G. Chang. 2014. Effects of weevil-infestation on seed dispersal of *Quercus aliena* handled by rodents. Acta Ecologica Sinica **34**:3937–3943.

Zhang, D., J. Li, Z. Wang, and X. Yi. 2016. Visual landmark-directed scatter-hoarding of Siberian chipmunks (*Tamias sibiricus*). Integrative Zoology **11**:175–181.

Zhang, H., Y. Luo, M. A. Steele, Z. Yang, Y. Wang, and Z. Zhang. 2013a. Rodent-favored cache sites do not favor seedling establishment of shade-intolerant wild apricot (*Prunus armeniaca* Linn.) in northern China. Plant Ecology **214**:531–543.

Zhang, H., M. A. Steele, Z. Zhang, W. Wang, and Y. Wang. 2014a. Rapid sequestration and recaching by a scatter-hoarding rodent (*Sciurotamias davidianus*). Journal of Mammalogy **95**:480–490.

Zhang, H., Y. Wang, and Z. Zhang. 2011. Responses of seed-hoarding behaviour to conspecific audiences in scatter- and/or larder-hoarding rodents. Behaviour **148**:825–842.

Zhang, H., C. Yan, G. Chang, and Z. Zhang. 2016. Seed trait-mediated selection by rodents affects mutualistic interactions and seedling recruitment of co-occurring tree species. Oecologia **180**:475–484.

Zhang, M., Z. Dong, X. Yi, and A. W. Bartlow. 2014b. Acorns containing deeper plumule survive better: How white oaks counter embryo excision by rodents. Ecology and Evolution **4**:59–66.

Zhang, M., M. A. Steele, and X. Yi. 2013b. Reconsidering the effects of tannin on seed dispersal by rodents: Evidence from enclosure and field experiments with artificial seeds. Behavioural Processes **100**:200–207.

Zhengyi, W., P. H. Raven, and D. Hong, editors. 1999. Flora of China, vol. 4: Cycadaceae through Fagaceae. Science Press, Beijing, China, and Missouri Botanical Garden, St. Louis, MO.

Zollner, P. A. 2000. Comparing the landscape level perceptual abilities of forest sciurids in fragmented agricultural landscapes. Landscape Ecology **15**:523–533.

Zwolak, R., and E. E. Crone. 2012. Quantifying the outcome of plant-granivore interactions. Oikos **121**:20–27.

INDEX

Page numbers in italics refer to figures. The letter *t* following a page number denotes a table.

213, 357; and fire, 180, 358; pine barrens, 344; predation by rodents, 110, 182–86, *184*, 201–3, *202*, 251; selective pressure from rodents, 178–82, 201–3, *202*

Conotrachelus spp., 276t, 282

cork oak *(Q. suber)*, *330*; consumption and caching decisions, 53t, 222; dispersal of, *215*; distribution and range, 28, 330–31, *331*; industry, *330*, *331*; insect predation in, 276t, 332t–34t, *334*; as threatened, 329–36, 332t–34t

Corvidae, taxonomy and distribution, *214*, *215*. *See also* crows; jays; rook

cotyledonary petiole, *46*, 46–47, 104, 106–8, 109t, 204, 229

cotyledons, 38–39, *39*, *46*, 51, *105*, *106*. *See also* chemical gradients in cotyledon

cowbirds, 265t

crop, 217, 222–23

crossbills, 203

crowberry *(Empetrum nigrum)*, 13

crows, 13, 264t. *See also* jays

cupule, *39*, *106*

Curculio cardamomi, 277

Curculio elephas (chestnut weevil), 126, 276t, 279, 292, 296, 297, 312t, 332t

Curculio glandium, 126, 296, 297, 312t, 332t

Curculio glandium Marshall, 276t

Curculionidae spp., 277, *277*

Curculio nuncum, 297

Curculio occidentis, 276t

Curculio pellitus, 296

Curculio robustus, 294–95

Curculio spp. *See* weevils

Curculio venosus, 276t, 296, 297

Cyclobalanopsis breviradiata, 191t

Cyclobalanopsis glauca, 53t

Cyclobalanopsis glaucoides, 191t

Cyclobalanopsis multinervis, 191t

Cyclobalanopsis section, 29

Cyclobalanopsis stewardiana, 191t

Cyclobalanopsis subgenus, 24, 25, 26t, 30, 53t, 191t

Cydia spp., 277, *277*, 298

Cyllorynchites spp., 294

Cyrtepistomus spp., 282

deer. *See* Lyme disease; red deer; white-tailed deer

deer mouse *(Peromyscus maniculatus)*, 60, 89t, 111, 187, 360

defense of caches and middens, 180, 188, 255–60, 257t–59t

deforestation, 304–5, 345

dentate gyrus, 262, 267t, 269

desiccation, 228, 286, 288t, 289

diet breadths, optimal, 197

diplochory, 18, 360–61

directed dispersal, 13, 18, 171–74, 172t–73t, 210, 226, 229

dispersal: biogeographic variations, 362–63; by carnivores, 339; definitions, xiii, 4, 14; diplochory, 360–61; dispersal effectiveness hypothesis, 50, 51–55, 52t–53t, 71; by insects, 298–99; keystone dispersers, 357–59; list of variables in, 19; modes of, 4–15; research models, 15–19; seed dispersal effectiveness (SDE), 363–64; significance of, x, xiii, 3–4, 43, 362. *See also* caching; directed dispersal; dispersal by birds; dispersal by rodents; secondary dispersal

dispersal by birds: of conifers, 171, 203, 213, 357; directed dispersal, 13, 172t–73t, 174, 210, 226, 229; distance of, 215–21, *218*, 221, 225, 226, 231, 232–33, 344; and forest fragmentation, 138–40, 210, 234, 235, 241; long-distance dispersal, 17, 112, 210, 229–33, 358; and masting, 225; in Mediterranean oak forests, 333t, 335; multiple-prey loading, 217–21, *218*, *219*, *220*, 222–23; overview of, 209–12; rapidity of, 216, *217*; of weevils, 301. See also *specific species*

dispersal by insects, 298–99

dispersal by rodents: of American chestnut, 313–15; coevolution of, 203–4; of conifers, 186–87; directed dispersal, 172t–73t; distance of dispersal, 33, 42, 124, 165–71, *166*, *167*, 204–5, 270, 342, 344; and forest fragmentation, 310–13, 311t; and invasive oaks, 321–23, *322*, *323*; lower amounts in Mexico,

fenses, 256, 258t; cognition and cache
recovery in, 245, 254t, 264t; diet, 59;
directed dispersal by, 172t–73t, 226;
distance of dispersal, 216, 232–33; distribution of, 215; home range, 229–30;
and invasive species, 233, 319; as key
dispersal agent, 59, 211, 214, 335, 359;
numbers of acorns cached, 216; tri-trophic interactions, 224
Eurasian nuthatch (Sitta europaea), 211
Eurasian red squirrel (Sciurus vulgaris),
60, 100, 181, 186, 188, 192t, 193
European beech (Fagus sylvatica), 57, 190t,
212, 230
European bee wolf (Philanthus triangulum), 244–45
European crow (Corvis corone), 264t
European wood mouse. See wood mouse
evergreen broadleaf tree (Castanopsis
hystrix), 191t
evolution of oaks: coevolution of, 12–13,
201–7, 357; history of, 29–30, 38; and
oak migration, 230–33

Fagaceae family: evolution of, 29–30; taxonomy of, 23–24
Fagoideae subfamily, 23
fertilization, 38, 40
filbertworm moth (Cydia latiferreana),
274, 275
fire and fire suppression: and bear oak,
344; and conifers, 180, 358; and cork
oak, 332t; and forest degradation and
fragmentation, 304, 305, 306–7, 312t;
and Garry oak, 32, 329; prescribed
burns, 308, 309, 312t; and Q. oleoides,
337; and silviculture, 308
Florida scrub jay (Aphelocoma coerulescens), 214; caching defenses, 257t, 259t;
distribution of, 215; episodic memory
in, 261–62; as key dispersal agent, 214;
number of acorns cached, 216; tri-trophic interactions, 223
flowering, 36, 36–38, 37, 337
follicles, 25, 26t
food value: and decision making, 206–7;
.optimal diet breadths, 197

foraging theory, optimal, 196–97, 198
forest fragmentation: and cork oaks, 333t;
and dispersal by birds, 138–40, 210,
234, 235, 241; and dispersal by rodents,
310–13, 311t; effects of, 309–13, 310,
311t–12t; and Q. oleoides, 336
forests: composition and caching, 198; deforestation, 304–5, 345; degradation of,
304, 305–7; diversity of oak habitats,
31–32; loss of oak, 304–5; Mediterranean, 329–36, 332t–34t; oak migration
in, 230–33; silviculture, 307–9; and
wind dispersal, 7. See also forest fragmentation
fossil record, 29, 230
foxes, 339, 360
fox squirrel (Sciurus niger): body size, 183;
coevolution of, 205; cognition and
cache recovery in, 227, 244, 249–52,
250, 255; and conifers, 182–86, 184, 202;
dispersal by, 60, 190t; distribution of,
193; embryo excision by, 91; and forest
fragmentation, 311t, 313; summer
slowdown, 183–84; territoriality, 188
frosts, 315–16
frugivory, 9–11, 10, 16–17, 171
fungi: chestnut blight, 75, 307, 313–15; and
conifers, 358; and cork oak, 334; damage from, 284–86, 288t; ectomycorrhizal, 328, 336, 362; and Garry oak, 328;
invasive, 318; mycorrhizal, 361–62

game theory, 196
Garry oak (Q. garryana), 32, 47, 212, 215,
325, 325–29, 326
genetic analysis: Garry oaks, 328–29; seed
shadows, 352–56; weevils, 279
germination: and acorn pruning, 46–47,
104–8, 105, 204, 229; and climate
change, 315; of conifer seeds, 186–87;
dormancy detection, 45, 90, 119; and
dung beetles, 298–99; and epigeal cotyledons, 39; of invasive oaks, 322; and
jays, 228–29; morphological variations,
199, 200, 290; and nitrogen levels, 317;
in oak life cycle, 45–46, 46, 105–6; of
partially consumed acorns, 42, 112, 113,

pilfering, *164*, 164–65, 170–71, 174, 359; and seedling establishment, 174, 259–60, 360

motivation, as variable in scatterhoarding, 19, 194

mountain chickadee (*Poecile gambeli*), 258t, 266t

mutualism: ants, 298; as conditional, 17, 195–96, 198, 206, 358–59; dung beetles, 298–99; effect on dispersal, 18; and fungi, 328, 336, 362; and jays, 59, 223–24, 229, 299–301; and keystone species, 357, 358–59; research challenges, 18, 19; and secondary dispersal, 14

mycorrhizal fungi, 361–62

myrmecochory, 11, *12*, 187, 360

neurogenesis, 267t, 268–69

New World jays, 214

nitrogen, 110, 316–18

northern red oak (*Q. rubra*), 27; acorn size, 32, 41, 221–23, *223*; chemical gradients in, 111, 120–22, *121*, 204; cotyledonary petiole, 109t; dispersal of, 52t, 53t, 89t, 165–71, *166*, *167*, 172t–73t, 190t, 191t, 213, *215*; and ectomycorrhizal fungi, 362; embryo excision of, *92*, 92–94; and forest fragmentation, 311t; germination and caching decisions, 26, 44, 80–86, *83*, *85*, 89; germination and establishment of partially consumed acorns, 122, 122–24, *123*; insect predation of, 66, 88–89, 125, *125*; as invasive, 191t, *215*, 233, 318–23, *322*; longevity of, 306; masting by, 62–64, *64*, 66, 70, 70–71; and nitrogen levels, 317; tannin levels, 319; testing of Janzen-Connell model, 361; tri-trophic interactions, 300t, 301t

Notholithocarpus densiflorus, 206

nutcrackers (*Nucifraga* spp.): conifer dispersal by, 203, 213, 357; oak dispersal by, 213. *See also* Clark's nutcracker

nuthatches (*Sitta* spp.), 59, 210–11

oaks (*Quercus* spp.): distribution of, ix, 3, 23–28, *24*, 30–32; diversity in size, 31; diversity of, 3, 23, 30–32, *31*; evolutionary history of, 29–30, 38; life cycle, 35–47; migration of, 230–33; as research focus, ix, xiii, 19–20; as symbol, 304, *305*; taxonomy of, 23–29, 26t; uses, 304, 305. *See also* acorn consumption; acorns; dispersal; masting; red oaks (*Lobatae* section); white oaks (*Quercus* section)

oak wilt, 318

Ocotea endresiana, 171

Old World jays, 214

olfactory cues, x, 205, 244, 245, 251, 260, 261

opportunity costs, 19, 194

optimal density models, 160

optimal diet breadths, 197

optimal foraging theory, 196–97, 198

orangutan (*Pongo pygmaeus*), *10*, 11

Ord's kangaroo rat (*Dipodomys ordii*), 264t

Oriental cork oak. *See* Chinese cork oak (*Q. variabilis*)

ovary, 36, 37, *37*–38

ovules, 36, 37, *37*, 38, 40

painted spiny pocket mouse (*Heteromys pictus*), 289

Pallas's squirrel (*Callosciurus erythraeus*), 99, *100*, *101*

palm, 137–38, 189, 190t, 193, 290, 338, 339, 341–43

palm civet (*Paradoxurus hermaphroditus*), 339

Palmer's oak (*Q. palmeri*), 28

parids, hippocampus in, 264t

partial consumption, *112*, *114*; and acorn abundance, 118, 124; and acorn size, 42, 354–55; by birds, 111–12, 224, 359; and chemical gradients, 42, 60–61, 114–22, *121*, 204, 224, 354–55, 358–59; and germination/establishment, 42, 112, 113, *114*, 122, 122–24, *123*, 333t; by insects, 125, 359; and multi-seeded acorns, 290; observations of, 110–13; and pericarp thickness, 119; by rodents, 111–12, *114*, 202, 358. *See also* acorn pruning

Q. *aliena*, 42, 53t, 89, 191t, 199, 292, 300t

Q. *cassipes*, 284–86, 285t, 287t

Q. *chapmanii*, 215, 315

Q. *conspersa*, 97, 109t

Q. *costaricensis*, 343

Q. *crassifolia*, 96–99, 97, 109t, 284–86, 285t, 287t, 291

Q. *crassipes*, 96–99, 109t

Q. *crispula*, 294

Q. *crispula = mongolica*, 52t

Q. *dentata*, 298

Q. *deserticola*, 311t

Q. *dysophylla*, 109t

Q. *emoryi*, 215, 300t

Q. *gambelii*, 215

Q. *geminate*, 215

Q. *glaucoides*, 97, 109t, 284–86, 285t, 287t

Q. *inopina*, 215

Q. *insignis*, 32

Q. *laeta*, 97, 109t, 284–86, 285t, 287t, 312t

Q. *liebmannii*, 284–86, 285t, 287t

Q. *mexicana*, 97, 109t, 285t

Q. *microla*, 97

Q. *microphylla*, 104, 109t, 284–86, 285t, 287t

Q. *miyagii = salicina*, 41

Q. *myagii*, 25

Q. *myrtifolia*, 215, 315

Q. *obtusata*, 96–99, 97, 284–86, 285t, 287t

Q. *oleoides*, 190t, 193, 215, 310, 311t, 336–43, 337

Q. *peduncularis*, 109t

Q. *petrea*, 132, 215

Q. *rugosa*, 97

Q. *salicina*, 25

Q. *schottkyana*, 291, 291

Q. *scytophylla*, 109t

rabbits, 332t, 333t, 335

radicle, 39, 39, 45–46, 105–8, 113, 114

rain, 14, 202, 279, 295, 332t

raptors, 73

ratlike hamster (*Tscherskia triton*), 256, 257t

rats: cache defenses, 256, 257t–59t; cognition of, 245–46, 251, 264t; dispersal by, 191t, 193, 300t; distribution of, 193

ravens (*Corvus corax*), 13, 257t

re-caching, 138, 161, 187, 252, 257t, 259

red acouchy (*Myoprocta acouchy*), 190t

red-backed vole (*Myodes gapperi*), 67, 70

red-bellied woodpecker (*Melanerpes carolinus*), 59, 210–11

red-billed blue magpie (*Urocissa erythrorhyncha*), 264t

red crossbill (*Loxia curvirostra*), 203

red deer (*Cervus elaphus*), 332t–33t, 335

red fox (*Vulpes vulpes*), 339, 360

Red List, 345–46

red maple (*Acer rubrum*), 306

red oaks (*Lobatae* section): acorn pruning in, 47, 109; biogeographic variations, 362–63; and climate change, 315; cotyledonary petiole, 108–9, 109t; distribution of, 25, 26t; diversity of, 26, 31; evolution of, 30; leaf morphology, 25; limited dispersal in Mexico, 286–89, 288t; lipids in, 78–79, 79t, 120, 121, 121, 166; masting by, 62–64, 64, 65, 70, 70–71, 149–55, 157, 294, 356; multi-seeded acorns, 40–41, 126, 290; pollen tube development, 40; and re-caching, 143–44; seed shadows, 350–56, 351, 352, 353, 355; tannin in, 26t, 78–79, 79t; taxonomy of, 25–27; and weevils, 114, 115, 280–86, 281, 284, 285t, 287t, 290, 296. *See also* germination; germination schedules; *specific species*

red-rumped agouti (*Dasyprocta leporina*), 137–38

red spiny rat (*Maxomys surifer*), 191t

red squirrel (*Tamiasciurus hudsonicus*), 72, 74, 74, 100, 179, 179–81, 181, 188–89

red wood ant (*Formica yessensis*), 298

Reid's Paradox, 230–31

resource budget hypothesis, 56–58

resource tracking hypothesis, 56

ring-cupped oaks, 24, 25, 26t, 30, 53t, 191t

rock squirrel (*Otospermophilus variegatus*), 105, 289

rodents: and coevolution, 201–7; diet, 60, 74, 74, 178; distribution of, 177–78, 207; endozoochory, 187; hippocampus in, 262, 264t–67t, 268–69; as key dispersal

shagbark hickory (*Carya ovata*), 296, 311t

shelterwood harvesting, 308–9

Sherman's fox squirrel (*Sciurus niger shermani*), 183

short-snouted weevil (*Conotrachelus naso*), 274

short-tailed shrew (*Blarina brevicauda*), 67, 68, 70, 300t

shrews (*Sorex* spp.), 67–69, 70, 299, 300t

shrub oaks, 31

Siberian chipmunk (*Tamias sibiricus*): caching decisions, 89, 116; cognition of, 248, 252, 267t; dispersal by, 172t–73t, 191t, 192t, 193; distribution of, *193*; and nitrogen levels, 316

silviculture, 47, 307–9

single-leaf piñon pine (*Pinus monophylla*), *218*

small Japanese field mouse (*Apodemus argenteus*), 172t–73t, 192t

smoky shrew (*Sorex fumeus*), 68

sodium, 120, 121, *121*

soil moisture, 227

song sparrow (*Melospiza melodia morphna*), 265t

southern flying squirrel (*Glaucomys volans*), 67, 89t, 90, 100, 253–55

southern red oak (*Q. falcata*), 36, 52t

spatial memory and cues, 245–46, *246*, 253–55, 254t–55t, 260–61, 357

spiny pocket mouse (*Liomys salvini*), 338, 339–41

squirrels: biogeographic variations, 363; body size, 183; caching decisions, 44–45, 80–81, *80–90*, 84–86, 89t, 90, 96–99, *97, 98*, 119, 197–98, 204–5, 299; caching defenses, 256–59, 257t–59t; cognition and cache recovery in, 244, 245, *246*, 247, 248, 249–55, *250*; diet, 60, 74, *74*, 88; distribution of, 60, 93, *193*; and forest fragmentation, 311t; as key dispersal agent, 60, 192–93; and multi-seeded acorns, 41, 126, *127*, 290; partial consumption by, 111, *113, 114*, *202*, 358; phylogeny of, 100, *101*; and pilfering, 162–65, 188, 228, 251, 261;

population density of, 72; research overview, 190t–92t; secondary dispersal by, 186–87; selective pressure by, 179–82, 201–3, *202*; social structure, 188–89; spatial cognition, 253–55; territoriality, 188–89; tri-trophic interactions, 300t–301t; and weevils, 87–90, 118–19, 299. *See also* embryo excision; *specific species*

stamens, 36, 37

Steller's jay (*Cyanocitta stelleri*), 59, 214, *214, 215*, 216, 327, 328

stigma, 36, *37*, 38

stone oaks (*Lithocarpus* spp.), 24, 53t

stratification, 26, 44

striped field mouse (*Apodemus agrarius*), 191t, 256, 257t

styles, 25, 26t, 27, 36, *39*, 106

sublingual pouches, 217, *218*, 357

Sudden Oak Death, 318, 345

tagging methods: acorns and seeds, 81, *81*, 130–41, 130t, *131, 136, 137, 142, 143, 163, 163*, 240, 341; animals, 130t, 137–39, *138*, 234–41, *235*

tannic acid equivalence, 116–17, 117t

tannin: assaying, 116–17, 117t; decisions by birds, 222, 223–25, 299–301; decisions by rodents, 78–79, 116, 197, 198, 206–7, 339; as defense mechanism, ix, 78, 114–17, 204; defined, 78; digestibility and, 88, 115, 223, 299; gradients in, 42, 114–22, *121*, 197, 204, 224, 292–93, 354–55, 358–59; in invasive oaks, 319; in leaves, 298; and nitrogen, 317; in red oaks, 26t, 78–79, 166; in white oaks, 25, 26t, 78–79

tanoak (*Lithocarpus densiflorus*), 311t

tanoaks (*Notholithocarpus* spp.), 24, 206, 318

taproot, 46, 107, *108*

territoriality, 188–89, 225, *226*

threatened oaks: bear oak, 344; California oaks, 344–45; cork oak, 329–36, 332t–34t; Garry oaks, 325–29, *326*; overview of, 324; *Q. oleoides*, 336–43, *337*; Red List, 345–46

three-wattled bellbird (*Procnias tricarun-culata*), 171
thrushes, 74, *74*
ticks, 73–74, *75*
timbering, 304, 305, 306, 307–9
titmice, 59, 172t–73t, 210–11, 267t
tits, 211, 245, 254t, 264t–67t, 268
tortricid moth, 54, 277
traps, 66–69, *67*
tri-trophic interactions, 223–24, 299–301, 300t–301t, 328
tropical forests, 11, 16, 17, 336–43, *337*
tufted titmouse (*Baeolophus bicolor*), 59, 172t–73t, 210–11
turkey (*Meleagris gallopavo*), 58
turkey oak (*Q. laevis*), 111, 117, 117t

utility models, 196–98

valley oak (*Q. lobata*), 57, 215, 225, 345
variegated squirrel (*Sciurus variegatoides*), 338, *343*
veery (*Catharus fuscescens*), 74
Virentes section, 29
visual cues and cache recovery, 244
voles, 67, 190t

walnuts (*Juglans* spp.): and caching decisions, 199–201, 206; dispersal of, 5, 179, 192t, 234, 311t; as invasive, 234; masting by, 205
wasps (*Callirhytis* spp.), 244–45, 274, 275, 282
water dispersal. *See* hydrochory
water oak (*Q. nigra*), 111, 301t
weasels, 73, 339
weather, 14, 56–57, 58, 315–16, 332t
weevils, 275, 278; and acorn survival, *125*, 125–26; biogeographic variations, 363; and caching decisions, x, 87–89, 299; and chemical gradients, 113–14, *115*, 118, *121*, 289, 292–93; and climate change, 315; community structure, 295–97; dispersal of, 301; distribution of, 274, 276t; as food source, 88, 299; and forest fragmentation, 312t; impact of, x, 87, 277, *281*, 294–95; larvae, 65,

87, 126, *178*, 279; and latitude, 126, 285, 289, 363; life cycle, 87, 274, *278*, 278–80; and masting, 54, 65–66, 283, 289, 293–95, 296; in Mediterranean forests, 332t, 333t, 334t, 335; patterns of, 280–86, *281*, *284*, 285t, 287t; and pericarp, 42–43, 119, 293; pupa, *178*, 279; and *Q. oleoides*, 336; and rain, 278, 295, 332t; responses to, 289–95; and shelterwood harvesting, 308; specialization by, 296–97; tri-trophic interactions, 223–24, 299–301, 300t–301t
western fox squirrel (*Sciurus niger ru-fiventer*), 183, *193*, 244, 249–51, *250*, 252–53
western scrub jay (*Aphelocoma californica*), *214*; cache defenses, 256, 257t, 258t; dispersal by, 214, 216, 225; distribution of, *215*; hippocampus in, 265t; pilfering strategies, 261
white bark pine (*Pinus albicaulis*), 203, 357
white-breasted nuthatch (*Sitta carolinensis*), 59, 210–11
white-faced capuchin (*Cebus capucinus*), 338
white-footed mouse (*Peromyscus leucopus*): caching decisions, 89t; diet, 60; dispersal by, 190t, 256, 300t; and Lyme disease, 73–74, *75*; partial acorn consumption, 111; population density, 67, 70, 70–71, 314–15; and silviculture, 308
white-lipped peccary (*Tayassu pecari*), 338
white oak (*Q. alba*), 27, 36, 305; acorn size, 198, 221–23, *223*; chemical gradients in, 120–22, *121*; and climate change, 315; cotyledonary petiole, 109t; dispersal of, 53t, 89t, 172t–73t, 190t, 215; and ectomycorrhizal fungi, 362; and forest fragmentation, 312t; germination and caching decisions, 44, 71, 80–86, 89; germination/establishment of partially consumed acorns, *122*, 122–24, *123*; germination morphology, 45–46, 106–8, *107*; insect predation of, 54, 65, *87*, 88–89, 125, *125*, 276t, *281*, 294; and invasive species, 321–23, *322*, *323*; masting by, 62–64, *64*, 65, 70, 70–71,